Continent Formation, Growth and Recycling

Continent Formation, Growth and Recycling

Edited by

Paul J. Sylvester

Memorial University of Newfoundland
Department of Earth Sciences
300 Prince Philip Drive
St. John's, NF, Canada

2000
ELSEVIER
Amsterdam - London - New York - Oxford - Paris - Shannon - Tokyo

ELSEVIER SCIENCE B.V.
Sara Burgerhartstraat 25
P.O. Box 211, 1000 AE Amsterdam, The Netherlands

© 2000 Elsevier Science B.V. All rights reserved.

This work is protected under copyright by Elsevier Science, and the following terms and conditions apply to its use:

Photocopying
Single photocopies of single chapters may be made for personal use as allowed by national copyright laws. Permission of the Publisher and payment of a fee is required for all other photocopying, including multiple or systematic copying, copying for advertising or promotional purposes, resale, and all forms of document delivery. Special rates are available for educational institutions that wish to make photocopies for non-profit educational classroom use.

Permissions may be sought directly from Elsevier Science Global Rights Department, PO Box 800, Oxford OX5 1DX, UK; phone: (+44) 1865 843830, fax: (+44) 1865 853333, e-mail: permissions@elsevier.co.uk. You may also contact Global Rights directly through Elsevier's home page (http://www.elsevier.nl), by selecting 'Obtaining Permissions'.

In the USA, users may clear permissions and make payments through the Copyright Clearance Center, Inc., 222 Rosewood Drive, Danvers, MA 01923, USA; phone: (978) 7508400, fax: (978) 7504744, and in the UK through the Copyright Licensing Agency Rapid Clearance Service (CLARCS), 90 Tottenham Court Road, London W1P 0LP, UK; phone: (+44) 207 631 5555; fax: (+44) 207 631 5500. Other countries may have a local reprographic rights agency for payments.

Derivative Works
Tables of contents may be reproduced for internal circulation, but permission of Elsevier Science is required for external resale or distribution of such material. Permission of the Publisher is required for all other derivative works, including compilations and translations.

Electronic Storage or Usage
Permission of the Publisher is required to store or use electronically any material contained in this work, including any chapter or part of a chapter.

Except as outlined above, no part of this work may be reproduced, stored in a retrieval system or transmitted in any form or by any means, electronic, mechanical, photocopying, recording or otherwise, without prior written permission of the Publisher.
Address permissions requests to: Elsevier Global Rights Department, at the mail, fax and e-mail addresses noted above.

Notice
No responsibility is assumed by the Publisher for any injury and/or damage to persons or property as a matter of products liability, negligence or otherwise, or from any use or operation of any methods, products, instructions or ideas contained in the material herein. Because of rapid advances in the medical sciences, in particular, independent verification of diagnoses and drug dosages should be made.

First edition 2000

Reprinted from TECTONOPHYSICS, Vol. 322 (1/2), 2000

Library of Congress Cataloging in Publication Data
A catalog record from the Library of Congress has been applied for.

ISBN: 0-444-50622-5

⊚ The paper used in this publication meets the requirements of ANSI/NISO Z39.48-1992 (Permanence of Paper).
Printed in The Netherlands.

Special Issue

Continent Formation, Growth and Recycling

Selected contributions from the symposium "Continent Formation, Growth and Recycling", convened at the 1998 Annual Meeting, Geological Society of America, Toronto, Canada
26–29 October 1998

Edited by

Paul J. Sylvester
Department of Earth Science, Memorial University of Newfoundland, 300 Prince Philip Drive, St. John's, NF A1B 3X5, Canada

CONTENTS

Preface	vii
Coupled magmatism–mantle convection system with variable viscosity	
M. Ogawa	1
Early formation and long-term stability of continents resulting from decompression melting in a convecting mantle	
J. De Smet, A.P. Van den Berg and N.J. Vlaar	19
The diversity of tectonics from fluid-dynamical modeling of the lithosphere–mantle system	
B. Schott, D.A. Yuen and H. Schmeling	35
Fast mechanisms for the formation of new plate boundaries	
K. Regenauer-Lieb and D.A. Yuen	53
Growth and recycling of early Archaean continental crust: geochemical evidence from the Coonterunah and Warrawoona Groups, Pilbara Craton, Australia	
M.G. Green, P.J. Sylvester and R. Buick	69
Age constraints on recycled crustal and supracrustal sources of Archaean metasedimentary sequences, Eastern Goldfields Province, Western Australia: evidence from SHRIMP zircon dating	
B. Krapez, S.J.A. Brown, J. Hand, M.E. Barley and R.A.F. Cas	89
Nd isotopic evidence for Early to Late Archean (3.4–2.7 Ga) crustal growth in the Western Superior Province (Ontario, Canada)	
P. Henry, R.K. Stevenson, Y. Larbi and C. Gariépy	135
Episodic continental growth models: afterthoughts and extensions	
K.C. Condie	153
Quantifying Precambrian crustal extraction: the root is the answer	
D. Abbott, D. Sparks, C. Herzberg, W. Mooney, A. Nikishin and Y. S. Zhang	163
Continental emergence and growth on a cooling earth	
N.J. Vlaar	191

Preface

Continent formation, growth and recycling

About 30 years ago, two landmark papers concerned with growth of the continents appeared in the literature. They set the stage for a debate that continues today. One paper, by Hurley and Rand (1969), argued that the distribution of ages of rocks in the continents, corrected for intra-crustal processing (re-melting, metamorphism, erosion and re-deposition), reflects the growth rate of the continents. This notion was held widely by geologists in the late 1960s but, for the first time, Hurley and Rand quantified the available database on a global scale. Their calculations suggested that the continents have been extracted from the mantle progressively through time.

A second paper by Armstrong (1968) presented a fundamentally different view of continental development — one in which the present volume of continental crust formed in a 'big bang' early in Earth history, and has been modified by crustal additions and subtractions from/to the mantle ever since. Because the new additions are balanced by the subtractions, the net result is a constant volume of continental crust through time. Thus, according to the Armstrong view, calculations of the sort performed by Hurley and Rand give only a minimum estimate of the growth rate of continental crust.

In the late 1960s almost all geologists would have disagreed with Armstrong's premise that crustal recycling was important process. The crust was seen as indestructible, being simply much too buoyant to be dragged down into the mantle. Today however many geologists think that continents have been recycled, both as sediments carried on subducting oceanic slabs and as parts of lithosphere delaminated during continental collisions. Many would accept, therefore, the idea that the ages of rocks in the continental crust are a record of crust preservation rather than crust formation, and that Hurley and Rand-type calculations provide only the minimum continental growth rate. However, the devil is in the details and most geologists remain skeptical that crustal recycling has destroyed huge volumes of Archean continental crust, as Armstrong envisioned.

Herein lies the frontier of research on continental growth. Using the Hurley and Rand approach, and the modern database of crust ages, we know that at least half of the present volume of the continents existed by 2 Ga. But is there evidence in either in the crust or mantle for more extensive crustal development before 2 Ga, which we now do not observe? Obviously there are diverse approaches to the problem, including studies of cratonic geology and geochronology, mantle geochemistry, lithosphere/asthenosphere dynamics, and marine sedimentation and tectonics. Advances will probably require combinations of approaches.

With these thoughts in mind, scientists gathered in a special session of the 1998 Annual Meeting of the Geological Society of America in Toronto in October to discuss "Continent Formation, Growth and Recycling". Twenty-four talks were presented, ten of which have been developed into papers published in this special issue of *Tectonophysics*. The papers are grouped by subdiscipline: Ogawa, de Smet et al., Schott et al., and Regenauer-Lieb and Yuen present numerical models of asthenospheric melting and convection and lithospheric break-up and delamination. Green et al., Krapez et al., and Henry et al. report geologic, geochronologic, geochemical and isotopic data for some key Archean cratonic terranes. Condie, Abbott et al., and Vlaar discuss global models for crustal growth, emphasizing episodic magmatism, crustal and lithospheric thickness, and crustal isostasy, respectively.

Readers of the entire special issue will see that a definitive answer to the overall question of continental growth rates has not been reached but that much progress is being made on understanding the processes involved in continent development.

References

Armstrong, R.L., 1968. A model for the evolution of strontium and lead isotopes in a dynamic earth. Rev. Geophys. 6, 175–199.

Hurley, P.M., Rand, J.R., 1969. Pre-drift continental nuclei. Science 164, 1229–1242.

Paul J. Sylvester
Memorial University of Newfoundland,
Department of Earth Science,
300 Prince Philip Drive,
St. John's, Newfoundland, Canada A1b 3X5
E-mail address: pauls@sparky2.esd.mun.ca

Coupled magmatism–mantle convection system with variable viscosity

Masaki Ogawa *

Department of Earth Sciences and Astronomy, University of Tokyo at Komaba, Meguro, Tokyo 153, Japan

Received 11 January 1999; received in revised form 1 July 1999; accepted for publication 16 August 1999

Abstract

Numerical models are presented for the thermal and chemical evolution of the mantle when it is controlled by both magmatism and mantle convection. The mantle is modeled as a fluid with temperature- and pressure-dependent viscosity in a two-dimensional square box heated by incompatible radioactive elements that decay with time. The influence of phase transitions at 660 km between the upper and lower mantle is also included. Magmatism is modeled as permeable flow of magma produced upon pressure release partial melting of the convecting mantle materials. It is shown that the mantle may exist in one of two regimes. At first, when the internal heating rate is higher than a threshold, the mantle is on a regime where episodic, but active, magmatism takes place to make the mantle chemically stratified with the upper mantle occupied by magma residue (harzburgite) and the deeper part of the lower mantle occupied by basaltic materials. There is a chemical discontinuity along the 660 km phase boundary. Mantle convection occurs as a layered convection punctuated by flushing events and is sluggish except at the time of flushing events owing to the compositional buoyancy from the chemical stratification. The flushing events induce vigorous magmatic activity. As the internal heating rate falls below the threshold owing to the decay of radioactive elements, the thermal and chemical state jumps to another regime. In this regime, magma does not erupt, and mantle convection occurs as layered thermal convection. In spite of the stirring due to the thermal convection, the chemical discontinuity and a portion of the chemical stratification formed in the earlier regime remains for billions of years. The numerical models suggest that (a) mantle convection was much less active than expected from the strong internal heating, and chemically distinct reservoirs are formed in the early Earth, and once formed (b), the reservoirs have survived convective stirring for billions of years. © 2000 Elsevier Science B.V. All rights reserved.

Keywords: magmatism; mantle convection; mantle evolution; numerical model

1. Introduction

Plate motion and the upwelling plumes beneath hot spots are well understood as a part of thermal convection in the mantle at present. Many studies on the thermal history of the Earth have been carried out, based on an uniformitarian assumption that solid-state convection occurred as thermal convection in the past mantle, too (e.g. Schubert, 1979). A simple estimate of the activity of mantle magmatism and mantle convection in the early Earth (Davies, 1993) and numerical models of magmatism in a convecting mantle (Ogawa, 1997; Ogawa and Nakamura, 1998), however, suggest that mantle magmatism, that is partial melting and differentiation of mantle rock, strongly affected the style and vigor of mantle convection and the

* Fax: +81-3-3465-3925.
 E-mail address: masaki@chianti.c.u-tokyo.ac.jp (M. Ogawa)

thermal and chemical state of mantle in the Archean when the internal heating due to radioactive elements was considerably stronger than at present; hence, the uniformitarian assumption does not hold. Here, I further develop the earlier models of magmatism in a convecting mantle in Ogawa and Nakamura (1998) (hereafter, called Paper 1) to study the dynamics and the thermal and chemical state of the mantle in the early Earth.

The influence of mantle magmatism on mantle convection was first pointed out for plate motion by Oxburgh and Parmentier (1977). At a ridge, the original mantle material is differentiated into basalt and harzburgite, and oceanic plates are chemically buoyant. Chemical buoyancy is much weaker than the thermally induced negative buoyancy of old oceanic plates and hardly affects their subduction at present (Oxburgh and Parmentier, 1977). The chemical buoyancy, however, may have made subduction of oceanic plates difficult in the Archean mantle according to a simple estimate by Davies (1990). The vigor of ridge volcanism strongly depends on the potential temperature of the mantle. Oceanic crust becomes more than twice as thick as the present oceanic crust when the mantle temperature is higher than 1500°C (Mckenzie, 1984). An increase in the mantle temperature by 100–200°C, therefore, makes the chemical buoyancy of plate considerably stronger. The chemical buoyancy may have been comparable to the thermal buoyancy of old oceanic plates in the Archean mantle that was strongly heated by radioactive elements. Thermal convection may not be a good model of solid-state convection in the Archean mantle, and it is reasonable to ask whether the uniformitarian assumption holds in the early Earth.

2. Two regimes in the thermal and chemical state of the mantle

A problem with the simple estimate referred to above is that the influence of mantle convection on the magmatism is not fully considered (Davies, 1993). Mantle convection is the agent that induces mantle magmatism. Mantle convection is also one of the most important agents that determines the mantle temperature and, in general, the thermal and chemical state of the mantle and hence indirectly controls the vigor of mantle magmatism. It is also possible that mantle magmatism may significantly affect mantle convection. Mantle magmatism and mantle convection are likely to strongly influence each other in the strongly heated mantle in the early Earth and must be treated as a coupled system. In this section, I review the earlier numerical models of the coupled system in Paper 1.

2.1. Model description

The mantle is modeled by a two-dimensional square box uniformly heated by an internal heat source that is constant in time. The height of the box is chosen to be 2000 km to make the ratio of volume of the lower mantle to that of the upper mantle 2, the value for the Earth; the boundary between the upper and the lower mantle is at a depth of 660 km, as will be described below. The temperature is fixed at 300 K along the top surface boundary. The box is placed on the top of a heat bath, which is a model of the core. The temperature of the bottom boundary of the box is equal to the temperature of the heat bath that is calculated from the energy balance of the heat bath. The heat bath does not contain any internal heat source, and the heat flux at the boundary is calculated from the vertical temperature gradient just above the boundary.

The mantle material is modeled by an incompressible binary eutectic material that has Newtonian rheology with constant viscosity. The rheology is kept as simple as possible in Paper 1 to highlight the interactions between magmatism and mantle convection. The composition of the material is denoted as $A_\xi B_{1-\xi}$ where B stands for a garnet-rich material and A stands for an olivine-rich material (harzburgite). The eutectic composition is at $\xi \approx 0.1$ and corresponds to basaltic composition. The end member A is transformed into its high-pressure phase at a depth of around 660 km, and the phase boundary has negative Clausius–Clapeyron slope, as is the case for γ-spinel versus post-spinel transition (Ito and Takahashi, 1989). The end member B is gradually transformed into its high-pressure phase at a depth of around 660 km, as is the case for garnet versus perovskite transition (Irifune and Ringwood, 1993). The Clausius–Clapeyron slope of the phase transition is assumed to be zero for simplicity, though the slope is recently found to be positive (e.g. Akaogi and Ito, 1999). (The phase boundary of the end member A will be called the 660 km phase boundary, and the two phase transitions will be called the 660 km phase transitions hereafter.) The end member B is also transformed into its crustal phase at depths of less than 40 km, corresponding to the basalt-versus-eclogite transition. The end member B is denser than the end member A at all depths except at depths less than 40 km and, in some cases described below, in a narrow depth interval just beneath the 660 km phase boundary (Irifune and Ringwood, 1993). Other solid–solid phase transitions of mantle materials are neglected for simplicity. The solidus temperature of the convecting material is close to the solidus temperature of mantle materials (Takahashi, 1986; Ohtani and Sawamoto, 1987) in the upper mantle where partial melting occurs in the numerical models.

Mantle magmatism is modeled by a permeable flow of magma produced by pressure-release melting of the binary eutectic material in partially molten regions (Mckenzie, 1984). The relative velocity of the magma and the coexisting matrix is proportional to the density difference between the magma and matrix. The density inversion between magma and the coexisting matrix (e.g. Agee, 1998) is not included, and magma always migrates upward in the numerical model. Magma migration makes divergence of matrix velocity non-zero since both magma and matrix are incompressible. Chemical equilibrium is assumed to hold in partially molten regions to calculate the melt content, composition of the magma, and temperature.

Mantle convection is modeled by convection of the binary eutectic material driven by thermal and compositional buoyancy as well as the buoyancy coming from phase transitions. The compositional distribution changes with time owing to mass transport by convecting matrix and migrating magma, while temperature and magma distributions change owing to the energy transport by convecting matrix and magma, thermal diffusion, absorption and release of latent heat of the phase transitions, viscous dissipation in partially molten regions, and internal heating.

The basic equations are numerically solved by a finite difference method described in Ogawa (1997). The adopted mesh is uniform in horizontal direction but non-uniform in vertical direction with a higher resolution around the 660 km phase boundary and the top and bottom boundaries. The mesh size for the calculation of momentum and continuity equations is 16.5 km in the horizontal direction and 13.5 km on average in the vertical direction; the minimum mesh size in the vertical direction is 4.7 km. Twice this resolution is employed for the calculation of energy and mass transport equations. Among the successful benchmark tests of the adopted numerical code (see Paper 1), a benchmark calculation of layered convection in a chemically layered mantle is particularly important. The chemical diffusivity is negligibly small in the Earth's mantle, but numerical diffusion that is necessary to keep numerical stability in the calculation of mass transport equation smears out fine structures in chemical distribution in numerical models. I found that the effect of numerical diffusion on the chemical layering was not significant for more than 6 Gyr, even when the chemical density contrast was rather small across the chemical boundary and the boundary largely undulated with time owing to the convection. The change in the chemical contrast is, of course, negligible when the chemical density contrast is large, and the location of the chemical boundary is almost fixed. The effect of numerical diffusion on the overall structure of chemical distribution is not, therefore, significant.

Numerical calculations were carried out for a long period of time until the effect of artificial setting of the initial condition disappeared. Namely, the thermal and chemical state of the numerically modeled mantle described in Section 2.2 below is statistically steady.

2.2. Coupled magmatism–mantle convection system with constant viscosity

The main conclusion of Paper 1 is that there are two regimes in the thermal and chemical state

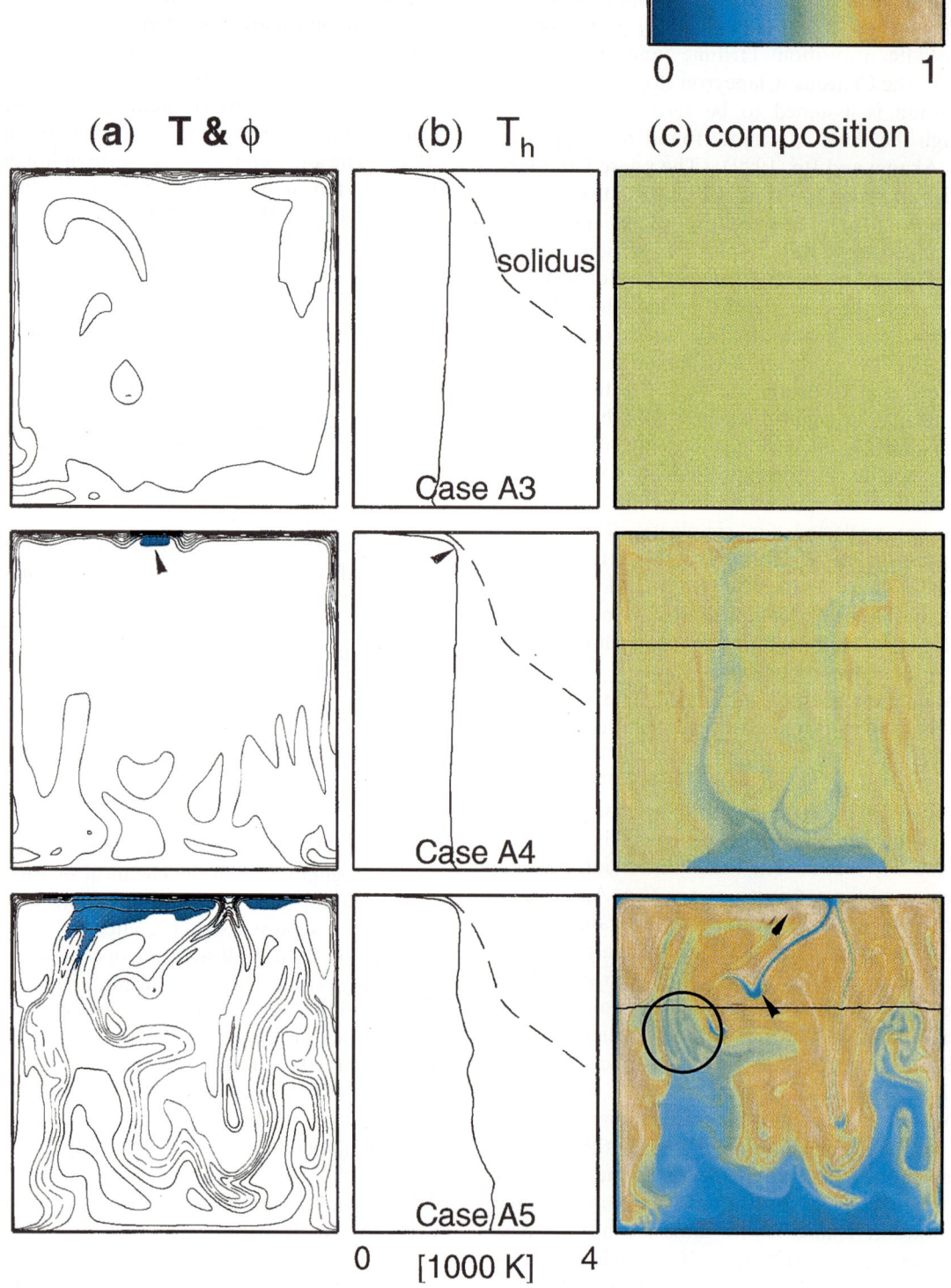

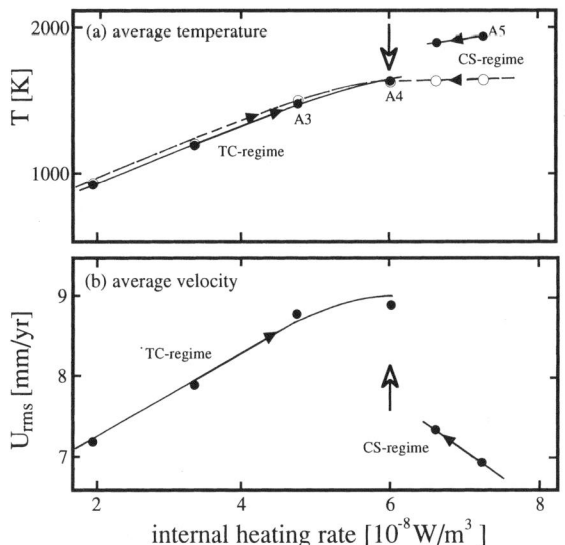

Fig. 2. (a) Average temperature in the entire mantle (solid line) and in the upper mantle (dashed line) and (b) root mean square velocity plotted against internal heating rate for the numerical models of the mantle shown in Fig. 1. The points denoted as A3–A5 show the results of Cases A3–A5 in Fig. 1, respectively. Notice that there is a bifurcation at an internal heating rate slightly higher than 6×10^{-8} W/m^3. After Ogawa and Nakamura (1998) with modification.

in the mantle controlled by the coupled system depending on the internal heating rate. An example of numerical results showing the two regimes is presented in Figs. 1 and 2. Here, the barrier effect of the 660 km phase transitions against mass exchange across the phase boundary (Christensen and Yuen, 1985; Irifune and Ringwood, 1993) is made negligibly weak by adjusting the Clausius–Clapeyron slope of the 660 km phase boundary and the depth interval for the 'garnet versus perovskite transition' to avoid the complications coming from the 660 km phase transitions and to make the effect of interaction between magmatism and mantle convection conspicuous.

When the internal heating rate is sufficiently low and the temperature is lower than the solidus temperature everywhere in the mantle (Fig. 1b for Case A3), magmatism does not occur at all, and the mantle is chemically homogeneous (Fig. 1c for Case A3). The solid-state convection in the mantle occurs as an almost steady thermal convection characterized by a cold thermal boundary layer along the top surface boundary, cold sinking plumes along the vertical side walls and isothermal core (Fig. 1a for Case A3).

When the internal heating rate is slightly higher (Case A4, see also Fig. 2a), the mantle temperature locally becomes equal to the solidus at its shoulder (see arrow in Fig. 1b; the average temperature is still below the solidus). Partial melting occurs in the small region indicated by the arrow in Fig. 1a. The generated magma that has the end member B-rich eutectic composition, that is basaltic composition, percolates to the surface. The cold basalt sinks, as shown by the pool of basaltic materials indicated by blue color in Fig. 1c for Case A4. The magma residue enriched in the end member A, however, has a slight tendency to remain in the upper mantle owing to its chemically induced positive buoyancy. The thermal convection in the mantle is, however, vigorous, and the once differentiated residue and a part of the basaltic materials are mechanically mixed, as can be seen from the blue and orange stripes in Fig. 1c for Case A4; the mantle still remains chemically homogeneous as a whole in spite of the chemical differentiation due to magmatism.

When the internal heating rate is further increased, however (Case A5), the mantle becomes chemically stratified. Basaltic materials occupy the

---

Fig. 1. Snap shots of (a) the distributions of temperature (contour lines) and melt content (color), (b) horizontally averaged geotherm (solid line) and solidus (dashed line), and (c) the compositional distribution in the mantle (color) and the location of the '660 km' phase boundary (solid line) obtained from the numerical models of coupled magmatism–mantle convection system of Paper 1. The color scales for (a) and (b) are shown at the top of the figure. In (a), the contour interval is 200 K, and the contour lines of 1000, 2000, and 3000 K are shown by a dashed curve. In (c), the color indicates the content of the residual end-member A (harzburgite). The primitive mantle materials are indicated by yellow (see the color in the figure for Case A3), the basaltic materials by blue, and the residue by orange. Cases A3–A5 correspond to A3–A5 in Fig. 2, respectively. The viscosity is constant, and the Rayleigh number defined with a temperature contrast of 1873 K is 1×10^7 for all cases. After Ogawa and Nakamura (1998) with modification.

deeper part of the lower mantle, and residue occupies the upper mantle and the upper part of the lower mantle (Fig. 1c for Case A5). Active magmatism takes place; partially molten regions sometimes extend to as deep as 400 km (Fig. 1a for Case A5). A large volume of basaltic materials and magma residue are newly added to the layer of basaltic materials and the layer of residue, respectively, each time magmatic activity takes place, as indicated by the arrow in Fig. 1c. Magmatic activity also causes mechanical mixing of the residue and basaltic materials produced by the earlier magmatic activities by inducing strong convective flow around partially molten regions, as can be seen from the fine structure encircled in Fig. 1c for Case A5. The chemical stratification develops as a result of the dynamic balance between convective stirring and new additions of basaltic materials and residue due to active magmatism. The chemical buoyancy that accompanies the stratification lowers the efficiency of heat transport by the solid-state convection and makes the horizontally averaged geotherm superadiabatic.

The thermal and chemical state of Case A5 is separated from the thermal and chemical state of Cases A3 and A4 by a bifurcation, as shown in Fig. 2, where the average temperatures in the upper mantle, T_{UM}, and the entire mantle, T_{WM}, and the root mean square velocity in the entire mantle, U_{rms}, are plotted against the internal heating rate. Here, the average implies both a spatial and time average. T_{UM}, T_{WM}, and U_{rms} all increase along the curve of 'thermal convection (TC)' regime indicated in Fig. 2 with increasing internal heating rate, q, for $q < 6 \times 10^{-8}$ W/m^3, but U_{rms} suddenly drops, T_{WM} suddenly jumps to a higher value, and T_{UM} becomes almost constant when q increases further [see the curve of the 'chemically stratified (CS)' regime in Fig. 2]. There is a bifurcation point at q slightly higher than 6×10^{-8} W/m^3, and the transition between the thermal and chemical state of Cases A3 and A4 and that of Case A5 is a 'phase transition'. A similar bifurcation is observed when the barrier effects of the 660 km phase transitions against the mass exchange across the 660 km phase boundary are strong, and mantle convection occurs basically as a layered convection, too, in Paper 1.

The key to understanding the bifurcation is the strong dependence of activity of mantle magmatism on mantle temperature (Mckenzie, 1984) and the chemical buoyancy that accompanies the chemical heterogeneity produced by mantle magmatism. When the mantle is on the TC regime but the internal heating rate is close to the bifurcation value, mild magmatism occurs (see Fig. 1a for Case A4). As the internal heating rate, q, is raised, the amount of basaltic materials and residue produced by the magmatism rapidly increases. Eventually, the chemical buoyancy of the basaltic materials and residue becomes comparable to the thermal buoyancy that drives the convection on the TC regime. The chemical buoyancy considerably weakens the thermally induced convection and lowers the efficiency of convective heat transport from deep mantle to the surface. The mantle temperature further increases, and the magmatism becomes more active, that is, it produces a larger amount of basaltic materials and residue. This positive feedback makes the TC regime unstable, and the thermal and chemical state in the mantle jumps to the CS regime; a bifurcation takes place. The bifurcation value is only slightly higher than the value of q for Case A4. Namely, the bifurcation value is well approximated as the value of internal heating rate at which the shoulder of horizontally averaged mantle temperature indicated by arrow in Fig. 2b for Case A4 touches the solidus curve if the thermal and chemical state of the mantle is on the TC regime. It is unlikely that this criterion for bifurcation value sensitively depends on the aspect ratio of the convecting box, though the bifurcation value itself may depend on the aspect ratio.

The reason for T_{UM} being almost independent of q on the CS regime is the efficient heat extraction by migrating magma. The internal heat source is strong enough to make the horizontally averaged temperature very close to the solidus curve in the uppermost mantle on the CS regime. When the temperature locally becomes equal to the solidus temperature in the uppermost mantle, magma is generated to efficiently extract heat from the partially molten part of the mantle. The temperature does not exceed the solidus temperature in the partially molten part. The average temperature, T_{UM}, therefore, cannot largely exceed the solidus temperature in the uppermost mantle, no matter how strong the internal heat source is, and becomes almost independent of the internal heating rate (see also Davies, 1990).

It is also shown in Paper 1 that the phase transition is 'reversible', provided that the 660 km phase transitions do not suppress the convective flow across the 660 km phase boundary. When a thermal and chemical state on the CS regime is taken as the initial condition and the internal heating rate, q, is decreased below the bifurcation value, the thermal and chemical state eventually falls on the TC regime shown in Fig. 1 for Cases A3 and A4, as indicated by the open arrow in Fig. 2. Namely, the chemical buoyancy that accompanies the chemical stratification on the CS regime is not strong enough for the stratification to survive convective stirring forever by itself if the effect of 660 km phase transitions is negligible and the viscosity is constant. (A comment is necessary here, however. A fairy long time, in the order of a billion years, was found to be necessary for the transition from the CS regime to the TC regime. The transition does not take place instantaneously).

Based on the bifurcation diagram in Fig. 2 and the snap shots in Fig. 1, the Archean mantle is suggested to have been on the CS regime in Paper 1: (a) Tectosphere, which is made of magma residue and hence is an example of chemical stratification in the mantle (Jordan, 1975), develops beneath continents older than 2 Gyr (Su et al., 1994; Stoddard and Abbott, 1996). In the numerical model of the coupled system, magmatism vigorous enough to induce chemical stratification in the mantle occurs only on the CS regime. (b) A study of Archean mid-oceanic ridge basalt suggests that the temperature in the Archean uppermost mantle was almost constant from 3.5 to 2 Gyr ago, in spite of the significant decay of radioactive elements during the period (Komiya, 1999; see also Abbott et al., 1994). This is also a feature of the CS regime. The constancy of the temperature in the uppermost mantle suggests that a significant portion of heat generated in the mantle is transported to the surface by migrating magma rather than by solid-state thermal convection in the mantle. Paper 1 also suggests that the thermal and chemical state of the mantle jumped from the CS regime to the TC regime in the early Proterozoic from the observation of the absence of tectosphere beneath the continents younger than 1.6 Gyr (Stoddard and Abbott, 1996).

3. Variable viscosity cases

To determine whether the above scenario of mantle evolution is the case, however, the CS regime must be shown to exist in more refined models of the mantle. In particular, a more realistic modeling of mantle rheology is crucial. Because of the assumption of constant viscosity, there is no lithosphere and, of course, no plate-like motion of the lithosphere in the numerical model of Section 2. Both the activity of mantle magmatism (Smrekar and Parmentier, 1996) and the vigor of convective stirring (Christensen and Hofmann, 1994) depend on whether or not the lithosphere and plate-like motion occur. As a first step toward this end, I include temperature- and pressure-dependent viscosity of mantle materials to reproduce the lithosphere in the numerical models. No weak zones (e.g. Zhong et al., 1998) and no yielding of lithosphere (e.g. Moresi and Solomatov, 1998) are included for simplicity here, and the lithosphere behaves as a stagnant lid rather than a moving plate. As a consequence, chemical differentiation at ridges and convective stirring by moving plates will not be included in the model. In particular, the mass exchange across the 660 km phase boundary due to subducting slabs that penetrate into the lower mantle (van der Hilst et al., 1997) will not be included in the model. However, we can still gain insights into the coupled system in the Earth's mantle by looking at both the numerical models presented in this section where there is a stagnant lithosphere and the numerical models in Section 2 where there is no lithosphere but, instead, the cold materials along the top surface boundary are involved in mantle wide convective circulation as the Earth's plate is involved in mantle convection (see Section 4 below).

3.1. Model description

A major difference between the model in this section and in Section 2 is that the viscosity depends on temperature, T, as well as depth, z, as

$$\eta = \eta_0 \exp[-E(T-T_0)+bz], \qquad (1)$$

where η_0, E, T_0 and b are constants taken to be 2×10^{21} Pa s, 0.0071 K^{-1}, 1573 K, and 1.7×10^{-6} m^{-1}, respectively. The value of E implies that a viscosity contrast of 10^4 arises when there is a temperature contrast by 1300 K, a typical temperature contrast across the lithosphere. If the convection is strictly thermal, therefore, it would occur beneath a highly viscous and stagnant lid that develops along the top surface boundary owing to this strong temperature dependence of viscosity (Ogawa et al., 1991; Solomatov, 1995).

The second major difference between the model presented here and the model of Section 2 is that the radioactive elements, which generate heat, are highly incompatible and decay exponentially with time. The partition coefficient of the elements is 100, the initial value of internal heating rate is 1.3×10^{-7} W/m^3, and the half-life of the elements is 1 Gyr. The initial value is slightly higher, and the half-life is shorter than the corresponding values appropriate for the Earth 10^{-7} W/m^3 and about 2 Gyr, respectively (Turcotte and Schubert, 1982). I assumed these values to search for a wider range of internal heating rate in a run where time integration is carried out for about 5 Gyr. I assumed that the heat bath beneath the mantle referred to in Section 2 also contains the same amount of radioactive elements to keep heat flow into the mantle from the heat bath; I found from several numerical experiments that the numerical results did not depend sensitively on the choice of internal heating rate in the heat bath provided that the heat capacity of the heat bath was comparable to that of the mantle.

The third major difference is that the problem is solved as an initial value problem in this section, while the calculation is continued until the thermal and chemical state in the mantle becomes statistically steady in Section 2. The initial condition is a chemically homogeneous and hydrostatic state with the initial temperature calculated as the minimum of the solidus temperature (see Fig. 3b below) and 2173 K. A rather high value is chosen for the initial temperature to simulate the hot origin of the Earth (Stevenson, 1981).

A problem in the numerical simulation of the coupled system with variable viscosity is the extremely heavy computational load required to solve the momentum and continuity equations; several million time steps are necessary to carry out the numerical simulation described below, and hence the momentum and continuity equations must be solved millions of times in a run. To reduce the computational load and make the calculation practical, I calculated the coefficients in the discretized momentum equation, which changes with time owing to the temperature dependence of viscosity, once for every five time steps when the time step calculated from the stability limit for the calculation of magma migration is much shorter than the time step calculated from the stability limit for the calculation of convective flow of matrix. I confirmed that the numerical results are still reliable by comparing the results calculated in this way with the results obtained when the coefficients of discretized momentum equation are renewed every time step.

3.2. Results

I present an example, called Case B hereafter, of numerical results in Figs. 3 and 4; Fig. 3 shows the snap shots of the distributions of temperature, melt, chemical composition, and internal heating rate, while Fig. 4 shows the plots of average temperature, root-mean-square velocity, magma eruption rate, and the flux across the 660 km phase boundary against time both obtained for Case B. The Rayleigh number defined with η_0 in (1) and the initial temperature contrast of 1873 K is 1×10^7. The Clausius–Clapeyron slope of the 660 km phase boundary is -0.7 MPa/K, considerably smaller in magnitude than the realistic value suggested to be -2 MPa/K (Bina and Helffrich, 1994) to -3 MPa/K (Akaogi and Ito, 1993). The barrier effect of the 660 km phase boundary has been suggested to be partly canceled out by the effect of positive slope of the garnet versus perovskite transition in the Earth's mantle (e.g. Akaogi and Ito, 1999), and hence, I calculated under the condition where the barrier effect was considerably weaker than expected from the Clausius–Clapeyron slope of the 660 km phase boundary alone. I confirmed by a numerical simulation that the convection occurs as a layered convection with flushing during the initial transient of elapsed time less than about 1 Gyr and then changes to a whole layer convection at the parameter values if the magmatism is turned off, and the convection occurs as a strictly thermal convection in the stagnant lid regime. The effect of garnet versus perovskite transition on mantle convection suggested by Irifune and Ringwood (1993) is neglected in Case B.

At the elapsed time, $t=0.04$ Gyr, a large portion of the upper mantle is partially molten (Fig. 3a) because of the assumed high initial temperature. Large volumes of basaltic materials enriched in radioactive elements and residue depleted in radioactive elements are formed (Fig. 3c and d). The basaltic materials sink to the bottom of the mantle, while a large portion of the residue remains in the upper mantle because of their chemical buoyancy. The residue could readily separate from the basaltic materials at the time of magmatic activity since the size of the residue was large, and the residue remained rather hot and hence remained at a low viscosity for a while after its formation. Though a small portion of residue is dragged into the lower mantle by the sinking basaltic materials, the residue eventually separates from the basaltic materials and uprise to the upper mantle.

During the subsequent period of $t=0.04$–0.97 Gyr, solid-state convection is layered, and a chemical discontinuity develops along the 660 km phase boundary (Fig. 3c and d). The upper mantle is occupied by residue depleted in radioactive elements, while the lower mantle is occupied by primitive materials and basaltic materials enriched in radioactive elements. (Here, any materials with the bulk composition close to the composition given in the initial condition are called 'primitive', even if the materials are actually a mechanical mixture of residue and basaltic materials on a spatial scale less than the resolution of the numerical simulation.) The discontinuity develops, owing to the homogenizing effect of convection within the upper and lower mantle. This process was first noticed by Weinstein (1992). Mass exchange across the 660 km phase boundary is suppressed by the phase boundary itself and the chemical buoyancy that accompanies the chemical layering; the barrier effect of the phase boundary alone does not completely suppress the mass exchange because of the assumed value of Clausius–Clapeyron slope of the phase boundary. Mild magmatism continues in the uppermost part of the convecting upper mantle, as can be seen from Fig. 3a and the small spikes in the curve of Fig. 4d. Because of the low efficiency of heat transport from the lower mantle to the surface in a layered mantle convection (e.g. Spohn and Schubert, 1982) and the strong internal heating in the early mantle, the temperature in the lower mantle gradually increases with time (Figs. 3b and 4a).

At $t=1.02$ Gyr, when the temperature in the lower mantle becomes high enough to induce partial melting at the shoulder of the horizontally averaged temperature indicated by the arrow in Fig. 3b for $t=0.97$ Gyr, a flushing event takes place. Primitive materials with a temperature of about 2500 K in the upper part of the lower mantle penetrate upward into the upper mantle (Fig. 3c and d), and a large portion of the upper mantle becomes partially molten (Fig. 3a). The entire mantle is convectively stirred (Fig. 3 for $t=1.08$ Gyr), and a large volume of basaltic materials and residue is formed (Figs. 3c and 4d). A high spike arises in the curve of heat flow (Fig. 4b), and the average temperature in the whole mantle drops, while the average temperature in the upper mantle rises (Fig. 4a) at the time of the flushing event.

Fig. 3. (a)–(c) Same as (a)–(c) of Fig. 1, respectively, for the case of temperature- and pressure-dependent viscosity and incompatible radioactive elements (Case B). The numbers at the base of (b) indicate the elapsed time in units of Gyr. (d) Distribution of the logarithm of the internal heating rate. See the scale bar at the top of the figure for the color scale of internal heating rate. The Clausius–Clapeyron slope of the 660 km phase boundary is -0.7 MPa/K, and the density inversion between basaltic material and primitive mantle material just beneath the 660 km phase boundary (Irifune and Ringwood, 1993) is not taken into account. The Rayleigh number defined with η_0 in Eq. (1) and the temperature contrast of 1873 K is 1×10^7.

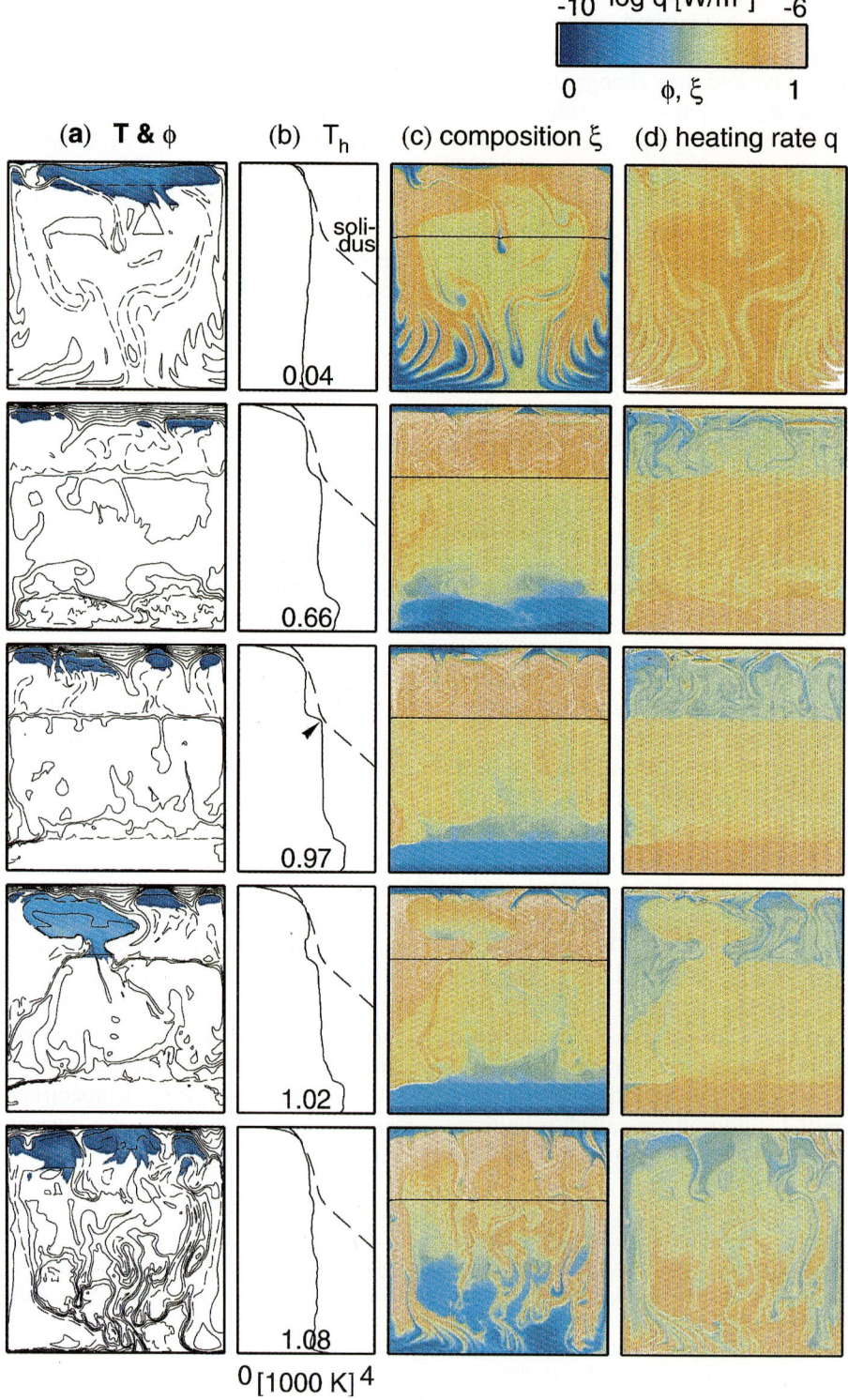

Fig. 3(a).

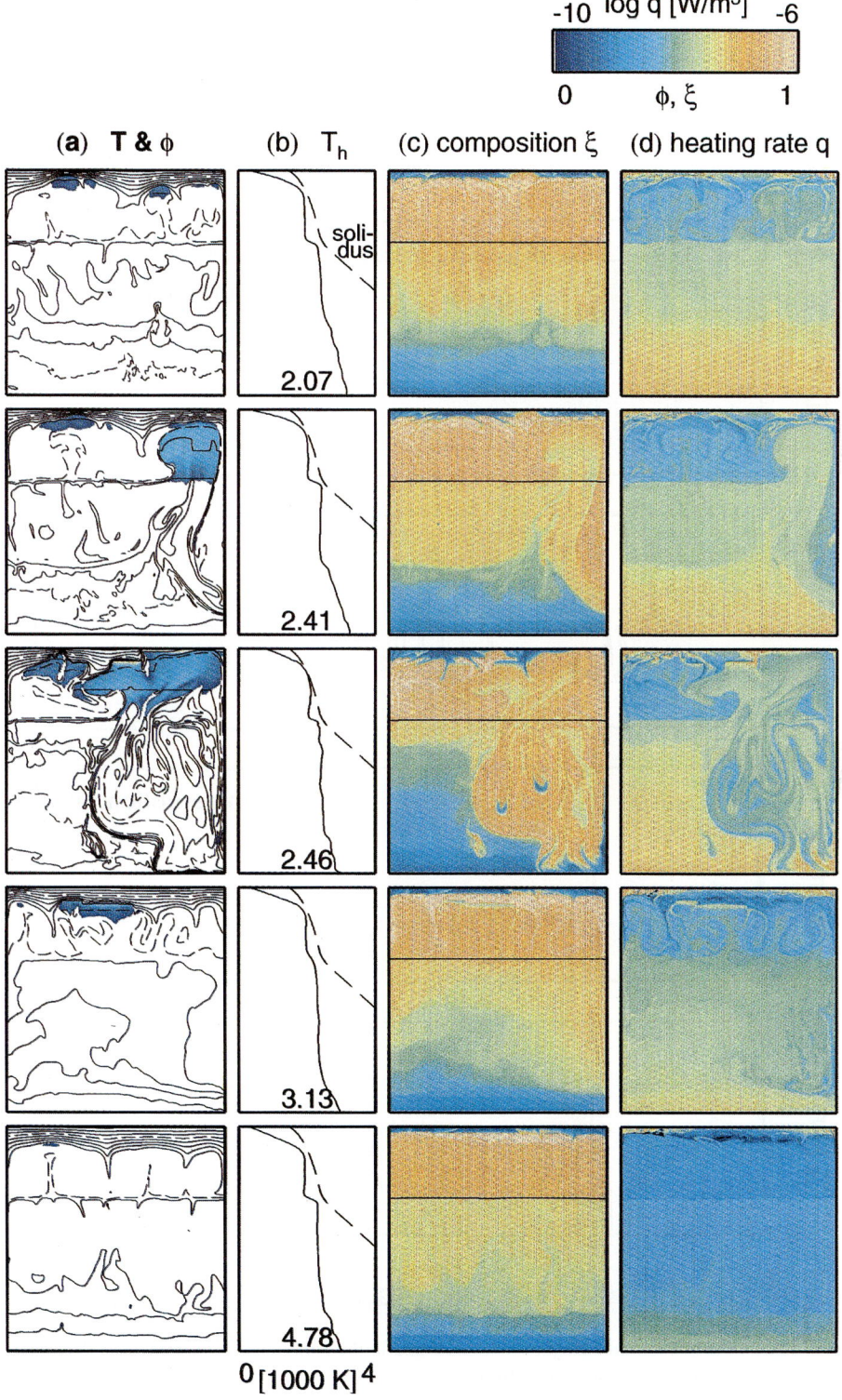

Fig. 3(b).

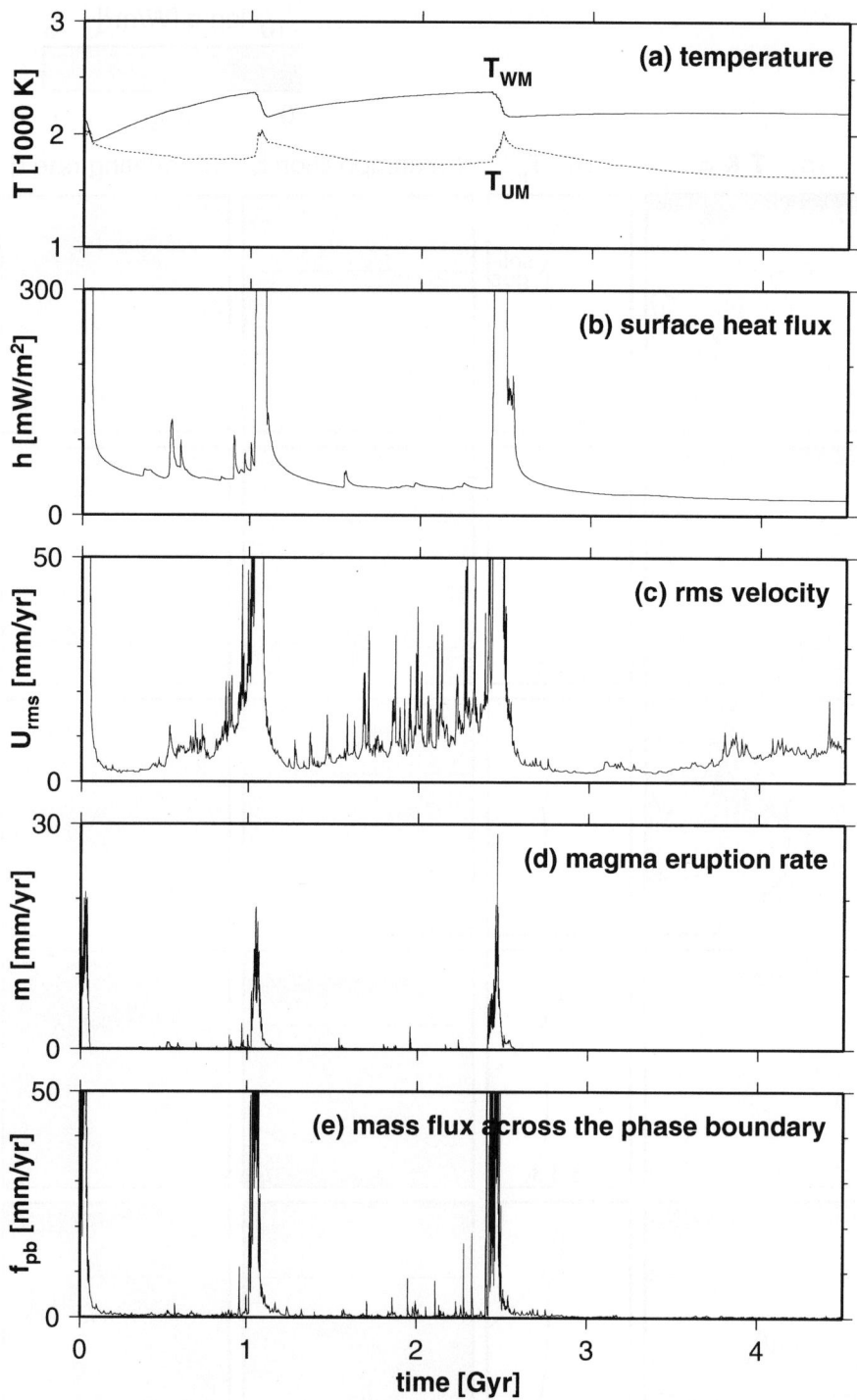

Fig. 4. (a) Average temperatures in the whole mantle (solid line), and in the upper mantle (dashed line), (b) surface heat flow, (c) root mean square velocity, (d) magma eruption rate, and (e) mass flux across the 660 km phase boundary plotted against time for Case B.

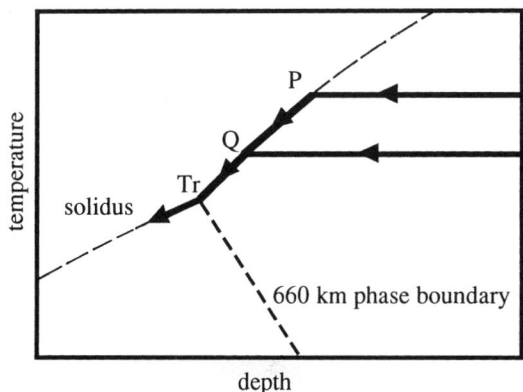

Fig. 5. Illustration of how temperature changes in two hot uprising mantle materials with different initial temperatures higher than the solidus temperature at the depth of the triple point 'Tr' at depth in the lower mantle. The temperature changes along the solidus curve above the depth level, Q, in both of the materials. The '660 km' phase transition, therefore, occurs at the same depth in the two uprising materials.

The flushing event at $t=1.02$ Gyr is a result of partial melting along the 660 km phase boundary. When material along the phase boundary becomes partially molten, the temperature and depth of the phase boundary are fixed at the temperature and depth of the triple point, 'Tr', illustrated in Fig. 5, respectively, and do not depend on the enthalpy of each material. Imagine, for example, two uprising mantle materials with different temperatures at depth in the lower mantle, as illustrated in Fig. 5. Partial melting occurs, and the temperature drops along the solidus curve in the materials above the points P and Q in the figure owing to the latent heat of melting. The 660 km phase transition occurs at the same depth, that is, the depth of the point, Tr, in the two uprising materials unless the end member, B, is exhausted by the time the materials arrive at the phase boundary. The barrier effect of the 660 km phase transition, therefore, disappears when partial melting occurs along the phase boundary. This is why partial melting along the 660 km phase boundary induces the flushing event in Figs. 3 and 4. The flushing event observed here is different from the flushing event due to the 660 km phase boundary alone found in Christensen and Yuen (1985) and referred to in Davies (1995).

When the turmoil induced by the flushing event subsides, again, the mantle becomes chemically layered with a chemical boundary along the 660 km phase boundary, mantle convection becomes layered, and mild magmatism occurs within the upper mantle (Fig. 3 for $t=2.07$ Gyr and Fig. 4d). The average temperature in the lower mantle increases with time, owing to the still strong internal heating (see the green to orange color in the lower mantle in Fig. 3d for $t=2.07$ Gyr), and a flushing event occurs again at $t=2.41-2.46$ Gyr, as can be seen from Figs. 3 and 4.

Because of the decay of the internal heat source (see the blue to green color in Fig. 3d for $t=3.13$ and 4.78 Gyr), however, the average temperature in the entire mantle does not rise significantly (Fig. 4a), and there is no further flushing event after the second flushing. Though partial melting occasionally occurs within the upper mantle (Fig. 3a for $t=3.13$ Gyr), the produced melt is cooled and solidifies within the lithosphere and does not rise to the surface, as can be seen from the absence of spikes in the curve of Fig. 4d at $t>2.46$ Gyr. Since chemical differentiation due to magmatism is negligible, both the upper and lower mantle become increasingly chemically homogeneous owing to convective stirring. Only the chemical contrast between the upper and the lower mantle and the chemical stratification in the uppermost mantle and just above the bottom boundary remain at $t=4.78$ Gyr (see Fig. 3c for $t=3.13$ and 4.78 Gyr). Because of the homogenization, the average velocity of convection gradually increases with time during the period of $t>2.46$ Gyr (Fig. 4c).

The active magmatism and the resulting development of chemical stratification before the second flushing strongly suggest that the mantle is on the CS regime of Section 2 at $t<2.46$ Gyr, while the absence of magma eruption, the convective homogenization, and the gradual increase in average velocity of mantle convection with time suggest that the mantle is on the TC regime after the second flushing event in Case B. Namely, the thermochemical state in the mantle of Case B is likely to have jumped from the CS regime to the TC regime just after the second flushing. (See the further discussion on the two regimes in Section 4.)

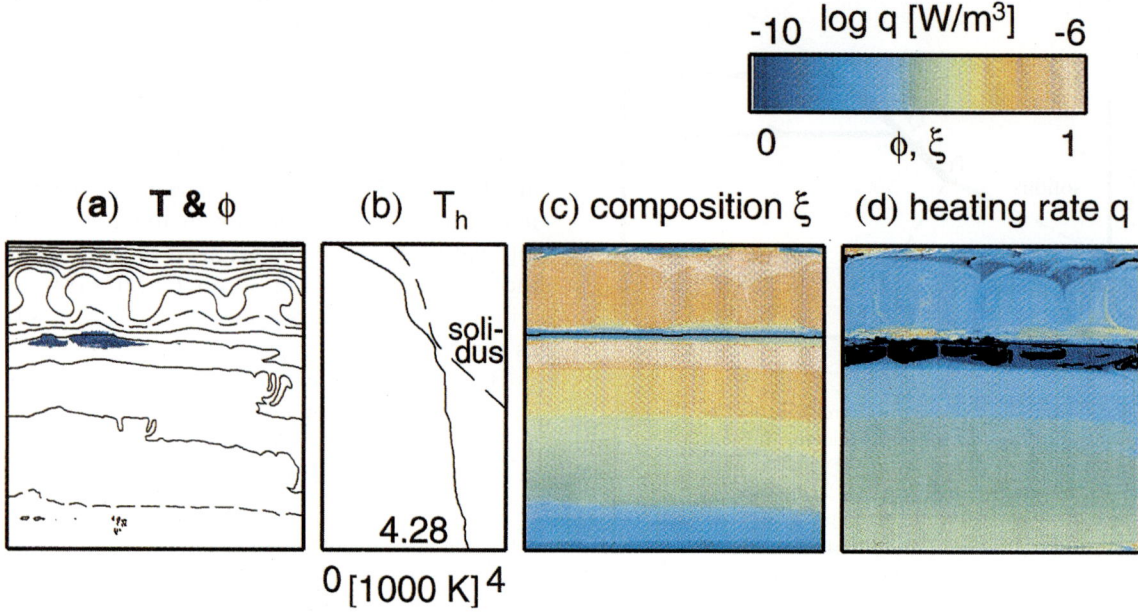

Fig. 6. Same as Fig. 3, but for Case C, where the Clausius–Clapeyron slope of the 660 km phase transition is −2.8 MPa/K and the density inversion between basaltic materials and primitive mantle materials at depths just beneath the 660 km phase boundary (Irifune and Ringwood, 1993) is taken into account. The thermochemical state in the mantle was observed to be on the CS regime at first and then was observed to jump to the TC regime as the internal heat source decayed in Case C, too. The snap shot was taken a long time after the jump from the CS regime to the TC regime.

I carried out a similar numerical simulation (hereafter, called Case C) with the Clausius–Clapeyron slope of the 660 km phase transition of −2.8 MPa/K, the value obtained for γ-spinel versus post-spinel transition by Ito and Takahashi (1989). The density inversion between basaltic material and primitive mantle material just beneath the 660 km phase boundary due to garnet versus perovskite transition (Irifune and Ringwood, 1993) is also included in Case C. I obtained the same overall features as in Case B shown in Figs. 3 and 4. Namely, I observed (a) chemically stratified mantle and layered convection punctuated by a flushing event, that is, the CS regime, when the internal heating is strong and (b) a transition from the CS regime to the TC regime that takes place owing to the decay of internal heat source. Some differences were, however, observed between the two results. First, the flushing event becomes more explosive and infrequent in Case C than in Case B and occurs only once in Case C. Second, a thin layer of basaltic materials develops along the 660 km phase boundary, as predicted in Irifune and Ringwood (1993), and a thin layer of highly depleted residue develops in the uppermost part of the lower mantle just beneath the basaltic layer in Case C, as shown in Fig. 6c. Because of these two thin layers, the compositional contrasts within the upper mantle and the lower mantle are larger in Case C than in Case B. Third, because of the stronger chemical buoyancy that accompanies the larger compositional contrast, mantle convection becomes much weaker in Case C than in Case B after the transition from the CS regime to the TC regime. In particular, the thermal convection in the lower mantle is almost completely suppressed by the chemical buoyancy in Case C, as can be seen from the almost horizontal isotherms and compositional distribution in Fig. 6a and c, respectively. A similar effect of the garnet-versus-perovskite transition is observed in the constant viscosity case of Paper 1, too, though the convection in the lower mantle is not completely suppressed (see fig. 10 of Paper 1).

4. Discussion

4.1. CS regime for the variable viscosity case

The CS regime originally identified for the coupled magmatism–mantle convection system with constant viscosity and without the 660 km phase transitions (Section 2) is most likely to exist for the coupled system with temperature- and pressure-dependent viscosity and the 660 km phase transitions, too, as suggested from the chemical stratification, the vigorous magmatism that accompanies flushing events, and the rather mild magmatism induced by the convective circulation within the upper mantle observed in Figs. 3 and 4 for $t < 2.46$ Gyr. This result is easy to understand. The chemical stratification on the CS regime observed in Section 2 (Fig. 1 for Case A5) develops as a result of a dynamic balance between chemical differentiation due to magmatism and stirring by solid-state convection. Convective stirring is much weaker when the viscosity strongly depends on temperature than when the viscosity is constant because the horizontal temperature contrast, and hence the thermal buoyancy in the mantle, becomes smaller as the temperature dependence of viscosity becomes stronger (e.g. Morris and Canright, 1984). Furthermore, the 660 km phase transitions impede mass exchange between the upper and the lower mantle. Chemical stratification, therefore, develops more easily as the viscosity depends more strongly on temperature, and the effect of 660 km phase transitions becomes stronger. Namely, the existence of the CS regime is a robust feature under changes in the magnitudes of dependence of viscosity on temperature and the strength of resistance to mass exchange between the upper and the lower mantle due to the 660 km phase boundary. The existence of the CS regime is likely a robust feature of the coupled system under the variation in the distribution of internal heat source coming from the incompatibility of radioactive elements, too. The implications for the Earth's mantle evolution suggested in Paper 1 and summarized in Section 2, therefore, still hold; the thermal and chemical state of the Earth's mantle is likely to have changed from the CS regime to the TC regime as the radioactive elements have decayed.

However, a caution is necessary here. (a) The basic equations that describe the coupled system are solved as an initial value problem in the numerical models described in Section 3. There is still a possibility that the thermal and chemical state observed in Fig. 3 for $t < 2.46$ Gyr is an artifact of the rather arbitrarily chosen initial condition. (b) The numerical simulations are carried out in a two-dimensional box with a small aspect ratio of 1. Both mechanical mixing and vigor of magmatism are likely to depend on the geometry of the convecting region. A numerical simulation of three-dimensional thermal convection (Schmalzl et al., 1996) suggests that the mechanical mixing is somewhat less efficient in three-dimensional space than in two-dimensional space. An analogy from numerical simulations of flushing events due to the 660 km phase boundary in a three-dimensional spherical shell (Tackley et al., 1993) suggests that the impact of flushing events on the thermal and chemical state of the mantle is overestimated in Figs. 3 and 4 because of the small aspect ratio and the spatial dimensionality of the model. It is important to see how the coupled system behaves in a wider box and in a three-dimensional spherical shell. (c) Plate motion is not included in the models. Moving plates and subducting slabs that penetrate into the lower mantle (van der Hilst et al., 1997) induce convective stirring that extends from the surface boundary to at least the shallower part of the lower mantle (Kellog et al., 1999). Though the CS regime occurs in Case A5 where convective stirring is mantle-wide, hard plates and slabs may induce an even stronger convective stirring than the solid-state convection with constant viscosity does in Case A5. (d) Continental crust, which is a major chemical reservoir in the Earth, is not included in the numerical model. If continental crust is included in Case B, for example, the transition from the CS regime to the TC regime probably occurs at an earlier time since a significant amount of radioactive elements will be removed from the model mantle by magma migrating upward to the continental crust and will not contribute to heating up the model mantle. Continental crust is, however, unlikely to affect the existence of the CS regime itself. In a manuscript in preparation (M. Ogawa), I will confirm that the 'CS regime' for a variable viscosity case suggested in Section 3 is not an artifact of the assumed initial condition and will discuss how the coupled system behaves in a wider box in two-dimensional space. Numerical simulations of the coupled system with plate motion in a three-dimensional space are an important issue that should be addressed in future.

4.2. Implications for crustal growth and mantle evolution

The thermal and chemical evolution of the mantle and the growth history of continental crusts heavily depend on the nature of mantle convection. The numerical model presented here, therefore, contains many implications for our understanding of mantle evolution and crustal growth though, of course, caution is necessary here because of the items (b)–(d) in Section 4.1.

First, it is often conjectured in the studies of growth of continental crusts that mantle convection was much more vigorous in the early Earth than it is today and that the vigor of mantle convection declined with time as the internal heating rate declined owing to the decay of radioactive elements (see Sylvester, 1998 for a review). The conjectured vigorous convection has been suggested to have induced rapid growth of continental crust in the early Earth, and the decline in the vigor of mantle convection has been suggested to have led to a decline in the rate of magmatic addition from the mantle to the continents (e.g. Armstrong, 1968). The conjectured vigorous convection in the early mantle has also been suggested to be at odds with the models of crustal evolution where the continental crusts have gradually grown throughout the Earth's history (Armstrong, 1991). The strong internal heating in the early mantle, however, probably made the early mantle fall on the CS regime where the mantle convection is much milder than expected from the strong internal heating (see Fig. 2b), as discussed above. Though there is still a large degree of uncertainty in the estimate of the vigor of mantle convection on the CS regime coming from the lack of plate motion in the numerical models, the models imply that the strong internal heating in the early mantle does not necessarily lead to vigorous mantle convection and hence to rapid growth of continental crusts.

Second, the numerical models presented here suggest that basaltic materials and magma residue easily separated from each other once they were formed by magmatism in the mantle of the early Earth because of the large size of residue formed by each of the magmatic activities on the CS regime (see Fig. 3 for $t=0.04$, 1.08, and 2.46 Gyr). Furthermore, the models suggest that basaltic materials and residue do not easily mix once they are separated in the Earth's mantle. In the model of Cases B and C, basaltic materials stagnate at depth in the lower mantle and significant mechanical mixing of the basaltic materials and the residue and/or primitive materials in the shallower part of the mantle takes place only at the time of mantle overturn induced by flushing events on the CS regime. Mechanical mixing of magma residue and basaltic materials is suppressed by (1) the compositional buoyancy of these materials, (2) strong temperature dependence of viscosity, and (3) the 660 km phase transitions. The mechanical mixing is, though not completely suppressed, difficult even in the model of Case A5 shown in Fig. 1 where only the agent (1) works. Only a small portion of the basaltic materials at depth in the lower mantle is mixed with residue and/or primitive materials in the shallower part of the mantle, as indicated by the circle in Fig. 1 and hence, basaltic materials, once formed, spend a long time, in the order of a billion years, at depth in the lower mantle. Namely, geochemical reservoirs of magma residue and basaltic materials are likely to have developed in the mantle of the early Earth that was probably on the CS regime, and the reservoirs are likely to have kept their identity for a long time, even after the suggested jump from the CS regime to the TC regime.

It is an important, but still open, issue to clarify whether the mechanical mixing between basaltic reservoir and residual reservoir was gradual, as modeled in Case A5 (Fig. 1), or episodic, as modeled in Cases B and C (Fig. 3), if the reservoirs did develop in the early mantle of the Earth, as suggested above. The mechanical mixing would be episodic if the 660 km phase boundary strongly impeded mass exchange between the upper and the lower mantle, and mantle convection occurred basically as a layered convection. When mantle convection is layered, heat builds up in the lower mantle (e.g. Spohn and Schubert, 1982), and the region along the 660 km phase boundary becomes partially molten. The partial melting would induce sporadic injection of basaltic materials from their reservoir at depth in the lower mantle to the residual reservoir in the upper mantle. In the Earth, however, plate motion is likely to have occurred throughout the history of the Earth since the Archean, as suggested from geologic observations of old continents (e.g. de Wit et al., 1992), and subducted slabs have probably penetrated through the 660 km phase boundary, inducing heat and mass exchange between the upper mantle and lower mantle (Christensen, 1996). The phase boundary is, however, still likely to have worked as a partial barrier against the convective flow across the phase boundary by suppressing the flow other than that induced by plate motion (Yuen et al., 1994). Heat transport by subducting slabs and the associated return flow may not have been efficient enough to prevent partial melting along the 660 km phase boundary and the resulting flushing events that cause vigorous mechanical mixing between the two reservoirs in the early Earth.

Acknowledgements

The author thanks the encouragement by T. Urabe at the computer center of Nagoya University. The numerical experiments were carried out on VPP-500 of the computer center of Nagoya University. The author also thanks A. Leitch and an anonymous reviewer for their comments on the earlier version of this manuscript. This work is financially supported by a Grant-in-Aid for Scientific Research #10640401 by the Ministry of Education and Culture of Japan.

References

Abbott, D., Burgess, L., Longhi, J., Smith, W.H.F., 1994. An empirical estimate of the earth's upper mantle. J. Geophys. Res. 99, 13835–13850.

Agee, C.B., 1998. Crystal-liquid density inversions in terrestrial and lunar magma. Phys. Earth Planet. Inter. 107, 63–74.

Akaogi, M., Ito, E., 1993. Refinement of enthalphy measurementof $MgSiO_3$ perovskite and negative pressure–temperature slopes for perovskite-forming reactions. Geophys. Res. Lett. 20, 105–108.

Akaogi, M., Ito, E., 1999. Calorimetric study on majorite-perovskite transition in the system $Mg_4Si_4O_{12}$–$Mg_3Al_2Si_3O_{12}$: transition boundaries with positive pressure–temperature slopes. Phys. Earth Planet. Inter. 114, 129–140.

Armstrong, R.L., 1968. A model for Sr and Pb isotope evolution in a dynamic Earth. Rev. Geophys. 6, 175–199.

Armstrong, R.L., 1991. The persistent myth of crustal growth. Austral. J. Earth Sci. 38, 613–630.

Bina, C.R., Helffrich, G., 1994. Phase transition Clapeyron slopes and transition zone seismic discontinuity topography. J. Geophys. Res. 99, 15853–15860.

Christensen, U., Yuen, D., 1985. Layered convection induced by phase transitions. J. Geophys. Res. 90, 10291–10300.

Christensen, U.R., Hofmann, A.W., 1994. Segregation of subducted oceanic crust in the convecting mantle. J. Geophys. Res. 99, 19867–19884.

Christensen, U., 1996. The influence of trench migration on slab penetration into the lower mantle. Earth Planet. Sci. Lett. 140, 27–39.

Davies, G., 1990. Heat and mass transport in the early earth. In: Newsom, H.E., Jones, J.H. (Eds.), Origin of the Earth. Oxford University Press, New York, pp. 175–194.

Davies, G., 1993. Conjectures on the thermal and tectonic evidence of the Earth. Lithos 30, 281–289.

Davies, G., 1995. Punctuated tectonic evolution of the earth. Earth Planet. Sci. Lett. 136, 363–379.

de Wit, M.J., Roering, C., Hart, R.J., Armstrong, R.A., de Ronde, C.J., Green, R.W.E., Tredoux, M., Peberdy, E., Hart, R.A., 1992. Formation of an Archean continent. Nature 357, 553–562.

Ito, E., Takahashi, E., 1989. Postspinel transformations in the system Mg_2SiO–Fe_2SiO_4 and some geophysical implications. J. Geophys. Res. 94, 10637–10646.

Irifune, T., Ringwood, A.E., 1993. Phase transformation in subducted oceanic crust and buoyancy relationships at depths of 600–800 km in the mantle. Earth Planet. Sci. Lett. 117, 101–110.

Jordan, T.H., 1975. Lateral heterogeneity and mantle dynamics. Nature 257, 745–750.

Kellog, L.H., Hager, B.H., van der Hilst, R.D., 1999. Compositional stratification in the deep mantle. Science 283, 1881–1884.

Komiya, T., 1999. The secular variation of the composition and temperature of MORB-source mantle. Ph.D. thesis, Tokyo Institute of Technology, Tokyo, Japan.

Mckenzie, D., 1984. The generation and compaction of partially molten rock. J. Petrol. 25, 713–765.

Moresi, L., Solomatov, V., 1998. Mantle convection with a brittle lithosphere: thoughts on the global tectonic styles of the earth and Venus. Geophys. J. Int. 133, 669–682.

Morris, S., Canright, D., 1984. A boundary-layer analysis of Benard convection in a fluid of strongly temperature-dependent viscosity. Phys. Earth Planet. Inter. 36, 355–373.

Ogawa, M., Schubert, G., Zebib, A., 1991. Numerical simulations of three-dimensional thermal convection in a fluid with strongly temperature-dependent viscosity. J. Fluid Mech. 233, 299–328.

Ogawa, M., 1997. A bifurcation in the coupled magmatism–mantle convection system and its implications for the evolution of the Earth's upper mantle. Phys. Earth Planet. Inter. 102, 259–276.

Ogawa, M., Nakamura, H., 1998. Thermochemical regime of the early mantle inferred from numerical models of the coupled magmatism–mantle convection system with the solid–solid phase transitions at depths around 660 km. J. Geophys. Res. 103, 12161–12180.

Ohtani, E., Sawamoto, H., 1987. Melting experiment on a model chondritic mantle composition at 25 GPa. Geophys. Res. Lett. 14, 733–736.

Oxburgh, E.R., Parmentier, E.M., 1977. Compositional and density stratification in the oceanic lithosphere — causes and consequences. J. Geol. Soc. Lond. 133, 343–354.

Schmalzl, J.G., Houseman, A., Hansen, U., 1996. Mixing in vigorous, time-dependent three-dimensional convecting and application to Earth's mantle. J. Geophys. Res. 101, 21847–21858.

Schubert, G., 1979. Subsolidus convection in the mantles of terrestrial planets. Ann. Rev. Earth Planet. Sci. 7, 289–342.

Smrekar, S.E., Parmentier, E.M., 1996. The interaction of mantle plumes with surface thermal and chemical boundary layers: Applications to hotspots on Venus. J. Geophys. Res. 101, 5397–5410.

Solomatov, V.S., 1995. Scaling of temperature- and stress-dependent viscosity convection. Phys. Fluid 7, 266–274.

Spohn, T., Schubert, G., 1982. Modes of mantle convection and the removal of heat from the earth's interior. J. Geophys. Res. 87, 4682–4696.

Stevenson, D.J., 1981. Models of the Earth's core. Science 214, 611–619.

Stoddard, P.R., Abbott, D., 1996. Influence of the tectosphere upon plate motion. J. Geophys. Res. 101, 5425–5433.

Su, W., Woodward, R.L., Dziewonski, A.M., 1994. Degree 12 model of shear velocity heterogeneity in the mantle. J. Geophys. Res. 99, 6945–6980.

Sylvester, P.J., 1998. Formation of the continents — dribble or big bang? Newslett. Geochem. Soc. 94, 12–25, January.

Tackley, P., Stevenson, D.J., Glatzmaier, G.A., Schubert, G., 1993. Effects of an endothermic phase transition at 670 km depth in a spherical model of convection in the Earth's mantle. Nature 361, 699–704.

Takahashi, E., 1986. Melting of dry peridotite KLB-1 up to 14 GPa: implications on the origin of peridotitic upper mantle. J. Geophys. Res. 91, 9367–9382.

Turcotte, D.L., Schubert, G., 1982. Geodynamics: Applications of Continuum Physics to Geological Problems. Wiley, New York. 450 pp.

van der Hilst, R.D., Widiyantoro, S., Engdahl, E.R., 1997. Evidence for deep mantle circulation from global tomography. Nature 386, 578–584.

Weinstein, S.A., 1992. Induced compositional layering in a convecting fluid layer by an endothermic phase transition. Earth Planet. Sci. Lett. 113, 23–39.

Yuen, D.A., Reuteler, D.M., Balachandar, S., Steinbach, V., Malevsky, A.V., Smedsmo, J.L., 1994. Various influences on three-dimensional mantle convection with phase transitions. Phys. Earth Planet. Inter. 86, 185–203.

Zhong, S., Gurnis, M., Moresi, L., 1998. Role of faults, nonlinear rheology, and viscosity structure in generating plates from instantaneous mantle flow models. J. Geophys. Res. 103, 15255–15268.

TECTONOPHYSICS

www.elsevier.com/locate/tecto

Early formation and long-term stability of continents resulting from decompression melting in a convecting mantle

J. De Smet *, A.P. Van den Berg, N.J. Vlaar

Department of Theoretical Geophysics, University of Utrecht, P.O. Box 80.021, 3508 TA Utrecht, The Netherlands

Received 19 January 1999; accepted for publication 20 September 1999

Abstract

The origin of stable old continental cratonic roots is still debated. We present numerical modelling results which show rapid initial formation during the Archaean of continental roots of ca. 200 km thick. These results have been obtained from an upper mantle thermal convection model including differentiation by pressure release partial melting of mantle peridotite. The upper mantle model includes time-dependent radiogenic heat production and thermal coupling with a heat reservoir representing the Earth's lower mantle and core. This allows for model experiments including secular cooling on a time-scale comparable to the age of the Earth. The model results show an initial phase of rapid continental root growth of ca. 0.1 billion year, followed by a more gradual increase of continental volume by addition of depleted material produced through hot diapiric, convective upwellings which penetrate the continental root from below. Within ca. 0.6 Ga after the start of the experiment, secular cooling of the mantle brings the average geotherm below the peridotite solidus thereby switching off further continental growth. At this time the thickness of the continental root has grown to ca. 200 km. After 1 Ga of secular cooling small scale thermal instabilities develop at the bottom of the continental root causing continental delamination without breaking up the large scale layering. This delaminated material remixes with the deeper layers. Two more periods, each with a duration of ca. 0.5 Ga and separated by quiescent periods were observed when melting and continental growth was reactivated. Melting ends at 3 Ga. Thereafter secular cooling proceeds and the compositionally buoyant continental root is stabilized further through the increase in mechanical strength induced by the increase of the temperature dependent mantle viscosity. Fluctuating convective velocity amplitudes decrease to below 10 mm a^{-1} and the volume average temperature of the sub-continental convecting mantle has decreased ca. 340 K after 4 Ga. Surface heatflow values decrease from 120 to 40 mW m^{-2} during the 4 Ga model evolution. The surface heatflow contribution from an almost constant secular cooling rate was estimated to be 6 mW m^{-2}, in line with recent observational evidence. The modelling results show that the combined effects of compositional buoyancy and strong temperature dependent rheology result in continents which overall remain stable for a duration longer than the age of the Earth. Tracer particles have been used for studying the patterns of mantle differentiation in greater detail. The observed (p, T, F, t)-paths are consistent with proposed stratification and thermo-mechanical history of the depleted continental root, which have been inferred from mantle xenoliths and other upper mantle samples. In addition, the particle tracers have been used to derive the thermal age of the modelled continental root, defined by a hypothetical closing temperature. © 2000 Elsevier Science B.V. All rights reserved.

Keywords: Archaean; continental roots; continent-formation; mantle convection; pressure-release melting; secular-cooling

---

* Corresponding author. Present address: Philips Centre for Industrial Technology, P.O. Box 218/SAQ p518, 5600 Eindhoven, The Netherlands.

E-mail addresses: smet@geo.uu.nl (J. De Smet), berg@geo.uu.nl (A. Van den Berg), vlaar@geo.uu.nl (A. Vlaar)

0040-1951/00/$ - see front matter © 2000 Elsevier Science B.V. All rights reserved.
PII: S0040-1951(00)00055-X

1. Introduction

The oldest continental shields have been stable for several billions of years. The origin of these regions and the reason for their long term stability is still debated. From seismological evidence (Jordan, 1975; LeFevre and Helmberger, 1989; Anderson, 1990; Woodhouse and Trampert, 1995; Polet and Anderson, 1995) and interpretation of the gravity field (Doin et al., 1996), and surface heat flow data (Pollack and Chapman, 1977; Nyblade and Pollack, 1993; Rudnick et al., 1998) for continental regions it has been concluded that continents are underlain by thick roots of compositionally distinct material. Analysis of mantle xenoliths from different continental regions (Jordan, 1979; Griffin et al., 1996) has shown that the continental roots consist of residual material from pressure release partial melting of mantle peridotite.

We have developed a numerical mantle convection model to investigate the formation of continents during the Archaean and their subsequent thermo-mechanical evolution since the late Archaean and Proterozoic.

In previous work (De Smet et al., 1998, 1999) we presented results which focused on the formation and early evolution of the first several hundred million years after the initiation of continental formation. It was shown that, after initiation, continental roots may have grown rapidly be addition from below of depleted peridotite in relatively small scale (± 50 km), hot and melt-producing diapiric upwellings which intermittently penetrate the growing continental root.

Here we present results of numerical modelling which deal with the long term evolution of continents from their formation in the early Earth until the present day. We also present an analysis of detailed temporal and spatial evolution of tracer particles which sample several physical quantities in the diapiric melting process.

Finally we have investigated the evolution of the 'thermal age' of evolving continental roots, defined as the elapsed time since a sample of mantle rock has cooled below a hypothetical closing temperature. The results show a vertical layering in thermal age related to the mechanism of growth from below. This layering is disturbed laterally by thermal rejuvenation by hot diapiric upwellings which overprint existing thermal age during the later evolution.

2. Model description

The model used here is identical to one of the models (Model A) described in more detail by De Smet et al. (1999). Important features of the model are: the use of a melting phase diagram for mantle peridotite with linear solidus and liquidus which practically limits occurrence of partial melting to the upper mantle above the transition zone at ca. 400 km depth. The mantle model is truncated at the upper to lower mantle boundary at a depth of 670 km. Thermal coupling between the upper mantle and the lower mantle and core is important in the thermal evolution models which operate on a time scale of several billion years where secular cooling of the Earth is significant. Therefore, we have extended the upper mantle model with a simple isothermal heat reservoir which accounts for the influence of lower mantle and core on the secular cooling of the upper mantle. We refer to De Smet et al. (1999) for a detailed discussion of the boundary conditions. The applied numerical modelling techniques (De Smet et al., 2000) consist of finite element solutions for the temperature T and the convective velocity field u. The degree of depletion F is defined as the mass fraction of melt extracted from the partially melted mantle material. Due to the occurrence of small (km) scale strength the F field is computed on a high resolution (sub-km scale) structured grid. Passive particle tracers are advected with the mantle flow in order to monitor the F, T, pressure p, and thermal age t_{close} for several hypothetical closing temperatures T_{close}.

2.1. Time convention

For elapsed model time we use Ma or Ga where the model evolution starts at $t=0$ Ma. In some discussions b.p. or Ga b.p. is used, meaning the time before present, that is, this is the age of the model measured from a present day situation back in time. We used for the present day model situation a model evolution time of 4000 Ma, that is, 0 Ma b.p. = 4000 Ma (end of the model computations) and 4000 Ma b.p. = 0 Ma (start of the model computations). See Section 3.4 for the application of this choice.

3. Modelling results

3.1. Model evolution up to 4 billion years

Fig. 1 shows the long-term evolution of the continental upper mantle model. The left-hand-side column shows snapshots of the compositional field from 0.7 to 4.0 Ga. On top of the depleted mantle the lower and upper crustal layers are represented as black and white areas, respectively. At corresponding times the lateral variations in the temperature fields, that is, $T-\langle T \rangle_{hor}$, are depicted in the right-hand-side column of Fig. 1. The extreme temperature values are not actually attained, which can be seen when the shown temperature variations at 2500 Ma are compared with the actual temperature field as given in Fig. 9 for 1500 Ma b.p. (the vertical scales are not identical for both figures). All frames for both the temperature and composition also contain instantaneous stream-lines of the convective velocity. Black and white lines correspond to clock and counter-clockwise flows, respectively.

Following the evolution in the F-field from 0.7 to 2.5 Ga we see that the total volume of depleted continental root increases. After 2.5 Ga, this increase of the depleted volume has stopped. Mantle differentiation has apparently ceased between 2.5 and 3.0 Ga.

The transition from root to mantle is set at the depth where $F \approx 10\%$. Up to ca. 2.0 Ga the compositional layering is relatively undisturbed and the root extends to depths varying between 150 and 300 km. Hot upwellings are sporadically penetrating the continental root after a model time of ca. 2 Ga. This is related to the increase in the occurrence of delamination of the continental root, which in turn results from the advancing cooling from the top down of the upper mantle system. Hot diapiric upwellings underneath old continental shields do occur at present (e.g., Yellowstone hotspot in North-America) or in the past in Africa (Ebinger and Sleep, 1998). A catastrophical large-scale overturn is, however, not observed during the investigated time-window. Depleted continental root delaminates by relatively small thread-like structures, which sometimes reach the upper to lower mantle transition depth. As a result, spirals of delaminated depleted material develop slowly as is observed in the F-field at 4.0 Ga.

From the decrease in the density of stream-lines in Fig. 1 with proceeding evolution it is concluded that the vigour of convection is gradually decreasing. Despite these sluggish convection rates, the compositional state of the upper mantle does show significant variations during the last 1 billion years of evolution shown.

The lateral variations of temperature in the right-hand-side column of Fig. 1 show that cold areas coincide with thick depleted continental roots. The maximum temperature is 2216°C at $t=$ 1.5 Ga and decreases to 2058°C at $t=4.0$ Ga. Most thermal anomalies have amplitudes of ca. 300°C, and the maximum positive thermal anomaly of the shown snapshots is 616°C at $t=2.0$ Ga.

Fig. 2 shows the evolution of several global quantities. The volume average temperature in Fig. 2a illustrates that secular cooling occurs at an almost constant rate. The drop in the volume averaged temperature due to cooling from the top is 340°C from 1740 to 1400°C.

Fig. 2b shows the evolution of the volume averaged degree of depletion $\langle F \rangle$. After the initial generation of the continental root, completed at ca. 50 Ma, three episodes follow during which the total volume of depleted material increases. The onset of these partial melting periods is marked with arrows. In spite of the overall secular cooling, upwellings apparently still cross the solidus. During the last 1.2 Ga of evolution the value of $\langle F \rangle$ remains constant because mantle differentiation has stopped after 2.8 Ga of evolution. If we assume that the initiation of continent formation in the model corresponds to 4.0 Ga b.p., the melt production stopped 1.2 Ga b.p., that is, the later Proterozoic. Several suggested continental growth curves (Windley, 1995) show a major surge of continental growth during the Mid and Late

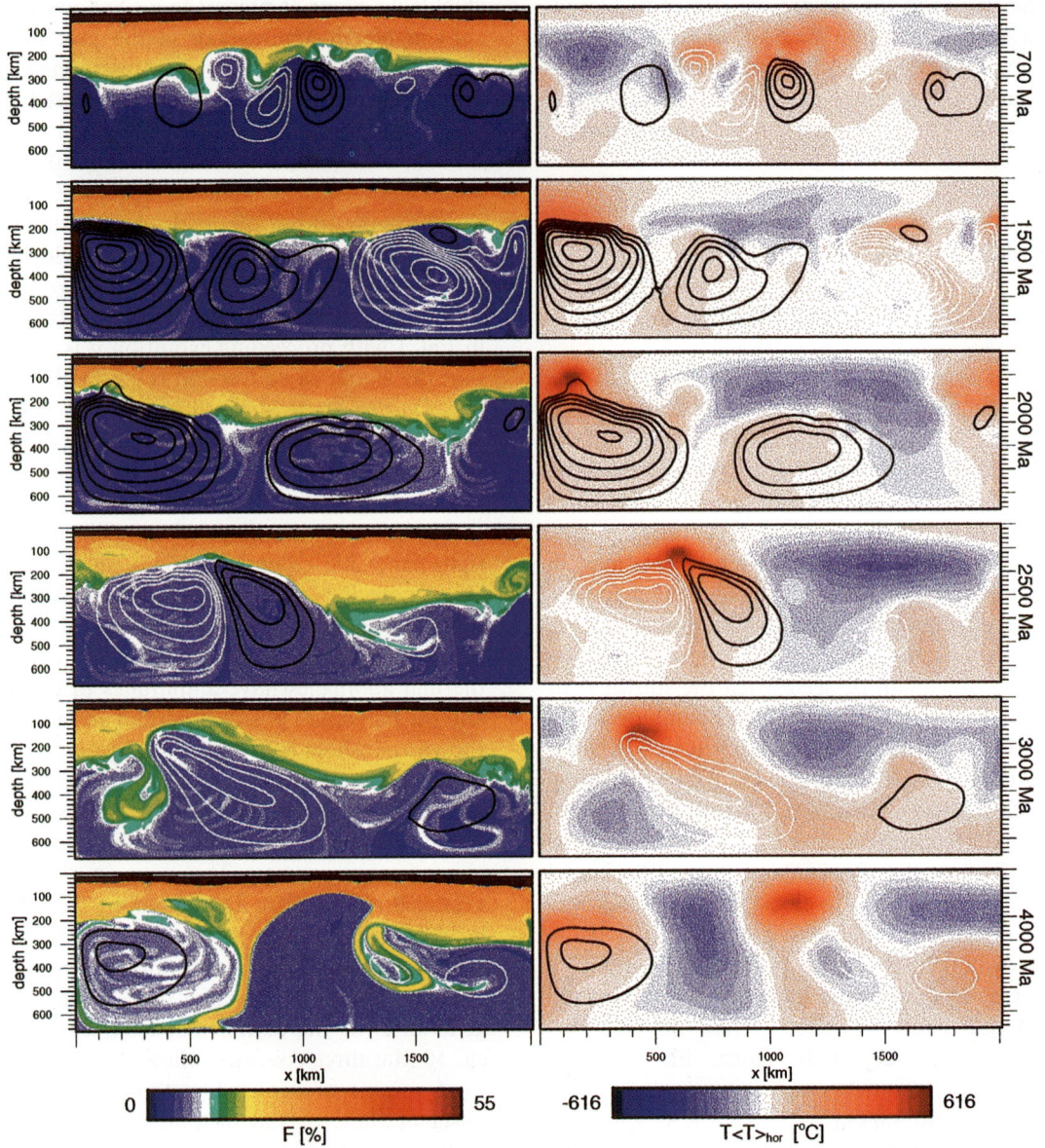

Fig. 1. The left-hand-side column shows the F-field. Evolution times are given at each frame. The white and black layers at the top are the upper and lower crustal layers, respectively. The white F-field below the crust corresponds to $9\% < F < 10\%$. The corresponding lateral variations of the temperature field (i.e., $T-\langle T\rangle_{hor}$) are shown in the right-hand-side column. The black and white contour lines in all frames indicate clock and counter clockwise flows, respectively. See Section 2.1 for the time use convention.

Archaean followed by a more slowly and steadily increase in crustal volume.

Fig. 2c illustrates that convection rates slowly decrease. Besides, the amplitude of the variations in the V_{rms} decrease with time. This is caused by the progressive cooling, which results in a significant increase in the overall viscosity.

Fig. 2d shows the averaged heat flow through the Earth's surface q_0. Starting at high values of ca. 120 mW m^{-2}, values decrease to ca.

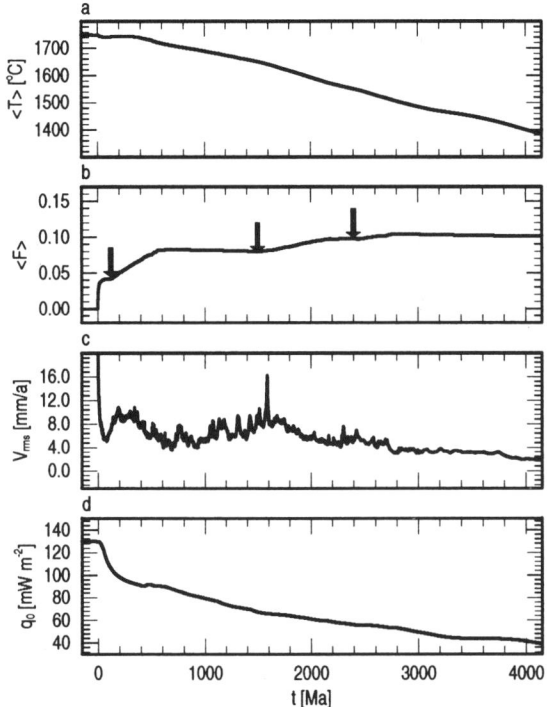

Fig. 2. Evolution of some volume averaged quantities. (a) Temperature. (b) Degree of depletion. (c) Root-mean-square velocity for which the off-scale part is shown in De Smet et al. (1998). (d) Average surface heat flow value.

40 mW m^{-2} after 4.0 Ga. Such a low surface heat flow value is reasonable for a present-day cratonic shield area (Chapman, 1986; Pollack et al., 1993). The steep decrease in q_0 during the early evolution of the model is followed by a much more steady decrease.

Fig. 2a suggests an almost constant cooling rate of the upper mantle model. On the other hand the surface heatflow decreases significantly during the 4 Ga evolution. From these curves it can be deduced that the surface heat flux is mainly produced by radiogenic heat sources, concentrated at shallow level which decrease due to radioactive decay with an effective half life time of ca. 2 Ga De Smet et al. (1999). A first order estimate of the heart flow component due to secular cooling of the upper mantle model follows from the drop in the volume averaged temperature of 340 K which represents a thermal energy decrease of $\Delta Q = 8 \times 10^{14}$ J m^{-2} in an average vertical column of unit cross section. Assuming a constant secular cooling rate in line with Fig. 2a this amounts to an average surface heatflux of 6 mW m^{-2} for the 4 Ga time window under consideration. This value corresponds well with interpretations of heat flow observations for the Canadian shield by (Jaupart et al., 1998) who report mantle heatflow contributions between 7 and 15 mW m^{-2}.

Fig. 3 shows several horizontally averaged quantities at four moments in time. Fig. 3a shows the horizontally averaged profiles for F. Due to delamination, vertical redistribution of depleted rock occurs. This results in an increase in $\langle F \rangle_{hor}$ in the 300–650 km depth range.

Fig. 3b shows the horizontally averaged temperature of the upper mantle model. Although the geotherm is below the solidus (dashed), partial melting events take place in upwellings associated with thermal anomalies that cross the solidus. After prolonged evolution, a thermal boundary layer (TBL) develops at the bottom. This is due to the thermal inertia of the heat reservoir of the lower mantle and core, and the applied impermeable lower to upper mantle transition, which buffers the cooling of the upper mantle. The decrease in temperature of the shallow model region is enhanced by the decrease in radiogenic heating, which is concentrated in the crustal layers.

Fig. 3c shows a dramatic decrease of the horizontally averaged convection rates from a maximum value of ca. 10 mm a^{-1} at $t = 1.2$ Ga to 2 mm a^{-1} at 4.0 Ga. Also the vertical variations in convective velocities diminish with proceeding evolution.

Fig. 3d shows the evolving viscosity profile. The strong increase in viscosity is the main cause for the decrease in V_{rms}. The decrease in averaged mantle temperature results in a less pronounced minimum viscosity in the asthenospheric layer underneath the growing MBL. The minimum value of the viscosity increases considerably by approximately two orders of magnitude from 5×10^{19} to 5×10^{21} Pa s over 3 Ga. The shallow region of the continental root is stabilized by the growth of the MBL at the top. The slowly evolving TBL at the bottom causes an inversion in the deeper upper mantle viscosity profile.

Fig. 4 shows the distribution of melt produc-

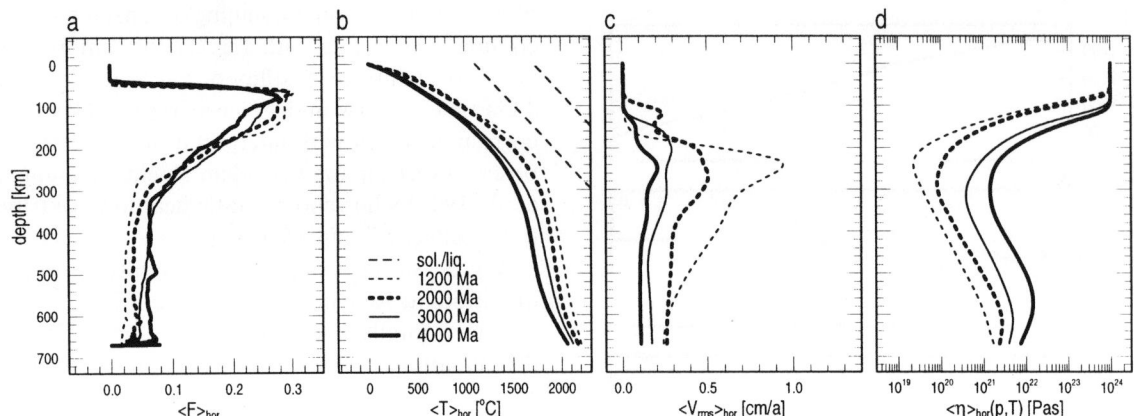

Fig. 3. Horizontally averaged profiles at several stages of the model evolution. (a) Degree of depletion. (b) Temperature. (c) Velocity root-mean-square. (d) Viscosity. The dashed curves in (b) are the solidus and liquidus equilibrium lines.

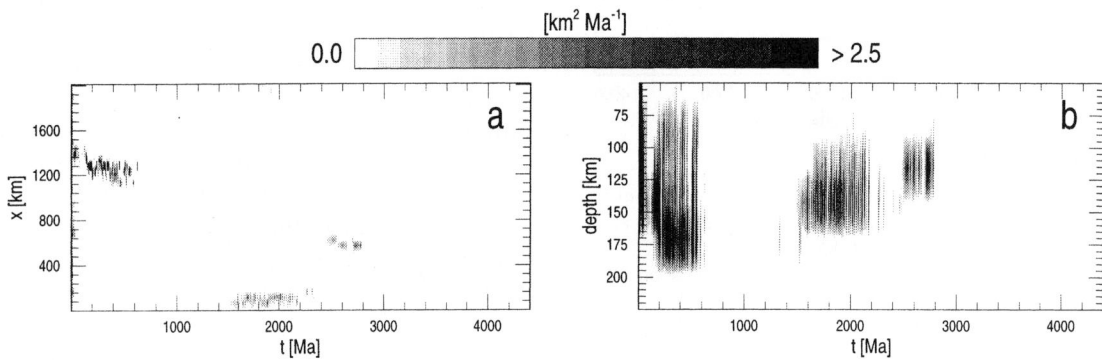

Fig. 4. The horizontal (a) and the one-dimensional depth (b) melt production distribution as a function of time. Note the three long distinct periods of partial melting (a+b), and the decrease in time of the depth range of melt production (b).

tivity versus time integrated over vertical columns (Fig. 4a) and horizontal rows (Fig. 4b). Up to 2.8 Ga three episodes of upper mantle differentiation can be recognized from 120–600 Ma, 1.5–2.2 Ga, and 2.5–2.8 Ga. Apparently, the continental upper mantle system experiences quiescent periods during which no intermittent melting occurs. During evolution the maximum depth of melting decreases [Fig. 4b] because of the decrease in mantle potential temperature with time. At the same time the growth of the MBL increases the minimum depth to which melt producing diapirs can rise.

3.2. Detailed dynamics of diapiric partial melting

In order to investigate the patterns of mantle differentiation in greater detail we applied a set of tracer particles. Each single tracer in the set tracks the histories of the degree of depletion F, the temperature T, and its position (x, z).

Fig. 5 shows several enlargements of a region where mantle differentiation occurs. The time-window from 401 to 500 Ma is part of the first period of intermittent melting events shown in Fig. 4. The positions of nine tracers are shown for eight snapshots, which are unevenly spaced in

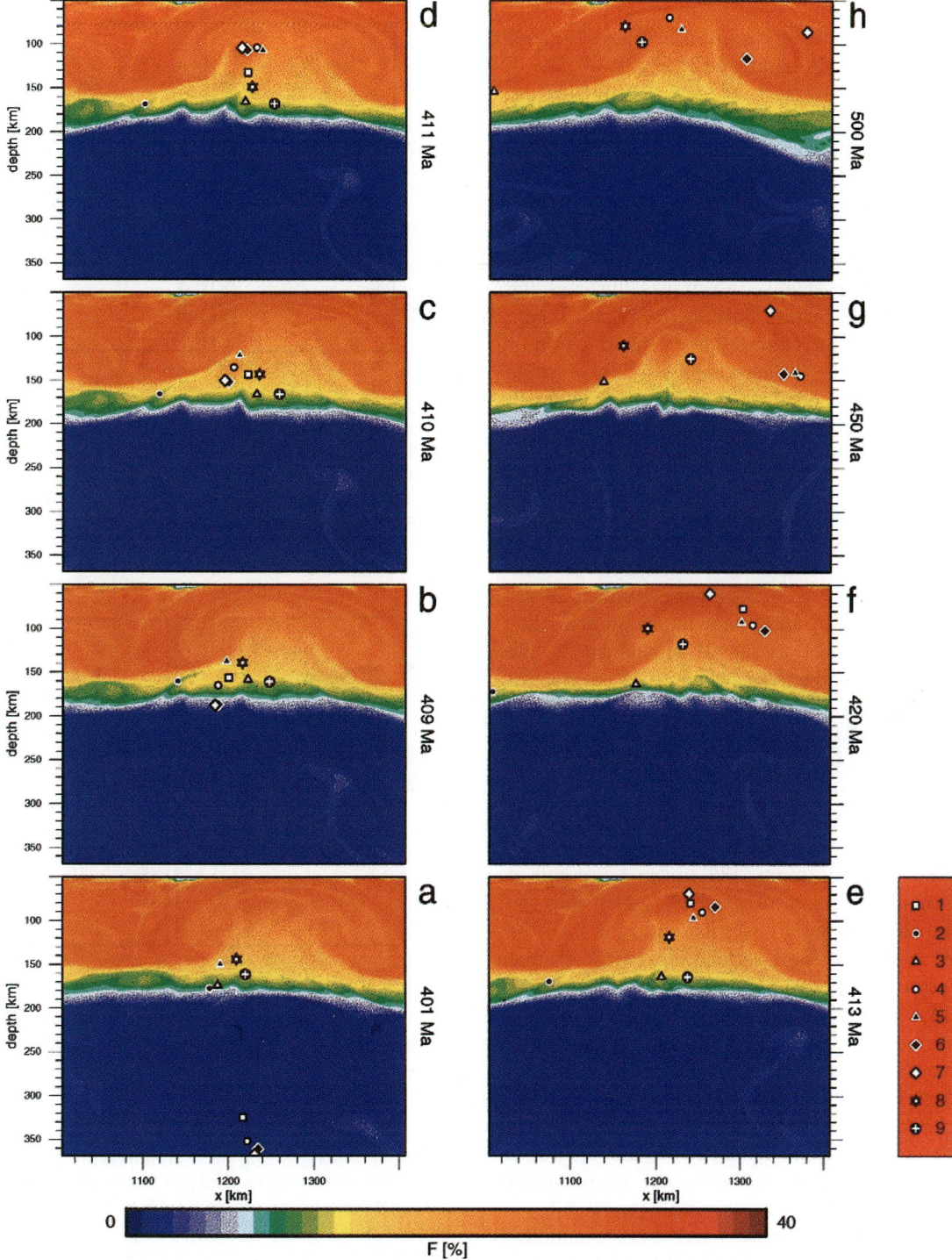

Fig. 5. Several enlargements of F-field snapshots illustrating the evolution of the continental root where small-scale melting diapirs impinge the depleted layer between 401 and 500 Ma. The material that melts comes from the undepleted deeper region as inferred from the flow pattern the nine tracers. This melting event corresponds to the first long period of the model evolution during which episodically melting occurs and melting starts at relatively large depths (see also Fig. 4). Note that the full available grid resolution was used in these contour plots. See Section 2.1 for the time use convention.

time. The relation between tracer symbols and corresponding numbers is given in the red legend at the bottom right-hand-side of Fig. 5.

In Fig. 5a four tracers (1, 4, 6, 7) are situated in the undepleted mantle at ca. 350 km depth at $t=401$ Ma. Following the tracer positions in time shows that they are part of a hot melt producing diapir that impinges on the continental root (Fig. 5b–h). They reach depths between 60 and 100 km and spread both laterally and vertically in the root. Their relative positions illustrate the vorticity of the mantle flow in the region where melting occurs. Furthermore, the central region of the mantle diapir, which exhibits a low F value during early ascend does not differentiate much further during the remaining ascend. This is illustrated by the F-values of tracers 1 and 4 in Fig. 5c and e.

The tracers 2 and 3 are in close proximity to each other at $t=401$ Ma (Fig. 5a) where they are part of the less depleted deeper part of the root. At $t=420$ Ma (Fig. 5f they have a much larger separation. During the time-window shown, they are mainly subject to lateral movement in the deeper part of the root where their degree of depletion is not altered.

Tracers 8 and 9 are subject to recurrent melting. Fig. 6a and b shows the (p, T, t)- and (p, F, t)-paths for tracer number 9 for the 401–500 Ma time-window. At point A the tracer has a low value of F and it slowly ascends to point B ($t=413$ Ma). At this point, the tracer represents a small mantle volume with $F=10\%$ which starts to rise much faster and crosses its solidus at $F=19\%$. Here, a relatively slow vertical movement around a depth of 125 km (4.2 GPa) is observed until $t=450$ Ma, when a further ascent to point D begins. Recurrent melting from C to D increases the degree of depletion to 26%. The uprise continues more slowly and without melting to point E after which the tracer descends to higher pressure (point F). It circulates back upward to point G, but is too cold to initiate another cycle of recurrent melting.

Three (p, T, t)-paths and corresponding (p, F, t)-paths are given in Fig. 7a and b, respectively. Each tracer is approximately situated in the centre of a melting diapir. They correspond to the

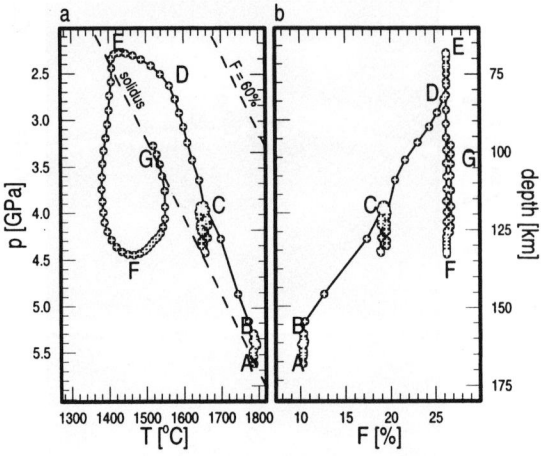

Fig. 6. (a) (p, T, t)-path for tracer that represents a mantle volume subject to recurrent melting. Time increment between symbols as indicated in the paths is 1 Ma. (b) Corresponding (p, T, t)-path. The symbols used in (a) and (b) corresponds to the tracer depicted with the same symbol in the contour plots for the F-field in Fig. 5.

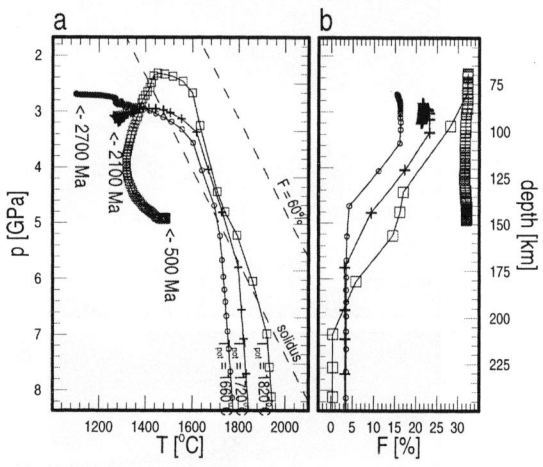

Fig. 7. (a) (p, T, t)-paths for three tracers. They correspond to tracers approximately positioned in the heart of different melt producing diapirs that impinge on the continental root. The final position of each tracer for the displayed time-window corresponds to the indicated time. Indicated times are model times (see Section 2.1). The elapsed model time between the symbols is 1 Ma. The hot and cold tracer correspond to the first and last period of melting, respectively. Tracer potential temperatures are indicated next to each (p, T, t)-path and the horizontal axis is the actual temperature. The time increment between symbols is 1 Ma. (b) (p, T, t)-paths corresponding to (a). The tracer path indicated with open squares in (a) and (b) is the tracer depicted with the same symbol in the contour plots for the F-field in Fig. 5.

three main periods of differentiation (see Fig. 4). The final position of each tracer for the displayed time-window corresponds to the indicated model time. The path marked with open squares corresponds to tracer 1 which is indicated with the same symbol in Fig. 5.

The subsolidus ascend of material occurs adiabatically and the corresponding potential temperatures T_{pot} are given in Fig. 7a. The mantle temperature decreases with time and partial melting therefore starts at shallower depth as time progresses, since the solidus is crossed at lower pressures. A hotter upwelling is capable of producing more melt over a larger depth interval and reaching higher degrees of depletion (Fig. 7b). This was also observed in the melt productivity shown in Fig. 4. When the upper mantle cools further, the MBL at the top of the mantle grows (Fig. 3d), which prevents the diapirs from penetrating the root to shallow levels. The presence of the MBL at the end of the melting path prevents further ascend after which the tracer becomes part of the stagnant layer and cools further by conductive heat loss. This is consistent with the (p, T, t)-paths derived for upper mantle samples described by Roermund and Drury (1998).

Highly depleted material is positioned directly underneath the crust during early evolution. After cooling from the top this material is trapped and can be brought to the surface during later evolution in the form of xenoliths. Bernstein et al. (1998) describes such xenoliths that contain highly depleted mantle material ($\sim 40\%$) originating from depths of 1–2 GPa.

After melting has stopped, the depleted material becomes part of the continental root. It maintains its attained F-value except for small effects of numerical diffusion, which causes the slow decrease in depletion observed at low pressures for the tracer path marked with crosses in Fig. 7b. This numerical diffusion is an artifact of the numerical method (De Smet et al., 2000). The two cold tracers in Fig. 7b are already modestly depleted ($F=3.5\%$) during upwelling before reaching the solidus. This is caused by the mixing of delaminated depleted material in the deeper mantle due to the finite amount of numerical diffusion, which acts as an artificial mixing mechanism in the numerical model.

3.3. Comparison with theoretical geotherms

We compare the global and local thermal state of the upper mantle model at $t=4$ Ga with theoretically derived continental geotherms for a present-day cratonic situation. The heat flow through the surface of the model has dropped to an averaged value of 41 mW m^{-2} at this point in the evolution. Surface heat flow measurements for cratonic shields with an age of >3.5 Ga range between 30 and 50 mW m^{-2} (Pollack et al., 1993). Since the model heat flow value is within this interval after 4 Ga of evolution, we compare this model state with the theoretical cratonic thermal model used by Chapman (1986) and Pollack et al. (1993). This model is thought to be valid for the relatively shallow part of the shield where conductive heat transport is dominant.

The model crustal thicknesses and heat flow values in combination with the incorporated radiogenic heat generation at $t=4$ Ga are used as the parameters for the theoretical geotherm computation as given by Chapman (1986). Note that the heat productivity in the two-dimensional convection models is based on the data also given by this author. In the model the exponential depth distribution of the heat productivity in the upper crust is simplified to an equivalent uniform value. We therefore give two theoretical geotherms: one corresponding to the exponential distribution of radiogenic heating according to Chapman (1986) and one with the equivalent uniform value. Furthermore, at large depths advective heat transport becomes more important and the theoretical geotherm deviates significantly from the results derived from the convection model.

Fig. 8a–c shows three temperature profiles derived from the convection model and Fig. 8d is the horizontally averaged state of the model. They can be compared to the corresponding theoretical geotherms defined by Chapman (1986). The thin solid lines are for an exponential distribution of radiogenic heating in the upper crust. The surface heat flow of the model are given in each frame. Lateral variations of the lower and upper crustal

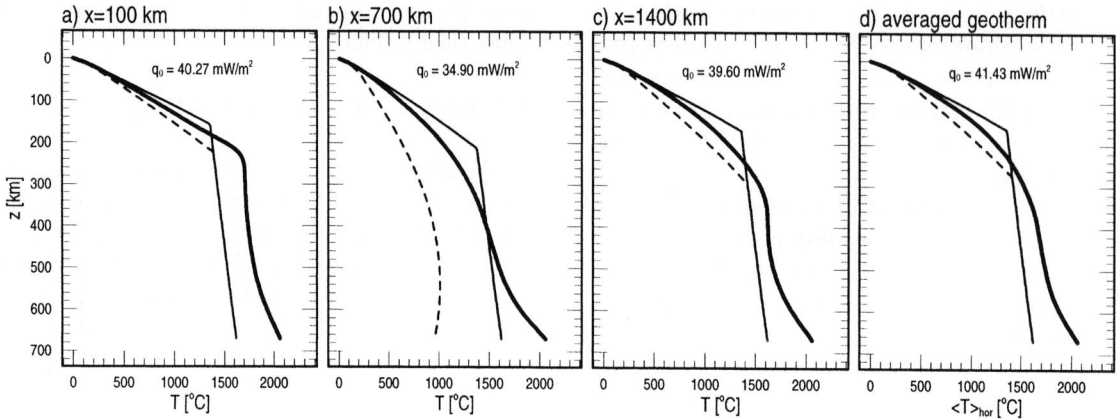

Fig. 8. (a–c) Geotherms derived from the convection model (thick solid line) at 4000 Ma at several x-locations as indicated above each frame. The last frame (d) shows the averaged temperature profile. The thin dashed and solid lines are theoretical continental geotherms as described by the model of Chapman (1986) where the heat flux at the surface (q_0) is obtained from the model. The deeper part of the theoretical curves are defined by an adiabatic geotherm corresponding to a present-day potential temperature of 1280°C (McKenzie and Bickle, 1988). The colder dashed geotherms corresponds to the uniform upper crustal radiogenic heat production at 4000 Ma as incorporated in the model. The solid thin line corresponding to an exponential distribution in the upper crust.

thicknesses in the model (see Fig. 1) are taken into account. It can be concluded from Fig. 8 that for shallow depths (<200 km) the geotherms derived from the convection model are in reasonable agreement with the continental geotherms according to Chapman (1986).

3.4. Age and compositional structure of continental roots

The mechanism of continental formation and evolution resulting from the mantle convection model presented here produces a specific structure of the continental root expressed in the thermally defined age and the composition. Samples of deep continental root material brought up as xenoliths and peridotite massifs have been used to derive information about (p, T)-conditions (Finnerty and Boyd, 1987; Boyd, 1987) age (Boyd et al., 1985; Boyd and Gurney, 1986; Pearson, 1997) and composition (Jordan, 1979; Griffin et al., 1984, 1996) of the upper mantle below continents. Here we present similar observables derived from the numerical modelling results.

The same set of tracers used in the previous section has been used to investigate the structure and evolution of the thermally defined age of the continental root. The thermal age t_{close} is defined here as the time elapsed since a rock sample cooled below a hypothetical closing temperature T_{close}. In the model calculations, the temperatures of a set of tracer particles is monitored at every time-step. When a tracer cools below the closing temperature the current model time value is stored. The closure time is reset when the closing temperature is exceeded during a later event. This thermal rejuvenation can occur by reheating during thermal events. Results are presented for several hypothetical closing temperatures ranging from 600 to 1200°C with intervals of 200°C. This is within suggested upper mantle ranges (Mezger et al., 1992). The state of the model after 4 Ga of evolution is assumed to correspond to the present-day situation, i.e., $t = 4$ Ga $= 0$ Ma b.p.

In the left-hand-column of Fig. 9 the evolution of the closure time is given for $T_{\text{close}} = 1200$°C for snapshots starting at 2.5 Ga b.p. and ending at 0 Ga b.p. with 0.5 Ga intervals. Note that the depth-scale is twice the x-scale to emphasize the details. Deep, white areas indicate that the temperature of the material is higher than T_{close}. White upper mantle areas just underneath the crustal layers have temperatures below the T_{close} from the onset of evolution. The corresponding temperature fields are given in the right-hand-side column.

At 2.5 Ga b.p. there is a shallow feature in the

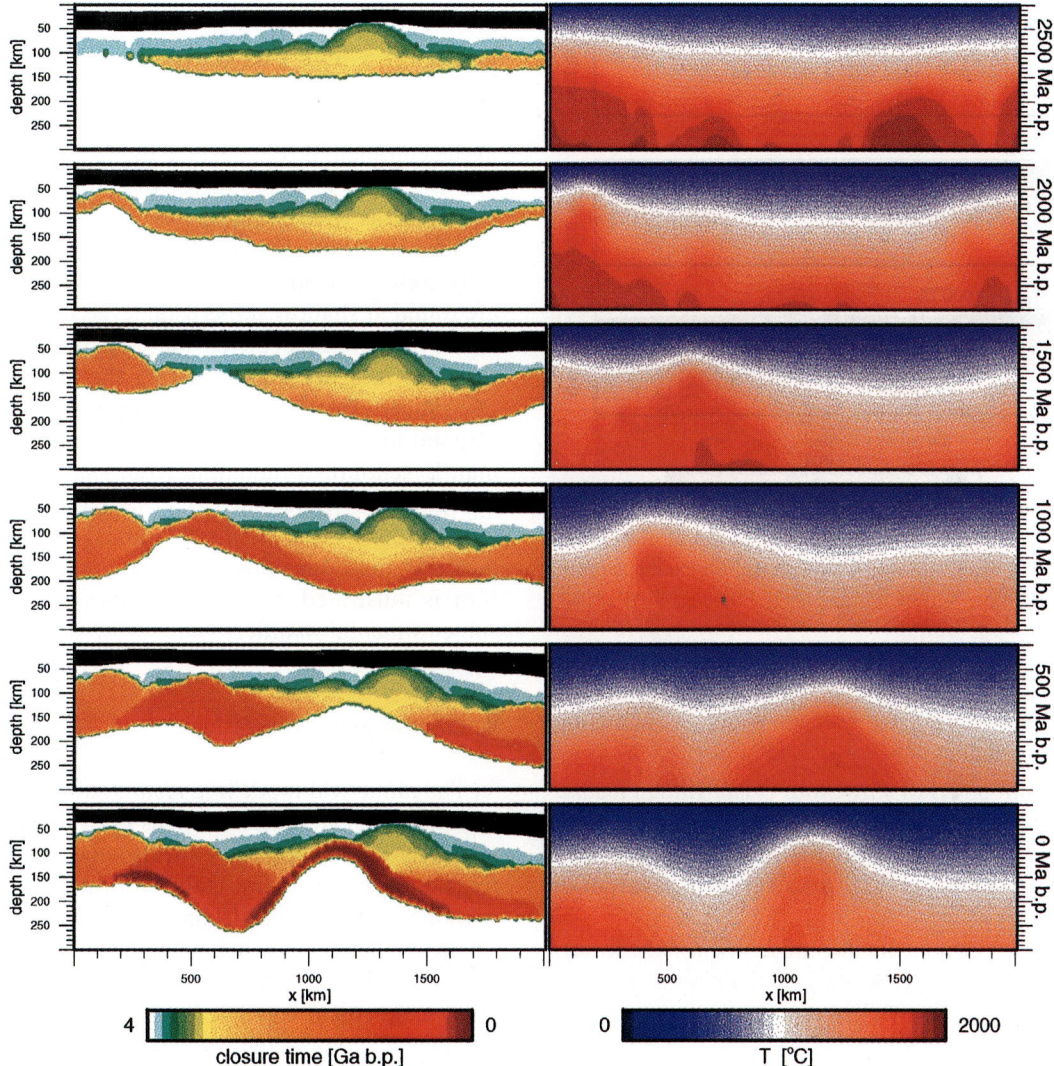

Fig. 9. Evolution of the closure time (left-hand-side column) for a closing temperature of 1200°C when 4000 Ma = 0 Ma b.p. is assumed. The white deep regions are still above the closure temperature. The transition from the coloured regions at the base to the deep, white part shows a very narrow band with old ages, which is an artifact of the post-processing and plotting procedure. This transition should be sharp from young age to no age (white). The corresponding temperature fields are depicted in the right-hand-side column. Note that the depth-scale is exaggerated twice with respect to the x-axis to emphasize the details. See Section 2.1 for the time convention use convention.

$1200 < x < 1400$ km range, which reaches the lower crust. This is the result of an older hot temperature anomaly, the remnant of a mantle diapir timed at 3.5 Ga b.p., no longer visible in the temperature field. A similar event is observed from 2.5 to 1.0 Ga b.p. at a slightly greater depth in the $100 < x < 300$ km range. The thermal anomaly that caused this feature is slowly disappearing as is illustrated by the temperature field at 2.5, 2.0 and 1.5 Ga b.p. This region on the 'left' closes within 1 Ga up to a depth of almost 200 km.

Over a depth range of $125 < z < 225$ km between $x = 1000$ and 1300 km rock reached temperatures $< 1200°C$ at 1.0 Ga b.p. During the next 500 Ma

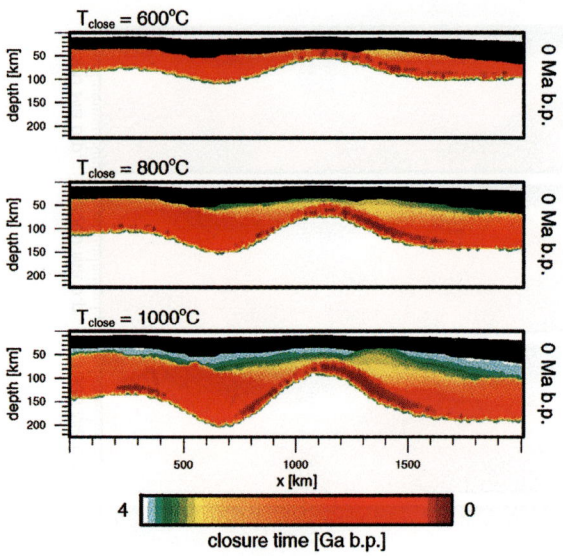

Fig. 10. The closure time fields at 0 Ma b.p. for closing temperatures of 600, 800 and 1000°C from top to bottom when 4000 Ma = 0 Ma b.p. is assumed. The depth-scale is exaggerated twice with respect to the x-axis. The white deep regions have not yet cooled enough and therefore have not sank below the closure temperature. See also the caption for Fig. 9.

of evolution, a hot upwelling resets these closure times after which the rock 'closes' again from 0.5 Ga b.p. on. A similar smaller scale process has been active between 2.0 and 1.0 Ga b.p. around $x = 600$ km.

Similar patterns are also observed for the lower range of T_{close}. Fig. 10 shows continental thermal age structures for different closing temperatures (600, 800 and 1000°C) at 0 Ma b.p. The effect of cooling from the top is reflected in the differences between the patterns of closure times at different closing temperatures.

Vertical profiles of the closure times and degree of depletion at 0 Ma b.p. are given in Fig. 11a at three horizontal positions identical to those given in Fig. 8.

Lateral variations in thermal history are illustrated by the differences between the frames in Fig. 11a1–a3. The sub-horizontal parts of these curves are the result of thermal rejuvenation, resetting the thermal age beneath the secularly growing MBL. For instance, the fast drop to greater depth of the t_{close}-curves in Fig. 11a2 at $t_{close} = 1.3$ Ga

b.p. is related to the accelerated shift of the isotherms to greater depth. This thermal shift starts between 1.5 and 1.0 Ga b.p., at $x = 700$ km as shown in the temperature frames of Fig. 9.

The stratification of the compositional layering in the continental root and its lateral variation are illustrated in Fig. 11b1–b3 for the same horizontal positions as those given in Fig. 11a1–a3. The detailed structure shown in these profiles is numerically well resolved. Maximum values of $F = 30\%$ are found at shallow sub-crustal levels corresponding to the oldest depleted material in the model that was in a hot mantle during the initial phase of continental formation. This material might correspond to the strongly depleted material found in mantle xenoliths originating from shallow depths as reported by Carswell et al. (1984) and Bernstein et al. (1998). The results of partial remixing of material that delaminated from the continental root is illustrated by the low values of F in the sub-continental mantle.

4. Discussion and conclusions

The numerical modelling results show that continental upper mantle which has been formed in a relatively short period during the Archaean remains gravitationally stable for at least 4 Ga. Although small scale delamination events occur, which result in remixing of depleted continental root material in the sub-continental mantle, no large scale collapse of the continental root occurs for the observed time window. This long term stability is due to the effect of the low density of the depleted residual peridotite. This is reinforced by the effect of the temperature dependence of the viscosity which results in the growth of the strong mechanical boundary layer during secular cooling. The thickness of the resulting continental root (200–250 km) corresponds to similar values reported from seismological observations (Woodhouse and Trampert, 1995) although larger values have been reported also (Durrheim and Mooney, 1994; Polet and Anderson, 1995).

Detailed analysis of the (p, T, F, t)-paths illustrate the nature of recurrent melting in mantle diapirs which penetrate the continental root from

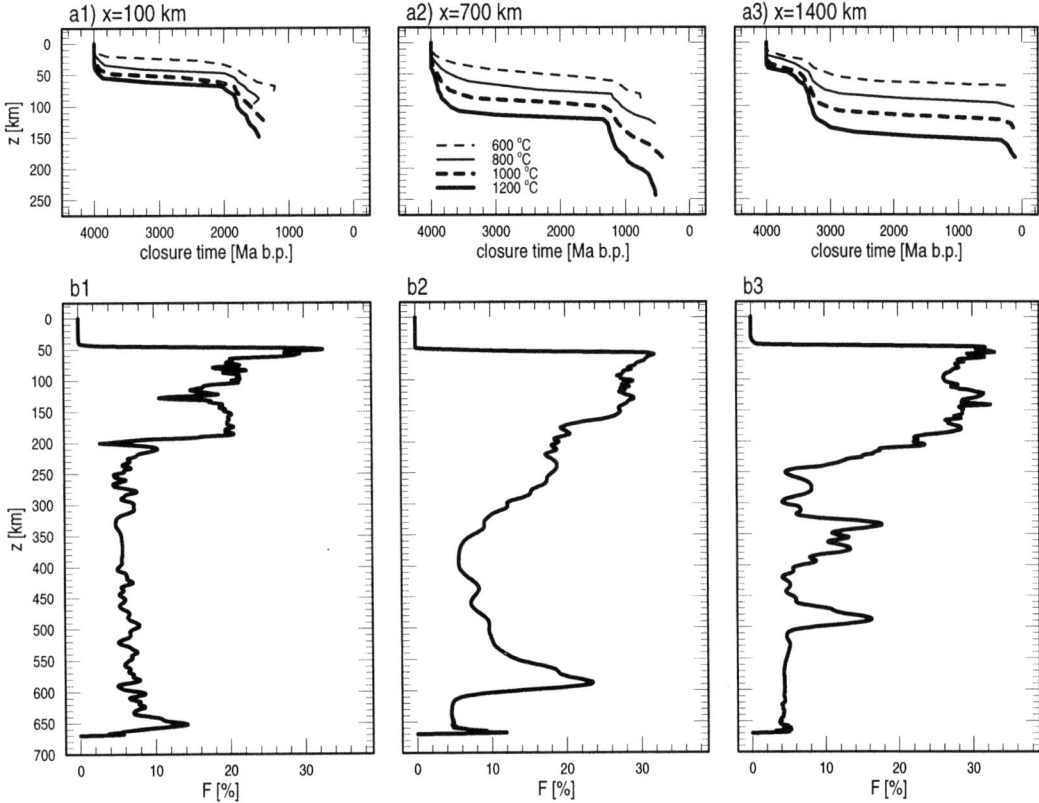

Fig. 11. Profiles of the F-field (b) and four closure times (a) at three x-locations identical to the locations used for the profiles as given in Fig. 8. The profiles correspond to $t=4000$ Ma $=0$ Ma b.p. for which the closure temperature fields are depicted in Figs. 9 and 10. The deeper region for which the closure time profile is not shown has temperatures higher than the indicated closing temperatures. The 'wiggles' in the F-profiles (b) are resolved by the numerical method.

below. The results show that partially melted material in the root can be entrained in diapiric upwellings, resulting in recurrent melting. The model (p, T, F, t)-paths show similarities with recent peridotite samples found in mantle xenoliths (Bernstein et al., 1998) and peridotite massifs (Roermund and Drury, 1998). The analysis of the thermal age of the model continental root reveals a definite layering in the thermal age with the oldest material of ca. 4 Ga situated at shallow depth directly beneath the crust. This reflects the mechanism of growth of continental lithosphere from below. Lateral variation of the thermal age increases with depth in the continental roots due to thermal rejuvenation by diapiric events in the weaker deeper parts of the root during later evolution.

Several limitations of the model should be considered. First the model configuration is limited to a completely continental upper mantle. This prevents lateral heat transport form the sub-continental mantle, which would enhance the secular cooling (Nyblade and Pollack, 1993; Pollack, 1997). At present, cratonic shields are not adjacent to oceanic upper mantle but are typically surrounded by Proterozoic continental upper mantle systems. Nevertheless, during early evolution in the Archaean those Proterozoic systems still had to be formed, and some interaction with the non-continental environment must have taken place. This requires more insight into non-continental Archaean mantle dynamics, which was probably completely different from today's situation (Vlaar et al., 1994).

An important limitation of the model is the restriction to the uppermantle domain. The lower mantle has been represented as an isothermal time dependent heat reservoir, which operates as a buffer in secular cooling. This is related to the assumption of fully layered convection, which may have prevailed especially during the earlier periods of the Earths evolution when the effective Rayleigh number for thermal convection must have been higher, thus favouring layered convection (Christensen and Yuen, 1985; Steinbach et al., 1993).

The strengthening effect of devolatilization on the rheology of the depleted mantle root has been neglected in the present model (Karato et al., 1986; Pollack, 1986; Karato and Jung, 1998). This would slow the convective cooling and increase the mechanical stability. Modelling results illustrating this effect are given by (De Smet et al., 1998).

Geotherms derived from the numerical modelling results after 4 Ga of secular cooling are in agreement with the theoretical present-day geotherms as given by (Chapman, 1986), at least for the shallow lithospheric regime, where conductive transport is dominant. Thermal evolution of the upper mantle shows a drop in average temperature of ca. 340°C and a decrease in surface heat flow from 100 to ca. 40 mW m^{-2} after 4 Ga. This is in reasonable agreement with estimates from whole mantle parameterized convection models (Jackson and Pollack, 1984).

The estimated small contribution of secular cooling of 6 mW m^{-2} to the surface heatflux compares well with the range of 7–15 mW m^{-2} given by (Jaupart et al., 1998), based on heatflow observations for the Canadian shield.

Acknowledgements

We thank Jean-Claude Mareschal and Paul Morgan for their insightful comments, which improved the manuscript. This research has been supported by the Dutch Science Foundation (NWO) and by NATO.

References

Anderson, D.L., 1990. Geophysics of the continental mantle: a historical perspective. In: Menzies, M. (Ed.), Continental Mantle. Clarendon Press, Oxford, UK, pp. 1–30.

Bernstein, S., Kelemen, P.B., Brooks, C.K., 1998. Depleted spinel harzburgite xenoliths in tertiary dykes from east Greenland: restites from high degree melting. Earth Planet. Sci. Lett. 154, 221–235.

Boyd, F.R., 1987. High- and low-temperature garnet peridotite xenoliths and their possible relation to the lithosphere-asthenosphere boundary beneath southern Africa. In: Nixon, P.H. (Ed.), Mantle Xenoliths. Wiley, pp. 403–412.

Boyd, F.R., Gurney, J.J., 1986. Diamonds and the African lithosphere. Science 232, 472–477.

Boyd, F.R., Gurney, J.J., Richardson, S.H., 1985. Evidence for a 150–200-km thick Archaean lithosphere from diamond inclusion thermobarometry. Nature 315, 387–389.

Carswell, D.A., Griffin, W.L., Kresten, P., 1984. Peridotite nodules from the Ngopetsoeu and Lipelaneng kimberlites Lesotho: a crustal or mantle origin. In: Kornprobst, J. (Ed.), Kimeberlites II: The Mantle and Crust–Mantle Relationships vol.IIB, Developments in Petrology. Elsevier Science, The Netherlands, p., 243.

Chapman, D.S., 1986. Thermal gradients in the continental crust. In: Dawson, J.B., Carswell, D.A., Hall, J., Wedepo, H.K.H. (Eds.), The Nature of the Lower Continental Crust vol. 24. Spec. Publ. Geol. Society, London, pp. 63–70.

Christensen, U.R., Yuen, D.A., 1985. Layered convection induced by phase transitions. J. Geophys. Res. 300, 10291–10300.

De Smet, J.H., Van Den Berg, A.P., Vlaar, N.J., 1998. Stability and growth of continental shields in mantle convection models including recurrent melt production. Tectonophysics 296, 15–29.

De Smet, J.H., Van Den Berg, A.P., Vlaar, N.J., 1999. The evolution of continental roots in numerical thermo-chemical mantle convection models including differentiation by partial melting. Lithos 48, 153–170.

De Smet, J.H., Van Den Berg, A.P., Vlaar, N.J., 2000. A characteristics based method for solving the transport equation and its application to the process of mantle differentiation and continental root growth. Geophys. J. Int. 140 (3), 651–659.

Doin, M.-P., Fleitout, L., McKenzie, D., 1996. Geoid anomalies and structure of continental and oceanic lithospheres. J. Geophys. Res. 101, 16119–16135.

Durrheim, R.J., Mooney, W.D., 1994. Evolution of the Precambrium lithosphere: seismological and geochemical constraints. J. Geophys. Res. 99, 15359–15374.

Ebinger, C.J., Sleep, N.H., 1998. Cenozoic magmatism throughout east africa resulting from impact of a single plume. Nature 395, 788–791.

Finnerty, A.A., Boyd, F.R., 1987. Thermobarometry for garnet peridotites: basis for the determination of thermal and compositional structure of the upper mantle. In: Nixon, P.H. (Ed.), Mantle Xenoliths. Wiley, pp. 381–402.

Griffin, W.L., Wass, S.Y., Hollis, J.D., 1984. Ultramafic xenoliths from Bullenmeri and Gnotuk Maars, Victoria, Australia: petrology of a subcontinental crust-mantle transition. J. Petrol. 25, 53–87.

Griffin, W.L., Kaminsky, F.V., Ryan, S.Y., O'Reilly, C.G., Win, T.T., Ilupin, I.P., 1996. Thermal state and composition of the lithospheric mantle beneath the Daldyn kimberlite field, YakutiaKimberlites and Structure of Cratonic Lithosphere. Tectonophysics 262, 19–33.

Jackson, M.J., Pollack, H.N., 1984. On the sensitivity of parameterized convection to the rate of decay of internal heat sources. J. Geophys. Res. 89, 10103–10108.

Jaupart, C., Mareschal, J.C., Guillou-Frottier, J., Davaille, A., 1998. Heat flow and thickness of the lithosphere in the Canadian shield. J. Geophys. Res. 103, 15269–15286.

Jordan, T.H., 1975. The continental tectosphere. Rev. Geophys. Space Phys. 13 (3), 1–12.

Jordan, T.H., 1979. Mineralogies densities and seismic velocities of garnet lherzolites and their geophysical implications. In: Boyd, F.R., Meyer, H.O.A. (Eds.), The Mantle Sample: Inclusions in Kimberlites and Other Volcanics. American Geophysical Union, Washington, DC, pp. 1–14.

Karato, S., Jung, H., 1998. Water, partial melting and the origin of the seismic low velocity and high attenuation zone in the upper mantle. Earth Planet. Sci. Lett 157, 193–207.

Karato, S., Paterson, M.S., FitzGerald, J.D., 1986. Rheology of synthetic olivine aggregates: influence of grain size and water. J. Geophys. Res. 91, 8151–8176.

LeFevre, L.V., Helmberger, D.V., 1989. Upper mantle p velocity structure of the Canadian shield. J. Geophys. Res. 94, 17749–17765.

Mezger, K., Essene, E.J., Halliday, A.N., 1992. Closure temperatures of the sm–nd system in metamorphic garnets. Earth Planet. Sci. Lett. 113, 397–409.

Nyblade, A.A., Pollack, H.N., 1993. A global analysis of heat flow from Precambrium terrains: implications for the thermal structure of Archaean and Proterozoic lithosphere. J. Geophys. Res. 98, 12207–12218.

Pearson, D.G., 1997. The age of continental roots. Workshop on Continental Roots. Harvard University & Massachusetts Instititute of Technology.

Polet, J., Anderson, D.L., 1995. Depth extent of cratons as inferred from tomographic studies. Geology 23 (3), 205–208.

Pollack, H., 1997. Thermal characteristics of the Archaean. In: de Wit, M.J., Ashwal, L.D. (Eds.), Greenstone Belts. Oxford University Press, Oxford, pp. 223–232.

Pollack, H.N., 1986. Cratonization and thermal evolution of the mantle. Earth Planet. Sci. Lett. 80, 175–182.

Pollack, H.N., Chapman, D.S., 1977. On the regional variation of heat flow, geotherms, and lithospheric thickness. Tectonophysics 38, 279–296.

Pollack, H.N., Hurter, S.J., Johnson, J.R., 1993. Heat flow from the Earth's interior: analysis of the global data set. Rev. Geophys. 31, 267–280.

Roermund, H.L.M., Drury, M.R., 1998. An ultra-deep ($d > 200$ km) orogenic peridotite body in Western Norway. In:Fall Meeting Abstracts vol. V21E-03. AGU, p., F971.

Rudnick, R.L., McDonough, W.F., O'Connell, R.J., 1998. Thermal structure, thickness and composition of continental lithosphere. Chem. Geol. 145, 395–411.

Steinbach, V., Yuen, D.A., Zhao, W., 1993. Instabilities from phase transitions and the timescales of mantle evolution. Geophys. Res. Lett. 20, 1119–1122.

Vlaar, N.J., Van Keken, P.E., Van Den Berg, A.P., 1994. Cooling of the Earth in the Archaean, consequences of pressure-release melting in a hotter mantle. Earth Planet. Sci. Lett. 121, 1–18.

Windley, B.F., 1995. The Evolving Continents. Wiley.

Woodhouse, J.H., Trampert, J., 1995. Global upper mantle structure inferred from surface wave and body wave data. In: Fall Meeting Abstracts vol. S42C-9. AGU, p., F422.

TECTONOPHYSICS

www.elsevier.com/locate/tecto

The diversity of tectonics from fluid-dynamical modeling of the lithosphere–mantle system

Bertram Schott [a,*], David A. Yuen [b], Harro Schmeling [c]

[a] *Department of Earth Sciences, HRTL, Uppsala University, Villav. 16, 75326 Uppsala, Sweden*
[b] *Minnesota Supercomputing Institute and Department of Geology and Geophysics, University of Minnesota, Minneapolis, MN 55415-1227, USA*
[c] *Institute of Meteorology and Geophysics, University of Frankfurt/Main, Feldbergstr. 47, 60323 Frankfurt/Main, Germany*

Received 15 March 1999; accepted for publication 24 September 1999

Abstract

Numerical modeling of the lithosphere has been carried out from many points of view. Most popular and prevalent is the use of the kinematic boundary conditions to move the lithosphere about in the dynamics. We have approached the numerical modeling of the lithosphere–mantle system by treating this as a fluid-dynamical system in which crustal and mantle rheologies and thermal and compositional buoyant forces play dominant roles in delivering the dynamics in a self-consistent manner. We have investigated the development of extensional and compressional tectonic regimes from variations of the crustal density, and compare their temporal evolutions. In comparing with kinematically driven thickening of the lithosphere, we find that the timescales are more self-consistently determined in a full fluid-dynamical treatment than in kinematic models. We emphasize the delicate nature of balancing the chemical buoyancy of the crust with the thermal negative buoyancy of the mantle lithosphere, as the modes and associated timescales of the delamination process are controlled critically by their relative importance. The role of an actively participating buoyant crust cannot by ignored in the overall lithosphere–mantle dynamics. © 2000 Elsevier Science B.V. All rights reserved.

Keywords: crustal recycling; geodynamics; numerical modeling; thermal–chemical convection

1. Introduction

Studies of tectonic processes due to lithospheric deformation have often been carried out within a regional framework involving kinematic boundary conditions (Willett et al., 1993; Beaumont et al., 1994); i.e. crustal–lithospheric flexure (Cloetingh and Burov, 1996; Ter Voorde et al., 1998), lithosphere extension (Buck and Poliakov, 1998; Regenauer-Lieb and Yuen, 1998), or compression (Buck and Sokoutis, 1994) have been investigated by either numerical or analogue models, assuming viscoplastic material properties. The kinematic models have been very successful in describing small-scale and subduction tectonics (Faccenna et al., 1999). They can usually well reproduce tectonic structures and are helpful in understanding deformation patterns over a lithospheric scale. However, kinematic models cannot portray faithfully the strong time dependence of the deformation processes and therefore do not result in a 'realistic' temporal model evolution.

Large-scale tectonics have already been studied

* Corresponding author.
E-mail address: bs@geofys.uu.se (B. Schott)

by Schmeling and Schott (1994), Buck and Poliakov (1998), Moresi and Solomatov (1998), and Trompert and Hansen (1998) using dynamical models of mantle convection and assuming a stick–slip plastic behavior of the lithosphere. However, the dynamical interaction with the underlying asthenosphere and upper mantle dynamics is often not taken into consideration (Batt and Braun, 1997). This mindset in the partitioning of the different dynamical regimes into submodels is due either to limited computational resources or to the lack of appreciation of the need to consider lithospheric–mantle interaction as a holistic, tightly coupled nonlinear system. One way of studying the coupled dynamics would entail treating the entire upper mantle–lithospheric system as a fluid-mechanical system with temperature- and pressure-dependent rheological properties and thermal and chemical buoyant forces. By treating the dynamical interior with this thermal–chemical convective approach, we do not need to supply the kinematic boundary conditions. However, this self-consistent approach would require us to deal with large viscosity contrasts between the cold lithosphere and the hot asthenosphere, and the large deformation due to mantle convection over a geological timescale of several tens of million years. While analogue models can handle finite strains of the order of 10, they cannot treat the viscosity contrasts and thermal gradients associated with heat and mass transfer in thermal–chemical convection in the lithospheric–mantle system. Large deformations due to thermal–chemical convection do not pose any difficulties for numerical models using several million markers to monitor the compositional field, and the method of finite differences with less than a 1 km spatial resolution along with a direct solver to tackle viscosity contrasts up to 10 orders in magnitude. Even viscoelasticity can be handled by finite difference spectral methods (Schmalholz and Podladchikov, 1999) at short wavelengths. However, in long-wavelength deformation and heat-transfer processes exceeding a few hundred kilometers, numerical finite difference methods represent a very useful and versatile tool for studying lithospheric dynamics. To resolve the relevant thermal–mechanical, time-and-length scales, down to the level of 1 km, the lithosphere–upper mantle system is best studied today in 2D. Due to the demands of computational resources, it is not possible to model all of the relevant scales in the coupled thermal–mechanical problem in 3D.

The major force driving the lithospheric dynamics is the negative buoyancy force of the relatively dense mantle–lithosphere, which may develop a gravitational instability, if it is not balanced properly by the positive buoyancy of the less dense crust (Schott and Schmeling, 1998). In this work we will illustrate that another major force acting on the lithosphere is the viscous drag exerted by subasthenospheric convection, which can trigger the gravitational instability of the mantle lithosphere, even if its negative buoyancy is balanced by the less dense crust. The purpose of this paper is to demonstrate the diversity in style of tectonics in a self-consistent thermal–chemical convective model for the lithospheric–mantle dynamics. We will focus on the effects from varying the crustal thickness in the model, and how this will cause two different tectonic regimes, compressive or extensional, depending on the initial state of the lithosphere and the subsequent lithospheric–mantle convective interaction. We will also discuss the role played by chemical buoyancy in crustal dynamics, and the simultaneous interaction with the thermal buoyancy. Finally, we will focus on the role played by an active low-density buoyant crust.

2. Extensional and compressional tectonic regimes

2.1. Modeling approach

To study lithospheric–mantle dynamics, we have employed a two-dimensional thermal–chemical convective model with a variable viscosity η depending on temperature T, depth z, second invariant of the deviatoric strain rate $\dot{e}_{II}$, and yield strength τ_{max}, or $\eta = \eta(T, z, \dot{e}_{II}, \tau_{max})$. A full description of this model can be found in Schott and Schmeling (1998) and also in Appendix A. The dynamics of the temporal evolution of this model are only governed by the internal buoyant forces, without the application of any external forces on the side, such as compression (Faccenna et al.,

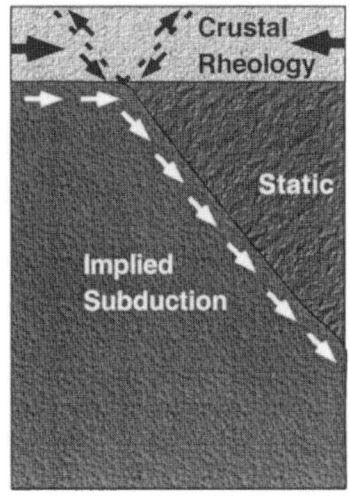
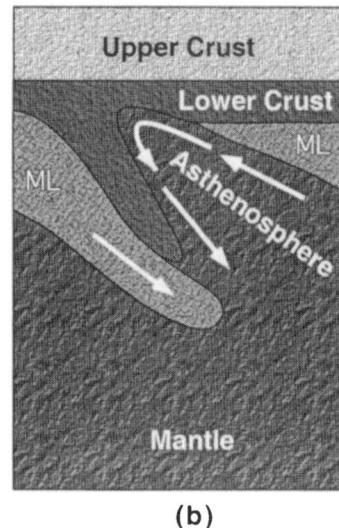

Fig. 1. A sketch of the differences between the kinematic models: Beaumont et al. (1996), left panel; our dynamical model, right panel.

1999). The lithosphere has only been decoupled from the left sidewall of the box by a low viscosity zone Schmeling and Jacoby (1981) modeled by a low pore pressure factor $\lambda = 0.03$ [see Eq. (A9) for definition]. There is therefore no need to be concerned with kinematic boundary conditions, which may be problematic in purely kinematically controlled models of lithospheric deformation (e.g. Beaumont et al., 1996). The fundamental difference between the two types of model is sketched schematically in Fig. 1. In our model we have full thermal–mechanical coupling, which includes the advection of the temperature field and its attendant influence on the viscosity, but also the positive feedback mechanism of viscous heating allowing the viscosity to be influenced by the shear heating. Other models (e.g. Beaumont et al., 1996; Marotta et al., 1998) coupled the viscosity loosely with the temperature by advection, and there is no feedback allowed from viscous heating. See Fig. A1 for a schematic comparison of the 'loosely' and the 'tightly' coupled thermal–mechanical models. The kinematic models do not exhibit any information about the intrinsic timescales of the deformation process, which is essential for obtaining some ideas about the characteristic timescales of geological evolutions. Any sort of positive feedback process will shorten the dynamical timescales (Larsen and Yuen, 1997; Regenauer-Lieb and Yuen, 1998). The combination of a temperature-dependent viscosity with viscous dissipation is one of the most effective positive feedback mechanisms in mantle convection, and is therefore included in our model.

2.2. Initial conditions

We have integrated as a function of time the full set of thermal–chemical convective equations with the finite difference method to study the combined lithospheric–mantle system. Technical details can be found in Appendix A. We started out with a lithospheric root of amplitude f_T (see Fig. 2), and this root represents the unsupported portion of length L of a subducting slab. The amplitudes f_T and f_C of the thermal and crustal roots are defined as the ratio of the maximum thickness to the minimum thickness at time $t=0$ (the initially unchanged thickness) of the thermal boundary layer and crust. The maximum thickness of the thermal boundary layer is 300 km in the center of the left panel in Fig. 2, the minimum thickness is 100 km outside the root, giving a value of $f_T = 3$. Likewise, the amplitude of the crustal root can be calculated, giving $f_C = 2.4$. A variation

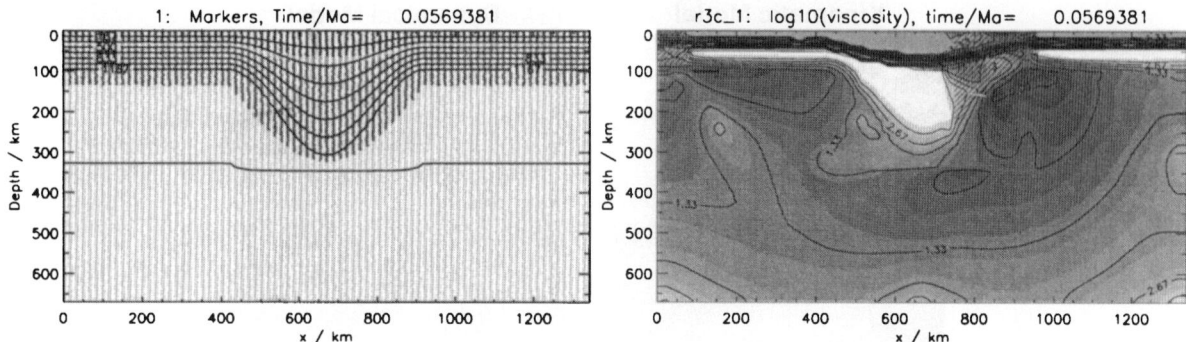

Fig. 2. Left: isotherms and chemical components. From top to bottom: upper crust, black; lower crust, light gray; mantle lithosphere, dark gray; asthenosphere, gray. Amplitude of the crustal root: $f_C=2.4$. Amplitude of the thermal root: $f_T=3$. Density difference between crust and mantle: $\Delta\varrho=340$ kg m^{-3}, therefore $f_C<f_T$ is required for a gravitationally unstable lithospheric root. Right: logarithmic scaled plot of the viscosity η [$\log_{10}(\eta/\eta_{\text{scale}})$, $\eta_{\text{scale}}=10^{21}$ Pa s], dark ≙ low viscosity, light ≙ high viscosity. Zones of weak brittle rheology (hatched gray) are introduced in 737 km $\leq x \leq$ 938 km and $x \leq 67$ km. Spatial resolution is 61 × 121 grid points, and 200 × 400 markers have been used.

of f_T from 3 to 4.3 corresponds to a slab length L changing from approximately 100 km to 165 km. The initial conditions are portrayed in Fig. 2 after a time of 0.06 Ma, allowing us to show the viscosity distribution due to non-Newtonian, temperature-dependent mantle rheology.

A rheologically weak zone is assumed at the right-hand flank of the lithospheric root (737 km $\leq x \leq$ 938 km), mimicking the relatively weak suture between two lithospheric plates after collision. Delamination of the lithospheric root starts in this relatively weak zone, eventually propagating to the left. The weakness is modeled by a low pore pressure factor ($\lambda=0.03$), please consult Appendix A for additional details. Likewise, the lithosphere is mechanically decoupled from the left model boundary ($\lambda=0.03$ for $x\leq 67$ km). These weak zones, which are effective at temperatures lower than 1000°C, are shown as hatched gray in Fig. 2.

We note that this system is two-component in nature, consisting of the crust and mantle, with $\varrho_{\text{crust}}/\varrho_{\text{mantle}}=0.9$, and the crust is also thickened to form a crustal root of amplitude $f_C=2.4$. The internal buoyancy forces therefore have both thermal and chemical origins. Hence, at times the middle lithosphere has a greater density than the underlying mantle by virtue of its being simply older. Therefore, a thickened mantle lithosphere would provide a negative buoyancy force, which would tend to pull the lithosphere deeper into the mantle. The crust is compositionally less dense than the mantle by around 10%, thus providing a positive buoyancy force, allowing the crust to float in the mantle. If the upper crust is mechanically strongly coupled to the mantle–lithosphere by a lower crust, which does not have too low a viscosity, the mantle–lithosphere cannot delaminate easily from the upper crust.

The average lithospheric density is less than the asthenospheric density in the case of an upper crust with $\Delta\varrho \approx 10\%$. In this case we classify the lithosphere as being inherently gravitationally stable, because an initially thickened mantle–lithosphere will not develop a gravitational instability. However, if additional forces, such as buoyancy from phase transitions or viscous drag from mantle convection, can help to overcome the positive buoyancy of the upper crust, the lithosphere can start to sink into the mantle.

The average lithospheric density is greater than the asthenospheric density in the case of an upper crust with $\Delta\varrho \approx 5\%$ only. This second case we classify as being inherently gravitationally unstable, because the negative buoyancy force of the mantle lithosphere is strong enough to overcome the positive force of the mantle lithosphere. Sinking of the mantle lithosphere starts at once from the initial state, driving a passive mantle circulation. The vertical density profiles located at

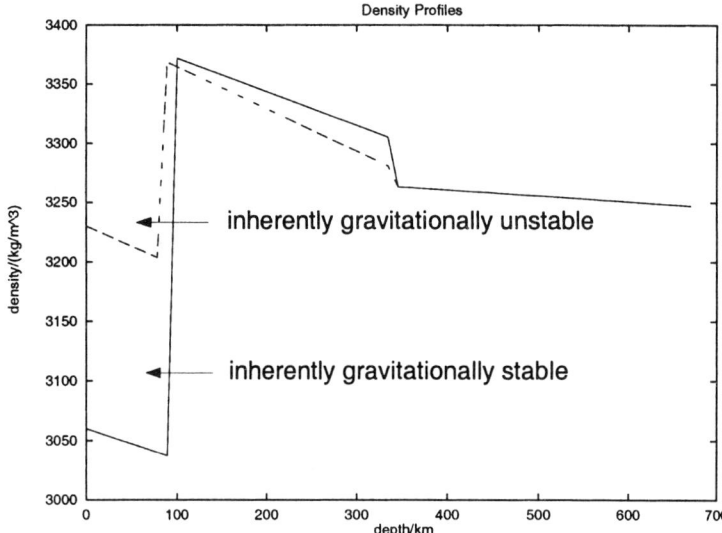

Fig. 3. Initial vertical cross-section of the density in the orogenic root being modeled. Solid line: inherently gravitationally stable density profile. Dashed line: inherently gravitationally unstable density profile.

the center of the thickened lithosphere are shown for both cases in Fig. 3.

The initial configuration associated with the solid line is inherently gravitationally stable, because the integrated density of the orogenic lithospheric root is less than the asthenospheric density. The low crustal density overcompensates the high density of the mantle lithosphere, leading to delamination in a compressive regime, if additional forces are applied. By contrast, the initial configuration associated with the dashed line is inherently gravitationally unstable, since the integrated density of the orogenic lithospheric root is greater than the asthenospheric density. The higher crustal density cannot compensate for the high density of the mantle lithosphere. This condition causes delamination to take place in an extensional regime without any additional forces.

2.3. Tectonic regimes: dynamical consequences from variations in crustal thickness

The fluid-dynamic approach to lithospheric deformation can basically provide information only about the state of the deviatoric stress. However, because of the incompressibility constraint, we can still glean some information about the tectonic regime by looking at the horizontal component σ_{xx}. Henceforth, we designate the tectonic regime to be compressive for $\sigma_{xx}<0$ and extensional for $\sigma_{xx}>0$ at shallow depths. These stress fields are produced self-consistently during the evolution of the model.

2.3.1. Compressive regime

In Fig. 4 we show how a compressive regime is able to develop. This comes from the interaction of an inherently stable lithosphere with the viscous drag induced by mantle convection. The viscous forces arise from the rising plume at the left boundary of the model. One can see the deformation suffered by the mantle through impingement of the plume by viewing the deformation of the initially vertically distributed markers in the right panel of Fig. 4. Initially the markers are uniformly distributed over the model area, with noise added to their positions to avoid numerical anisotropy. Not all of them are shown in the figures. The shown markers are selected in a way to give vertical stripes, when there is no deformation present. The deviation of the marker lines from the vertical becomes clearly visible with increasing material movement and is therefore a qualitative measure of the total material deformation. The strong

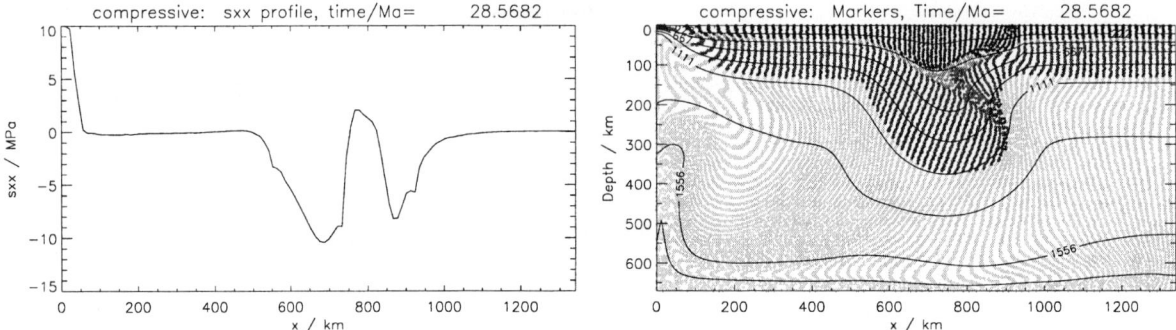

Fig. 4. Compressive tectonic regime. Left: deviatoric horizontal stress in approximately 50 km depth. Right: markers and isotherms. Delamination of the inherently stable lithosphere takes place only after the upwelling rises at the left boundary.

deviation of the pattern from the vertical in the upper left-hand corner (right panel) of Fig. 4 indicates that the mantle material is flowing from left to right, thus allowing the hotter, less viscous and lighter material from below to rise. This horizontal flow exerts a drag on the lithosphere, allowing the thickened mantle lithosphere to sink into the mantle. This sinking takes place in spite of the upper crust having a lower density by 10% than the mantle lithosphere. This left-to-right movement of the mantle lithosphere acts like a conveyor belt, which serves to compress the crustal material in the area where the mantle–lithosphere subducts. This compression is indicated by the negative horizontal deviatoric stress σ_{xx} at a depth of $z \approx 50$ km in the interval $x \approx 500$ km to $x \approx 1000$ km. The hill with positive σ_{xx} values in the center is caused by the buoyancy of the thick crustal root with a width of about 100 km. We note that the sinking mantle lithosphere is not opening any gap to allow the hot and less dense asthenospheric material to rise to shallow depths.

Formation of the Red Sea. Although the upwelling in Fig. 4 has only reached a depth of about 300 km according to the $T = 1550°C$ isotherm, the plume is exerting 'strong' forces on the lithosphere by viscous drag. This situation in Fig. 4 can be likened to the continental rifting between the present African and Arabian plate, forming the Red Sea. The opening was probably initiated along a relatively weak zone in the African plate due to the pan-African orogeny (Ghebreab, 1998), and was driven both by the slab pull from the subduction of the present Arabian plate under Asia in the Zagros region and by the Afar plume. There is still no definite timing on the lithospheric extension. However, our model results predict that the African lithosphere might have been severely stretched about 40 Ma ago, when the Afar plume might have reached the transition zone from the lower to the upper mantle (400 km to 670 km depth) (Schilling et al., 1992). Later on, about 15 Ma ago (Orihashi et al., 1998), the plume head probably impinged at the base of the already stretched lithosphere. If viscous drag from mantle convection is the only force acting on the lithosphere, and no additional extensional forces are present, then plumes may need to reach the lithospheric base to sever it.

We note that an upwelling plume of course always requires some downwelling, if the system is to conserve the mass. The point is that the starting plume 'has' a sublithospheric downwelling, corresponding to a stable lithosphere. When the ascending plume comes closer to the surface, the lithosphere becomes part of the downwelling. In this way the plume triggers the lithospheric gravitational instability, initiating the subduction process.

Dating of the rifting process is often conducted by dating syn-rift magmatic rocks, however, the extension of the lithosphere and hence the rifting process may have started much earlier. The early extension of the lithosphere by the viscous mantle drag cannot be accompanied by any decompression melting of hot asthenosphere, because there is no hot plume head underplating the lithosphere, which could lift it up by its buoyancy. It is therefore difficult to uncover signs of the early extensional deformation in the geological record and to date it using geological methods. However, we can estimate the time span between the onset of extension and major volcanic activity to be around 10 to 20 million years from our model results.

2.3.2. Extensional regime

Different from the compressional regime, the extensional regime develops from an inherently gravitationally unstable lithosphere driving a passive mantle flow, with just enough force for the mantle–lithosphere to sink. Hence, the total deformation of the sublithospheric mantle is small initially and the initially vertical marker stripes do not deviate much from the vertical. However, when the mantle lithosphere starts to sink, the movement of the mantle–lithosphere behaves like a conveyor belt again. But now the viscous mantle does not permit the lithosphere to move from the left to the right arbitrarily fast. This blocking effect allows a gap to open in the mantle–lithosphere (Fig. 5) at a depth of around 100 km at $x \approx 900$ km, thus allowing hot and less dense asthenospheric material to rise up to shallow depth. This near surface extension from the hot intrusion is reflected by the positive horizontal deviatoric stress σ_{xx} at $x \approx 900$ km. To the left of the extended region, the surface stress regime remains still compressive, due to the left-to-right movement of the mantle lithosphere (Fig. 5) described above. Buck and Sokoutis (1994) also showed by analogue experiments that extension and compression can occur close to each other at the same time.

2.4. Comparison of the temporal evolution

In Fig. 6 we plot the time history of the descent of the lithospheric root, as defined by the isotherm of $T = 1000°C$, for the extensional and compressional models. One notices the threshold effect of this nonlinear process, which means that it does not take off until some critical depth is reached. Then there is a fast timescale of the order of one to two million years for the lithospheric root to reach an asymptotic depth in the deep upper mantle. We note that these dynamical timescales are self-consistently determined by the balance between the rheology and the buoyant forces, in contrast to kinematic models with imposed boundary conditions (e.g. Giunchi et al., 1996). The delamination and descent of the lithospheric root last about 10 million years. The maximum sinking speed of the descending root is about 20 cm a^{-1} in the extensional regime and about 5 cm a^{-1} in the compressive regime.

The characteristic descent velocity of the lithospheric root is due mainly to truncating the viscosity at a minimum value of $\eta_{\min} = 10^{20}$ Pa s in

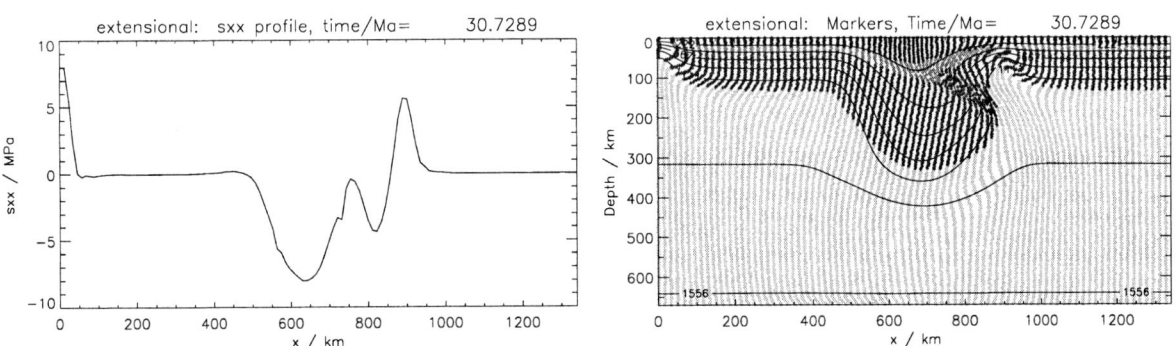

Fig. 5. Extensional tectonic regime. Left: deviatoric horizontal stress in ~50 km depth. Right: markers and isotherms. Delamination of the mantle–lithosphere starts from the inherently unstable state without being driven by sublithospheric convection. Asthenosphere rises into the gap.

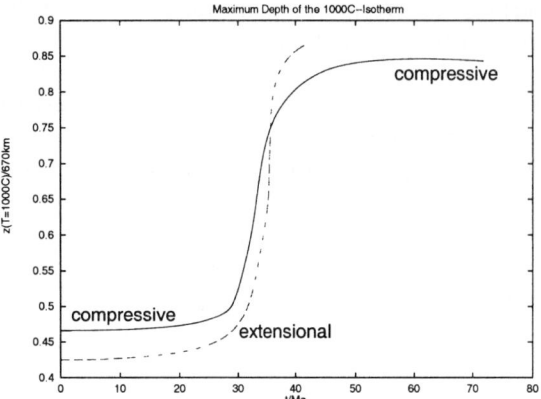

Fig. 6. Temporal evolution of the maximum depth of the lithospheric root (the $T=1000°C$ isotherm) for the two regimes, compressive (solid line) and extensional (dashed line), respectively.

the region surrounding the descending root. However, the strong acceleration of the sinking slab is a general feature of a gravitationally unstable layer with power law rheology (Yoshioka et al., 1995; Conrad and Molnar, 1997), which can be observed in dynamical models. While this lower bound has been chosen for numerical reasons, it is well known that with power law rheology and viscous heating, descending slabs would weaken and suffer detachment with increasing descending velocities due to positive feedback effects (Larsen et al., 1995; Yoshioka et al., 1995). Thus the rheologies employed are very important in modelling the dynamics of descending roots.

Although the rates of (current) plate motion are (the most) important observational constraints for geodynamic models, our numerical studies, and for example the analogue and numerical models of Becker et al. (1999), show that the assumption of constant plate or subduction velocities may not hold even over short timescales, $O(10^6 \, a)$. Assuming a constant boundary condition may therefore be valid only for 100,000 years or even less. One would need to apply adjoint models to the pure forward model, where the adjoint computes the sensitivities and derives in a least-square sense the best initialization and parameters consistent with observations from the present time (Thacker and Long, 1988).

This model is very sensitive to the combination of compositional density stratification and the underlying thermal structure. We emphasize that a change in crustal thickness and/or lithospheric temperature distribution can change the status of gravitational stability from stable to unstable or vice versa. A horizontally well-stratified lithosphere with a low density crust would be metastable, because any small additional forces (i.e. from viscous mantle drag) can disrupt the balance and initiate the delamination of the mantle–lithosphere, leading to its subduction.

Another dynamical effect, which is not present in kinematic models, arises from the isostatic rebound of the thickened, less dense crust. This subsurficial rebound process is illustrated in Fig. 7, where we plot the viscosity and temperature fields and the associated stream function for the compressive regime (left panel) and the extensional regime (right panel), showing the isostatic rebound of the thickened crust. Fig. 8 shows a zoom into the upper central part of the stream function plot represented in Fig. 7.

In Fig. 7 we can readily discern the large extent of the crustal flow coming from isostatic rebound of the crust, giving rise to a local change in the stress field from compressional to extensional. Such an internal change cannot be controlled by any externally prescribed boundary conditions. The thickened crust is driven upwards by its own positive compositional buoyancy, which is struggling against the negative thermal buoyancy of the cold mantle lithosphere. Hence the total buoyancy forces are changing with time and with changing angle of the descending root.

From Fig. 9 we observe that most of the viscous heating is associated with the descending root (Schott et al., 1999), but a non-negligible amount of dissipation is produced by the isostatic rebound of the crust (Schott et al., 2000). A very high spatial resolution, of the order of 1 to 2 km, is required to capture the details of viscous heating between the crust and mantle.

Both types of model show the possibility of crustal subduction. Here the crust recycling is incomplete, however, most of the lower crust is subducted in the vicinity of the delaminating slab.

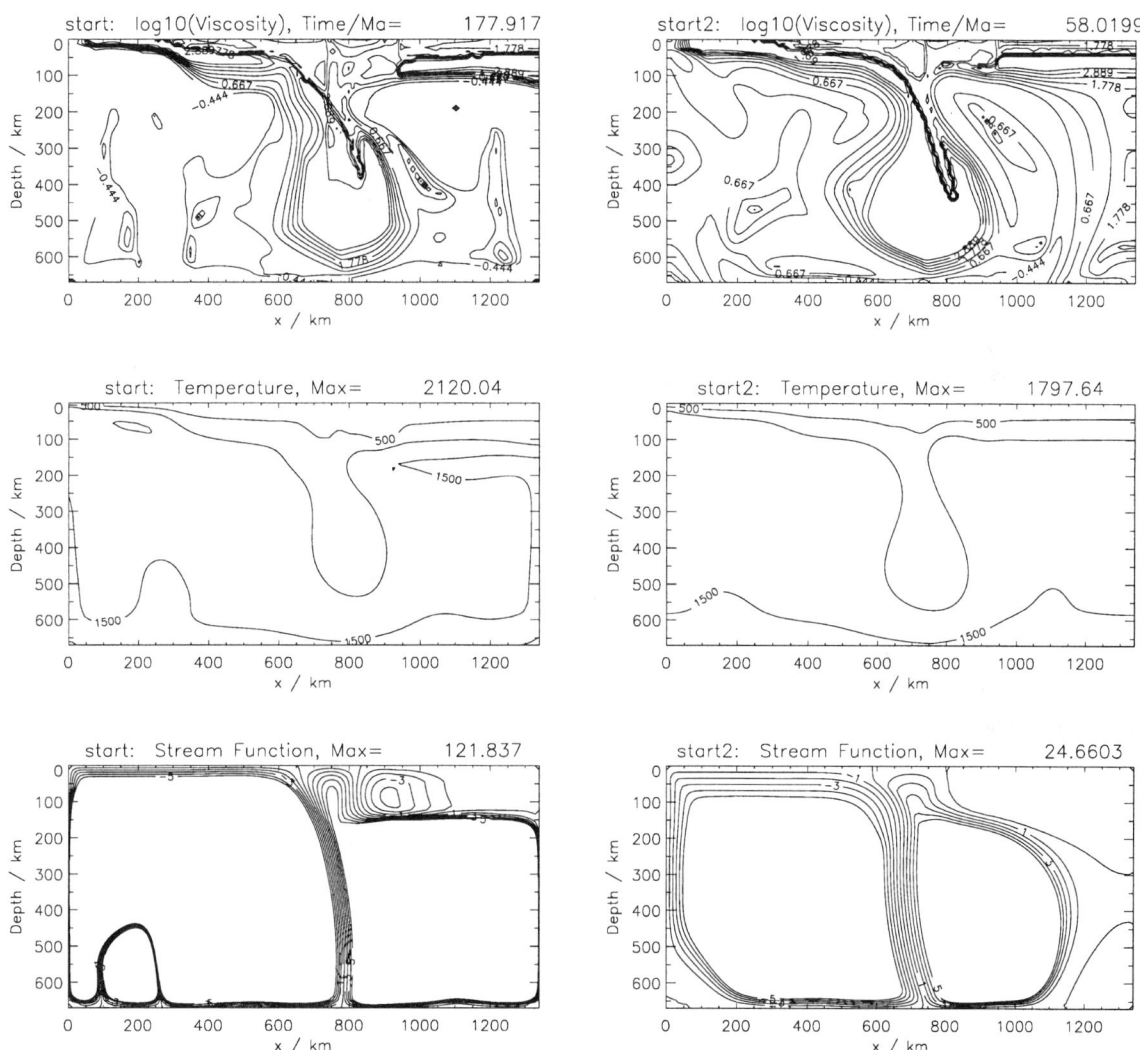

Fig. 7. Contour plot of the viscosity η [$\log_{10}(\eta/\eta_{scale})$, $\eta_{scale} = 10^{21}$ Pa s], top panel, temperature ($T = 500°C$, $T = 1000°C$, and $T = 1500°C$ isotherms), middle panel, and stream lines ($-5 < \Psi < 5$, $\Delta \Psi = 1$), bottom panel, for the compressive regime, left column, and the extensional regime, right column, respectively. Note that for both cases the strongly curved stream lines at shallow depth (0 km $< z <$ 100 km) at $x \approx 900$ km for the compressive regime and $x \approx 750$ km for the extensional regime indicate the upward flow of crust driven by compositional buoyancy forces, in other words, the isostatic rebound of the thickened crust.

3. Thickening of the lithosphere by kinematics

The thermal history of sedimentary basins has received great impetus through the kinematic stretching model of McKenzie (1978). The effects of a finite duration of rifting (Jarvis and McKenzie, 1980) and depth-dependent thinning (Royden and Keen, 1980) were incorporated. A finite strength during rifting was proposed (Braun and Beaumont, 1989). All of these models follow the

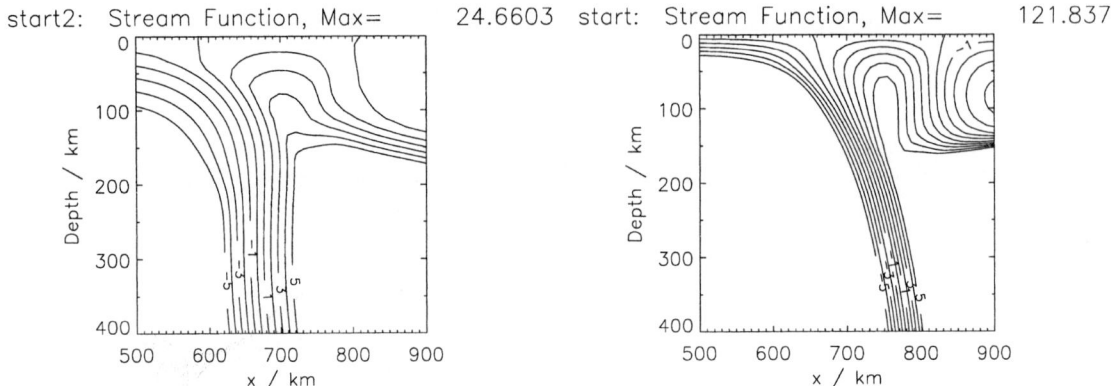

Fig. 8. Zoom into stream function Ψ in the subduction area. The central, almost vertical flow is subducting the lithosphere. The upwards oriented streamlines in the right upper corner indicate the isostatic rebound of the thickened crust.

same pattern. A new crustal configuration is produced by imposing extensional movements, resulting in a mass redistribution in the lithosphere. We will now consider the temporal evolution of the lithospheric thickening driven by prescribed boundary conditions, but including the dynamics of the combined lithospheric–mantle system.

To induce subduction flow, we have prescribed the stream function Ψ on a vertical profile at $x = 300$ km in the depth range between $z = 0$ km and $z = 300$ km. The stream function decreases linearly from $\Psi(z = 0) = 0$ to $\Psi(z = 300 \text{ km}) = -100$ along this line. The imposition of this boundary condition thickens the lithosphere in a weak zone modeled by a low pore pressure factor of $\lambda = 0.03$ [see Eq. (A9) for definition], which is assumed to lie between $x = 800$ km and $x = 900$ km, and is confined to the base of the lithosphere. While the imposed boundary condition provides a force contributing to the slab-pushing process, the thickened lithosphere exerts slab-pulling forces on the lithosphere. When the thickness of the lithosphere exceeds a critical amplitude $f_{T,\text{critical}}$, it can then become gravitationally unstable.

Fig. 10 shows the maximum thickness or amplitude f_T of the mantle lithosphere (the $T = 1000°C$ isotherm) as a function of time for three different numerical experiments. Their temporal evolutions are shown in Fig. 10, where Exp.1 marks the first experiment, Exp.2 the second experiment, and Exp.3 the third experiment.

In the first experiment, we have maintained the driving boundary condition for all time. Since it is mainly driving the lithospheric thickening, f_T increases linearly with time even for a very large f_T ($f_T \approx 4$), corresponding to the slab reaching 150 km in depth below an unthickened lithosphere. The influence from the boundary condition is so dominant that a typical gravitational instability does not get a chance to develop. However, in the next two experiments we will demonstrate that, in an unconstrained system, a gravitational instability of the thickened mantle–lithosphere can develop for a lithospheric amplitude $f_{T,\text{critical}} < 4$. From Schott and Schmeling (1998), the critical amplitude $f_{T,\text{critical}}$ of the unconstrained system can be estimated to be $f_{T,\text{critical}} \approx 3$ from their fig. 11. 'Critical' means here that the mantle–lithosphere

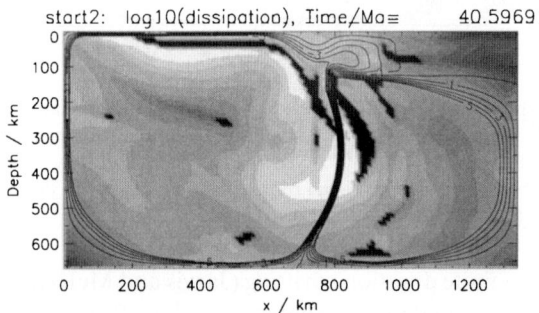

Fig. 9. Viscous heating Φ for the extensional regime. Lighter shading indicates higher values. At $t \approx 40$ Ma the subducting slab has reached a depth of approximately 500 km.

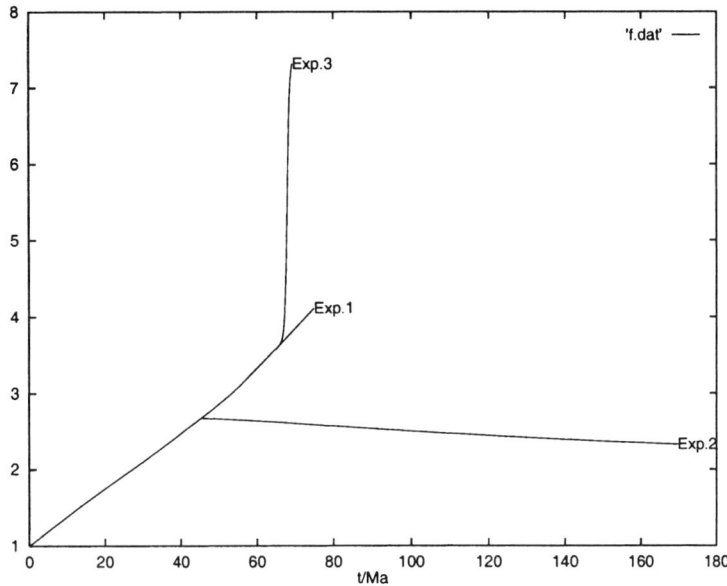

Fig. 10. The temporal evolution of the maximum depth of the lithospheric root f_T.

is gravitationally unstable for $f_T \geq f_{T,\text{critical}}$. Therefore, we will now determine how the system develops, when the driving mechanism is turned off around the critical lithospheric amplitude $f_{T,\text{critical}}$ estimated from the unconstrained model. Two lithospheric amplitudes f_T around the critical value $f_{T,\text{critical}}$ have been employed in this sensitivity analysis.

In the second experiment the driving boundary condition is switched off at a lithospheric thickening of $f_T \approx 2.5$, corresponding to the slab reaching a depth of around 75 km below an unthickened lithosphere. We observe from Fig. 10 that the lithospheric thickness slowly decreases with time due to thermal diffusion, the negative buoyancy force of the mantle–lithosphere decreases with time, thus showing that $f_T \approx 2.5$ lies in the subcritical regime ($f_T < f_{T,\text{critical}}$).

In the third experiment we turn off the driving boundary condition for a lithospheric thickening of $f_T \approx 3$, corresponding to the slab reaching a depth of about 100 km below an unthickened lithosphere. In this case we see from Fig. 10 that the lithospheric thickness increases rapidly with time, driven by the greater negative buoyancy force of the mantle lithosphere. The value of $f_T \approx 3$ is obviously slightly supercritical.

We interpret these results to be the consequence of the surface boundary condition on the velocity, which does not allow the horizontal portion of the mantle lithosphere to accelerate. If the sinking part of the mantle lithosphere should accelerate due to the development of a gravitational instability, then the horizontal part of the mantle lithosphere should follow suit, providing there is no severe internal deformation in the mantle lithosphere between the sinking and the horizontal parts. Therefore, studying the dynamics of the lithosphere–mantle system by applying velocity boundary conditions (Marotta et al., 1998) is problematical, and may yield results which are not applicable in natural systems without such dynamical constraints. The amplitudes required for a gravitationally unstable mantle–lithosphere may be overestimated by a factor of around 2. We note that the speed of sinking [slope of $f_T(t)$] in experiment 3 (unconstrained for $f_T > f_{T,\text{critical}}$) is much greater than the speed of sinking in experiment 1 (totally constrained), the velocities differ by some orders of magnitude. Therefore, the timescales for the development of gravitational instabilities may be underestimated by a few orders of magnitude if the shortening velocity from the imposed boundary condition is small.

4. Discussion and concluding remarks

We have illustrated in this paper the possibilities for many different styles of tectonics, which can develop from using a self-consistent treatment of the lithosphere–mantle system in which the role of chemical buoyancy must be modeled at the same time as thermal buoyancy. We have found that the lithosphere is very sensitive to the density difference between the crust and the mantle. A mere 5% difference in density separates the border between the gravitational stability or instability of the lithosphere. If the crustal density is 5% less dense than the mantle, the lithosphere can sink deep into the mantle, whereas a 10% less dense crust can prevent the lithosphere from descending into the mantle. In a model where the crustal density remains unchanged, the same effect can be simulated by varying the crustal thickness.

With a judicious choice of realistic lithospheric and mantle rheological parameters, we have found it possible to model the delamination and detachment of a lithospheric root in a dynamically self-consistent way (Schott and Schmeling, 1998; Schott et al., 1999). The mode of delamination, be it extensional ($\Delta\varrho=5\%$ and no plume) or compressive ($\Delta\varrho=10\%$ and plume), depends critically on the initial crustal buoyancy (thickness and density) and the proximity of ascending plumes.

The compressive regime develops from an inherently gravitationally stable state of the lithosphere. In this situation the buoyancy force of the less dense crustal root is able to overcome the negative buoyancy of the cold mantle lithosphere. Additional forces due to stresses from sublithospheric convection would cause a style of delamination marked by shortening. In this compressive regime delamination leads to thick crustal roots without any upwelling by the hot asthenosphere. This may result in crustal melting due to the internal heating by radioactivity. Therefore, in compressive regimes the magmatic genesis comes from crustal sources and not directly from the mantle or contamination from mantle melts. However, we note that the time scales needed for a significant temperature rise in the crustal root due to radioactive decay are on the order of 100 Ma, which is much longer than the time span over which the crust remains largely thickened in our model.

The extensional regime is derived from an inherently gravitationally unstable situation of the lithosphere. Here the upward buoyancy force of the less dense crustal root is not able to overcome the downward driving force of the cold lithosphere. Delamination can evolve without the agency of any additional forces. The extensional regime allows hot asthenosphere to rise and approach shallow depths beneath a thin crust. Thus pressure-released partial melting of asthenospheric and crustal material may occur in a region of extensional deviatoric horizontal stress. In this area magmatic rocks with mantle origins are likely to be found.

The extensional stresses in the crust do not cause stretching of the mantle–lithosphere (Liu and Shen, 1998), however, the dynamics of the mantle–lithosphere are strongly influenced by the crustal buoyancy, by allowing a gap to open in the mantle–lithosphere into which the hot asthenosphere can rise. This phenomenon is by no means the kind of lithospheric thinning Liu and Shen (1998) studied, but rather represents an opening of the mantle–lithosphere driven by its own negative buoyancy.

Intense interaction of the sublithospheric convective flow with the lithosphere can also induce delamination, even if the lithosphere is gravitationally stable. Hence, such vigorous interaction would likely promote the delamination process in the case of a gravitationally unstable lithosphere.

We have found from numerical experiments in which the shortening boundary conditions are imposed that the thickness of the mantle lithosphere: (1) increases linearly for a constant driving term; (2) increases superexponentially (Conrad and Molnar, 1997) if the driving mechanism is switched off at a supercritical amplitude; and (3) decays if the driving mechanism is turned off at the subcritical level. These results would argue for the sensitivity of tectonic results to kinematically imposed surface boundary conditions.

For thermal–chemical convection, each initial condition elicits a different response, unlike in 'simple' fluid mechanics, where initial conditions are 'pure', i.e. a linear stratification of the compositional field. There is a strong interaction between the thermal and chemical boundary layers in thermal–chemical convection, which may need to be described in terms of a local density Rayleigh number $R\varrho = Rc/Ra$. We have therefore used simple but reasonable initial conditions, which allow us to model the general evolution of the system with reasonable effort, like CPU time, etc. However, a systematic study of the lithosphere–mantle system will be needed for a better future understanding of large-scale tectonic processes.

Lithospheric–mantle dynamics have such an extremely complex phenomenology that their global modeling has not been attempted successfully. The accumulation of a wealth of data of increasing accuracy has been accompanied by a number of papers which, by breaking the global problem into self-contained subproblems, have sought an approach of piecewise modeling. It is not clear whether such a decomposition into subprocesses will yield a correct description of the global phenomenon, since the sum of subprocesses may not be the same as the whole process due to the essentially nonlinear nature of lithospheric–mantle material properties and interactions. Our work has argued for the need to consider sometimes the global problem of continental lithospheric evolution in the best way possible with the current computational capability.

Acknowledgements

We thank Ms. Joye Branlund for stimulating questions at Maxwell's, Dr. Klaus Regenauer-Lieb for his concerns, and Professor Chris Beaumont for stimulating philosophical conversations. We also thank Professor Paul J. Sylvester for his good comments and Dr. Ross Boutilier for a helpful review. This research has been supported by the Geosciences Program of the Department of Energy, the Geophysics Program of the National Science Foundation, the Deutsche Forschungsgemeinschaft and the Swedish NFR.

Appendix A: Physical and numerical 2D model for lithospheric–mantle dynamics

A.1. Governing equations

In contrast to some models of the lithosphere and related orogenic process (Bird, 1979; Houseman and Molnar, 1997; Conrad and Molnar, 1997), which do not include the equations of the heat and chemical transport, our model is dynamically and thermally self-consistent, i.e. driving forces develop as a consequence of thermal and chemical buoyancy and are not imposed by boundary conditions.

The time-dependent geodynamical process is described by the conservation equations of mass, momentum, energy, and chemical composition in a two-dimensional Cartesian geometry. The equations for the conservation of mass and momentum are combined, resulting in Eq. (A1) for the stream function Ψ with variable viscosity η and both thermally and compositionally induced buoyancy forces. We use the dimensionless form of the equations governing the model. The Prandtl number for the mantle is infinite and the extended Boussinesq approximation is used (Christensen and Yuen, 1985; Schmeling and Marquart, 1991). The resulting system of nonlinear partial differential equations consists of the biharmonic equation:

$$\left(\frac{\partial^2}{\partial z^2} - \frac{\partial^2}{\partial x^2}\right)\eta\left(\frac{\partial^2 \Psi}{\partial z^2} - \frac{\partial^2 \Psi}{\partial x^2}\right) + 4\frac{\partial^2}{\partial x \partial z}\eta\frac{\partial^2 \Psi}{\partial x \partial z}$$
$$= Ra\frac{\partial T}{\partial x} + Rc\frac{\partial C}{\partial x} \qquad (A1)$$

and the time-dependent equation for the heat transport:

$$\frac{\partial T}{\partial t} + (\mathbf{v} \cdot \nabla)T = \nabla \cdot (\kappa \nabla T) + H + Di\left[\frac{\Phi}{Ra} - v_z(T + T_0)\right] \qquad (A2)$$

Thermal-Mechanical Coupling

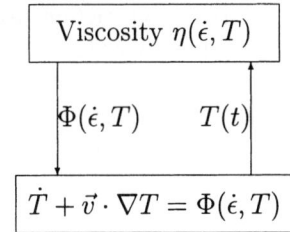

Fig. A1. Difference in thermal coupling for the case without viscous heating, left (case A), and with viscous heating, right (case B), which gives the positive feedback.

where the velocity v is related to the stream function Ψ by:

$$v = \frac{\partial \Psi}{\partial z} e_x - \frac{\partial \Psi}{\partial x} e_z. \quad (A3)$$

Here T denotes the temperature and T_0 is the surface temperature, κ is a depth-dependent thermal diffusivity (Schmeling and Bussod, 1996), H the internal heat production, C the chemical field, x and z the lateral position and the depth. In the biharmonic Eq. (A1) the thermal and chemical Rayleigh numbers are defined by $Ra := \varrho_0 g \alpha \Delta T h^3 / \kappa \eta_0$ and $Rc := \Delta \varrho g h^3 / \kappa \eta_0$, respectively, and $Di := \alpha g h / c_p$ is the dissipation number. The dissipation function Φ is defined by $\Phi = 2\eta \dot{\epsilon}_{ij} \dot{\epsilon}_{ij}$ with $\dot{\epsilon}_{ij}$ being an element of the strain rate tensor.

As an equation of state we used $\varrho(T, C) = \varrho_0 [1 - \alpha T + (\Delta \varrho / \varrho_0) C]$, where ϱ_0 is the reference density, $\Delta \varrho$ is the density difference due to the chemical composition, and α is the thermal expansivity. ΔT is the temperature difference and h is the distance between the bottom and top, respectively. η_0 is the scaling viscosity, the acceleration of gravity g points in the negative z-direction, and c_p is the heat capacity at constant pressure. Different chemical components are used to distinguish between the upper crust, the lower crust, and the mantle. Their distribution is described by the chemical field C, which solves the transport partial differential equation:

$$\frac{\partial C}{\partial t} + (v \cdot \nabla) C = 0. \quad (A4)$$

The chemical field C is calculated from the distribution of the markers that are advected with the flow, carrying information about density and/or rheology. Mantle–lithosphere and asthenosphere are compositionally identical. The Rayleigh number used in this paper is around 10^6 and the magnitude of the chemical Rayleigh number is of the same order.

Fig. A1 shows the 'tightly' coupling nature of the equations in this scheme, versus the 'loosely' coupled nature of that by Beaumont et al. (1996) and also by Marotta et al. (1998). In both types of coupling, the feedback from the temperature equation on the momentum equation is due to the temperature T, which affects the temperature-dependent density and rheology, and hence the buoyancy forces and the material properties.

With viscous heating (Fig. A1 right, case B) there is an additional strong nonlinear coupling between the momentum equation and the temperature equation by the viscous dissipation Φ, which affects the temperature and therefore all of the temperature-dependent material properties.

A.2. Rheology

The ductile rheology of the crust is assumed to be independent of the temperature and strain rate. The viscosity of the upper crust is $\eta_{UC} = 10^{23}$ Pa s, that of the lower crust varies from $\eta_{LC} = \eta_{UC} = 10^{23}$ Pa s to $\eta_{LC} = \eta_{UC}/10^3 = 10^{20}$ Pa s. This simplification takes into account the mobility of the ductile lower crust and the stiffness of the brittle upper crust. The mantle rheology is that of the dominating mineral, dry olivine, according to temperature- and pressure-dependent dislocation creep with a power law exponent of $n = 3.5$ (Cooper and Kohlstedt, 1987; Bussod and Christie, 1987; Bai et al., 1991; Karato and Wu, 1993). Significant anisotropy found in seismic waves (Montagner, 1996) in the upper mantle indicates that dislocation creep is the dominating deformation mechanism, which is consistent with a nonlinear rheology (Karato and Wu, 1993).

We use a superposition of two creep laws, $\dot{e} = \dot{e}_1 + \dot{e}_2$, see Bai et al. (1991). Each component is described by an empirical exponential law:

$$\dot{e}_i = D_i \Delta \sigma^{n_i} \qquad (A5)$$

where $\Delta \sigma$ is the difference between the maximum and minimum stress, and D_i is a diffusion term with:

$$D_i = A_i P_{O_2}^{m_i} \exp\left(-\frac{E_i + PV_i}{RT}\right) \qquad (A6)$$

which depends on the oxygen fugacity P_{O_2}, which is given by:

$$\log_{10}(P_{O_2}/\text{atm}) = F_1 + F_2/T + F_3(P-1)/T. \qquad (A7)$$

A.2.1. Summary of the rheological parameters

A_i is a constant, P_{O_2} is the oxygen fugacity, m_i is an exponent, E_i is the activation energy, P is the pressure, V_i is the activation volume, R is the gas constant, and T is the temperature. The olivine rheology used has been described by Schmeling and Bussod (1996). The parameters used are listed in Table A1 and apply to olivine (Karato and Wu, 1993) buffered by orthopyroxene. They have been measured by Bai et al. (1991).

The ductile rheology is combined with a quasi-brittle rheology to give an effective rheology. The brittle behavior of the material can be described by the Mohr–Coulomb law (Ranalli, 1995):

$$\tau_{max} = (A\sigma_n + B)\lambda \qquad (A8)$$

with the empirical constants $A = 0.6$ and $B = 60$ MPa found by Byerlee (Brace and Kohlstedt, 1980) for shallow depths. σ_n is the normal stress on the weakest existing fracture plane which does not allow frictional sliding as long as the shear stress τ is less than τ_{max}. For simplicity, the normal stress σ_n is assumed to be equal to the lithostatic pressure $P_{lith}(z)$, and hence it is directly proportional to the depth. The pore pressure factor λ is

Table A1
The parameters used for calculating the ductile mantle viscosity after Bai et al. (1991) and Karato and Wu (1993)

Symbol	Value and unit	Meaning
A_1	2.1×10^{-17} Pa$^{-n_1}$ atm$^{-m_1}$ s^{-1}	scaling factor
A_2	5.2×10^{-16} Pa$^{-n_2}$ atm$^{-m_2}$ s^{-1}	scaling factor
E_1	540 kJ mol^{-1}	activation energy
E_2	540 kJ mol^{-1}	activation energy
$V_1 = V_2$	15×10^{-6} m^3 mol^{-1}	activation volume
$n_1 = n_2$	3.5	power law exponent
F_1	9	factor in oxygen fugacity
F_2	2.5738×10^8	factor in oxygen fugacity
F_3	0.092	factor in oxygen fugacity
m_1	0.02	exponent in oxygen fugacity
m_2	0.23	exponent in oxygen fugacity

defined as:

$$\lambda := 1 - \frac{P_{\text{por}}}{P_{\text{lith}}} \Rightarrow \lambda \in [0, 1] \quad (A9)$$

where P_{por} is the pore pressure and P_{lith} is the lithostatic pressure. Eqs. (A8) and (A9) enable modeling the reduction of τ_{max} due to water on pre-existing fractures.

In the continuum mechanics approach the Mohr–Coulomb law is implemented by a 'Mohr–Coulomb viscosity':

$$\eta_{\text{MC}} = \frac{\tau_{\text{max}}}{2\dot{e}_{\text{II}}} \quad (A10)$$

where $\dot{e}_{\text{II}}$ denotes the second invariant of the strain rate tensor and τ_{max} is taken from Eq. (A8). The total effective viscosity η used in the biharmonic Eq. (A1) is then calculated from this 'Mohr–Coulomb viscosity' η_{MC} and the ductile viscosity η_{duc} by the relation:

$$\frac{1}{\eta} = \frac{1}{\eta_{\text{duc}}} + \frac{1}{\eta_{\text{MC}}} \quad (A11)$$

where, depending on position, η_{duc} is given by the olivine rheology mentioned above or by η_{UC} or η_{LC}. In cold parts $\eta \approx \eta_{\text{MC}}$ because η_{duc} is high. Weakening by small λ values would lead to very low viscosities, but due to numerical reasons, η is not allowed to become smaller than a given lower limit of η_{min} (here $\eta_{\text{min}} = 10^{19}$ Pa s). The crust is Newtonian isoviscous with $\eta_{\text{UC}} = 10^{23}$ Pa s and $\eta_{\text{LC}} = 10^{22}$ Pa s. The ductile mantle rheology is non-Newtonian, $n = 3.5$, dry olivine, laboratory data.

A.3. Numerical approach

The fourth-order equation for the stream function [Eq. (A1)] is solved by a finite difference technique on an equidistant grid in a rectangular box of aspect ratio 2 with a resolution of 61×121 grid points. The temperature equation is solved by an ADI scheme using a four times higher spatial resolution (241×481 grid points). The different compositional components are represented by 80,000 markers, which are advected with the flow.

References

Bai, Q., Mackwell, S.J., Kohlstedt, D.L., 1991. High temperature creep of olivine single crystals: mechanical results for buffered samples. J. Geophys. Res. 96, 2441–2463.

Batt, G.E., Braun, J., 1997. On the thermomechanical evolution of compressional orogens. Geophys. J. Int. 128, 364–382.

Beaumont, C., Fullsack, P., Hamilton, J., 1994. Style of crustal deformation in compressional orogens caused by subduction of the underlying lithosphere. Tectonophysics 232, 119–132.

Beaumont, C., Ellis, S., Hamilton, J., Fullsack, P., 1996. Mechanical model for subduction–collision tectonics of alpine-type compressional orogens. Geology 24 (8), 675–678.

Becker, T.W., Faccenna, C., O'Connell, R.J., Giardini, D., 1999. The development of slabs in upper mantle: insights from numerical and laboratory experiments. J. Geophys. Res. 104 B7, 15,207–15,266.

Bird, P., 1979. Continental delamination and the Colorado plateau. J. Geophys. Res. 84 B13, 7561–7571.

Brace, W.F., Kohlstedt, D.L., 1980. Limits on lithospheric stress imposed by laboratory experiments. J. Geophys. Res. 85, 6248–6252.

Braun, J., Beaumont, C., 1989. A physical explanation of the relation between flank uplifts and the break-unconformity at rifted continental margins. Geology 17, 760–764.

Buck, W.R., Poliakov, A.N.B., 1998. Abyssal hills formed by stretching oceanic lithosphere. Nature 392, 272–275.

Buck, W.R., Sokoutis, D., 1994. Analogous model of gravitational collapse and surface extension during continental convergence. Nature 369, 737–740.

Bussod, G.Y., Christie, J.M., 1987. Experimentally deformed lherzolite at hypersolidus conditions: an experimental example of diffusion-controlled power-law creep. In: Int. Union Geod. Geophys., Comptes Rendus 19th Assembly, 196.

Christensen, U.R., Yuen, D.A., 1985. Layered convection induced by phase transitions. J. Geophys. Res. 90, 10,291–10,300.

Cloetingh, S., Burov, E.B., 1996. Thermomechanical structure of European continental lithosphere: constraints from rheological profiles and EET estimates. Geophys. J. Int. 124, 695–723.

Conrad, C.P., Molnar, P., 1997. The growth of Rayleigh–Taylor-type instabilities in the lithosphere for various rheological and density structures. Geophys. J. Int. 129, 95–112.

Cooper, R.F., Kohlstedt, D.L., 1987. Rheology and structure of olivine–basalt partial melts. J. Geophys. Res. 91, 9315–9323.

Faccenna, C., Giardini, D., Davy, P., Argentieri, A., 1999. Initiation of subduction at Atlantic-type margins: insights from laboratory experiments. J. Geophys. Res. 104 B2, 2749–2766.

Ghebreab, W., 1998. Tectonics of the Red Sea region reassessed. Earth-Sci. Rev. 45, 1–44.

Giunchi, C., Sabadini, R., Boschi, E., Gasperini, P., 1996. Dynamic models of subduction: geophysical and geological evidence in the Tyrrhenian Sea. Geophys. J. Int. 126, 555–578.

Houseman, G.A., Molnar, P., 1997. Gravitational (Rayleigh–Taylor) instability of a layer with non-linear viscosity and convective thinning of continental lithosphere. Geophys. J. Int. 128, 125–150.

Jarvis, G., McKenzie, D., 1980. Sedimentary basin formation with finite extension rates. Earth Planet. Sci. Lett. 48, 42–52.

Karato, S., Wu, P., 1993. Rheology of the upper mantle: a synthesis. Science 260, 771–778.

Larsen, T.B., Yuen, D.A., 1997. Ultrafast upwelling bursting through the upper mantle. Earth Planet. Sci. Lett. 146, 393–400.

Larsen, T.B., Yuen, D.A., Malevsky, A., 1995. Dynamical consequences on fast subducting slabs from a self-regulating mechanism due to viscous heating in variable viscosity convection. Geophys. Res. Lett. 22, 1277–1280.

Liu, M., Shen, Y., 1998. Crustal collapse, mantle upwelling, and Cenozoic extension in North America Cordillera. Tectonics 17 (2), 311–321.

Marotta, A.M., Fernandez, M., Sabadini, R., 1998. Mantle uprooting in collisional settings. Tectonophysics 296, 31–46.

McKenzie, D.P., 1978. Some remarks on the development of sedimentary basins. Earth Planet. Sci. Lett. 40, 25–32.

Montagner, J., 1996. Surface waves on a global scale — influence of anisotropy and anelasticity. In: Boschi, E., Ekstrom, G., Morelli, A. (Eds.), Seismic Modelling of Earth Structure, first ed., Editrice Compositori, Bologna, Italy, section 3.2.1

Moresi, L.N., Solomatov, V., 1998. Mantle convection with a brittle lithosphere: thoughts on the global tectonic styles of the earth and venus. Geophys. J. Int. 133, 669–682.

Orihashi, Y., Al-Jailani, A., Nagao, K., 1998. Dispersal of the Afar plume: implications from the spatiotemporal distribution of the Late Miocene to Recent Volcanics, Southwestern Arabian Peninsula. Gondwana Res. 1 (2), 221–234.

Ranalli, G., 1995. Rheology of the Earth. second ed., Chapman and Hall, Boston.

Regenauer-Lieb, K., Yuen, D., 1998. Rapid conversion of elastic energy into plastic shear heating during incipient necking of the lithosphere. Geophys. Res. Lett. 24 (14), 2737–2740.

Royden, L., Keen, C., 1980. Rifting process and thermal evolution of the continental margin of eastern Canada determined from subsidence curves. Earth Planet. Sci. Lett. 51, 343–361.

Schilling, J.-G., Kingsley, R., Hanan, B., McCully, B., 1992. Nd–Sr–Pb isotopic variations along the Gulf of Aden: evidence for Afar mantle plume — continental lithosphere interaction. J. Geophys. Res. 97 B7, 10,927–10,966.

Schmalholz, S., Podladchikov, Y.Y., 1999. Buckling versus folding: importance of viscoelasticity. Geophys. Res. Lett. in press.

Schmeling, H., Bussod, G.Y., 1996. Variable viscosity convection and partial melting in the continental asthenosphere. J. Geophys. Res. 101, 5411–5423.

Schmeling, H., Jacoby, W., 1981. On modelling the lithosphere in mantle convection with nonlinear rheology. J. Geophys. 50, 89–100.

Schmeling, H., Marquart, G., 1991. The influence of second-scale convection on the thickness on continental lithosphere and crust. Tectonophysics 189, 281–306.

Schmeling, H., Schott, B., 1994. On the dynamics of lithospheric rootsDFT workshop on modeling of orogenic processes at lithospheric scale, March 21–23, Berlin. Terra Nostra 3/94, 139–141.

Schott, B., Schmeling, H., 1998. Delamination and detachment of a lithospheric root. Tectonophysics 296 3/4, 225–247.

Schott, B., Yuen, D.A., Schmeling, H., 1999. Viscous heating in heterogeneous media as applied to the thermal interaction between the crust and mantle. Geophys. Res. Lett. 26 (4), 513–516.

Schott, B., Yuen, D.A., Schmeling, H., 2000. The significance of shear heating in continental convergent processes. Phys. Earth Planet. Inter. 118, 273–290.

Ter Voorde, M., van Balen, R.T., Bertotti, G., Cloetingh, S.A.P.L., 1998. The influence of stratified rheology on the flexural response of the lithosphere to (un)loading by extensional faulting. Geophys. J. Int. 134, 721–735.

Thacker, W.C., Long, R.B., 1988. Fitting dynamics to data. J. Geophys. Res. 93, 1227–1240.

Trompert, R., Hansen, U., 1998. Mantle convection simulations with rheologies that generate plate-like behavior. Nature 395, 686–689.

Willett, S., Beaumont, C., Fullsack, P., 1993. Mechanical model for the tectonics of doubly vergent compressional orogens. Geology 21, 371–374.

Yoshioka, S., Yuen, D.A., Larsen, T.B., 1995. Slab weakening: thermal and mechanical consequences for slab detachment. Island Arc 40, 89–103.

TECTONOPHYSICS

www.elsevier.com/locate/tecto

Fast mechanisms for the formation of new plate boundaries

K. Regenauer-Lieb [a,*], D.A. Yuen [b]

[a] *Institut für Geophysik, ETH-Hönggerberg HPP P6.1, 8093 Zürich, Switzerland*
[b] *Minnesota Supercomputer Institute and Department of Geology and Geophysics, Univ. Minnesota, 1200 Washington Avenue, Minneapolis, MN 55415, USA*

Received 2 February 1999; accepted for publication 24 September 1999

Abstract

We demonstrate the important role played by non-linear feedback in promoting the fast breakup of the continental lithosphere and the formation of new plate boundaries. We compare the difference in fluid-dynamic and elasto-visco-plastic approaches and emphasize the advection of stresses in finite-element models with large strain Lagrangian formulation. Two types of instabilities were found at multiple time-length scales ranging from 100 years and km scale to 0.1 Myr and 100 km scale. One mechanism relies on the thermo-mechanical elasto-plastic energy conversion, and the other stems from the void–volatile interaction. Elasticity takes on a guiding role in the propagation of shear zones, closely followed by a thermal wave, which is again followed by a wave of volatiles. This last mechanism consolidates the fault zones and establishes the plate boundary for a geological time-scale. Using a fully coupled thermal–mechanical approach, we obtain for the first time a two-dimensional solution, which shows that the constant velocity boundary conditions can lead to episodic avalanche-type tectonic phases within <0.1 Myr. © 2000 Elsevier Science B.V. All rights reserved.

Keywords: continuum mechanics; fluid dynamics; instabilities; plate tectonics; shear zones

1. Introduction

The theme of the session was focused on continental evolution. This process develops at different time-scales, ranging from magmatic underplating O (Myr) to secular cooling O (100 Myr–Gyr). Here, we wish to address several basic mechanisms which can admit to fast time-scales or the cut of the lithosphere without invoking the conventional idea of plate tectonics over 100 million years with the steady-state cycle of the creation and destruction of plates. Instead, we have focused on local instability mechanisms that rely on non-linear feedbacks. We also address the problem of how these fast thermal time-scale mechanisms, based initially on adiabatic plastic deformation, can be linked into the long plate tectonic scale. For this task, we use a volatile controlled feedback mechanism that is not governed by thermal time-scales. Thus, the goal of this paper is to provide a self-consistent thermal–mechanical model of nucleation of plate boundaries and to provide an explanation of the subsequent long lifetime.

The fundamental basis of our proposed mechanism, can be cast into a four-part feedback process diagram involving momentum, rheology, viscous heating and microstructural damage (grain size and volatiles). Fig. 1a and b show the thermal–

* Corresponding author. Tel.: +41-1633-2058; fax: +41-1633-1065.
E-mail addresses: klaus@msi.umn.edu (K. Regenauer-Lieb), davey@krissy.msi.umn.edu (D.A. Yuen)

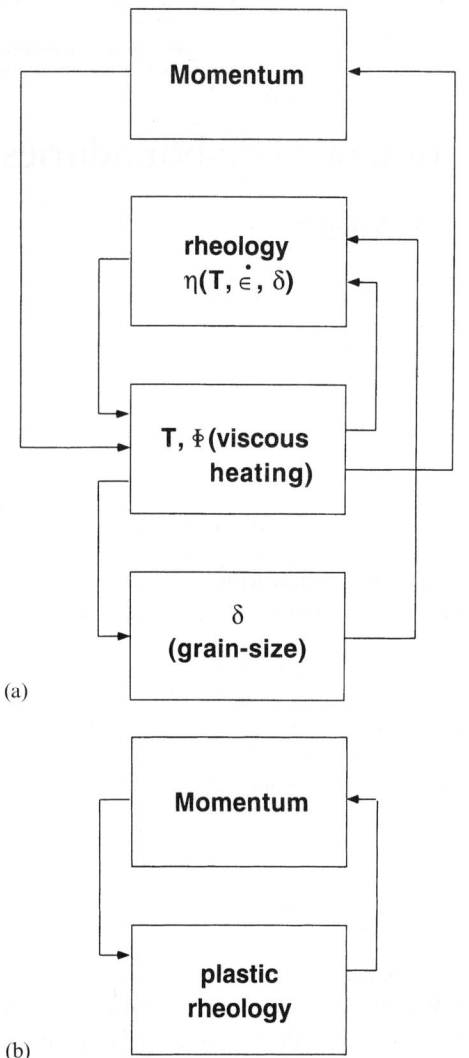

(a)

(b)

Fig. 1. Schematic diagram showing the feedback processes that can be expected during plate tectonic deformation. The first diagram (a) shows the complete set of feedback processes comprising momentum, rheology, viscous heating, and grain size or volatiles and the second model (b) a simplified version which neglects thermal and microstructural feedback components and couples a simple plastic rheology to the momentum equations.

mechanical and (other microscale, volatiles and grain size) feedback mechanisms. For clarity, we separate out the individual mechanisms and discuss the complete model (Fig. 1a) and a stipulated mechanical-momentum feedback mechanism through plastic rheology, as shown in Fig. 1b. The problem has drawn serious attention in the last few years from a fluid dynamic point of view. A Bingham type of fluid, i.e. a fluid with a lower yield stress for activating flow, or alternatively, a viscous fluid with an upper limiting yield stress, was found to be necessary for the creation of plate tectonics with mantle convection codes (Moresi and Solomatov, 1998; Tackley, 1998; Zhong et al., 1998). Although these models now emerge with almost realistic-looking plates, including subduction and spreading, they miss an essential element: they fail to produce transform faults over intermediate time-scales of a few million years. For now, we will leave aside fluid dynamic models and turn instead to elasto-plasticity. We will return to the fluid dynamic approaches in the discussion section.

Only very few authors have addressed the problem from the perspective of continuum mechanics and plasticity theory. We start with the simplest model in Fig. 1b. This model has been used extensively by a group of authors who place emphasis on the brittle part of the lithosphere (Davy et al., 1995; Brune and Ellis, 1997; Buck and Poliakov, 1998). In these models, brittle faults are built, owing to the rapid feedback of momentum and rheology. The advantage of this approach is that shear zones nucleate pervasively. They also appear similar to that which is observed on the brittle surface of the Earth. The disadvantage of the Coulomb approach is that it cannot be extended to the deeper part of the lithosphere and certainly is not applicable to the strongest part of the lithosphere that deforms in most cases by aseismic, ductile faulting. Ultramafic bodies uplifted and exposed to the surface give compelling evidence for these high-temperature mylonitic shear zones in the uppermost mantle (Hobbs and Ord, 1988; Obata and Karato, 1995).

As an alternative to the Coulomb plastic model, a von Mises plastic model has been suggested (Chery et al., 1991; Regenauer-Lieb and Petit, 1997; Regenauer-Lieb and Yuen, 2000), which is a remedy to the problem pointed out. This approach initially still stays within the same feedback scheme, as shown in Fig. 1b, but it produces far fewer faults than the Coulomb approach. As a rule, ductile faults do not rupture seismically, if other feedback processes are neglected. However, with a von Mises elasto-viscoplastic model, thermal mechanical feedback becomes important as soon as mantle faults propagate into the depth where temperatures rise above 900–1000 K. This is due to the coupled shear heating and the temperature-sensitive flow. Besides the thermal weakening aspect, structural damage during deformation needs to be considered.

For this subject, we move into the regime shown in Fig. 1a. It remains an open question which mechanism controls the further evolution of shear zones after they have been nucleated by the mechanism depicted in Fig. 1b. Therefore, we look at both the temperature and the volatile evolution within the same shear zone and decide which of the mechanisms takes over mechanical control and at which point in time. Although, in general, the feedback is subtle and keeps the fault zone narrow, both mechanisms can, in rare cases, cause catastrophic failure. One is known as thermal runaway (Gruntfest, 1963) and the other as ductile or creep fracture (Dodd and Baiy, 1987). Fracture mechanical interpretations will also be discussed. This feedback theme is the main focus of this paper on providing a potentially fast, time-scale mechanism for the development of continental lithosphere, its severage, and finally its ingestion into recycling. The episodic, autocatalytic type of behavior can be metaphored as a 'solid turbulent phenomenon' (Kagan, 1992).

2. Theoretical approach

The key to fast nucleation of ductile shear zones in the uppermost mantle lies in the ability of a highly non-linear material response of dunite, which becomes important at high stress levels. This non-linearity is similar to that observed in a brittle Coulomb material but is described by a smoother von Mises elasto-visco-plastic yield potential, Φ:

$$\Phi = \frac{1}{k_{(T,e)}^2} \sigma_{ij}^d \sigma_{ij}^d - 1 = 0 \qquad (1)$$

where $k_{(T,e)}$ is the temperature- and strain-rate-dependent flow stress and σ_{ij}^d is the deviatoric part of the stress tensor. A simplified approach to yielding of the lithosphere is used, where the ductile rheology is described by a single flow law, which describes low-temperature plasticity and a power law for higher temperatures. Alternatively, the Peierls stress creep law can be used for describing the low-temperature plasticity connecting the brittle- to the high-temperature creep regime (Poirier, 1985). For the problem of fast nucleation of fault zones, the mantle is considered elastic if the relaxation time is longer than 10^4 years and visco-elastic below this, thus defining an initial yield stress. In the visco-elastic regime, the additive strain rate decomposition is made in which:

$$e_{tot} = e_{visc} + e_{el} \qquad (2)$$

and the indices tot, visc and el indicate total, viscous, and elastic strain rates, respectively.

The flow law is derived from a power law fit to the low temperature visco-plastic Peierls stress mechanism of dunite, which defines the flow stress, k, as a function of temperature and strain rate:

$$k_{(T,e)} = k_0 \left[1 - \left(\frac{RT}{Q} (\ln e_0 - \ln e_{visc}) \right)^{0.5} \right] \qquad (3)$$

where Q (500–550 kJ/mol) is the activation energy, e_0 (10^{12}–5×10^{10} s^{-1}) the reference strain rate, and k_0 the so-called Peierls stress. At absolute zero, Eq. (3) gives the value of k_0, which is between 8.5 and 9 GPa. This value decreases with water content (Evans, 1984). It is obvious that there is an upper limit to the ductile yield stress, which is very much lower than this ideal value and also lower than the brittle fracture stress. We use Goetze's criterion (Goetze, 1978), repeatedly cited in the literature. This criterion is applicable when the flow stress from Eq. (1) is equal to the high-temperature fracture stress, $k_{(f)}$, which is in the notation used here:

$$k_{(f)} = \frac{1}{\sqrt{3}} (\sigma_1 - \sigma_3): \text{ for } k_{(f)} < k_{(T,e)} \qquad (4)$$

where σ_3 is the least principal stress and σ_1 the maximum principal stress (positive if compressive). The high-temperature stress therefore increases linearly with depth and, in the case of a tensile tectonic regime, intersects the Peierls stress at 0.5 GPa.

A Lagrangian framework (DT/Dt) is used in the finite element modeling. In the energy equation, we consider both conduction and shear heating. Advection is already included in the Lagrangian approach:

$$\frac{DT}{Dt} = \frac{c_k}{\rho c_p} \nabla^2 T + \frac{\chi}{\rho c_{cp}} \sigma_{ij}^d e_{ij}^{\text{visc}} \tag{5}$$

where c_k is the heat conductivity (3.4 W m^{-1} K^{-1}), ρ the density (3300 kg m^3), c_p (1340 J kg^{-1} K^{-1}) the specific heat and χ (0.9) is the efficiency of conversion of mechanical work into heat.

This approach gives a fully coupled set of thermal–mechanical equations giving rise to momentum, rheology, and shear heating feedback mechanisms described in Fig. 1a. For a second model, an additional feedback due to the presence of volatiles is considered, which has recently been suggested (Regenauer-Lieb, 1999). For this case, the following yield criterion replaces Eq. (1), and the other Eqs. (2)–(5) are still applicable:

$$\Phi = \frac{1}{k_{(T,e)}^2} \sigma_{ij}^d \sigma_{ij}^d + f \cosh\left(\frac{\sigma_{kk}\delta_{ij}}{2k_{(T,e)}}\right) - 1 - f^2 \tag{6}$$

where $\sigma_{kk}\,\delta_{ij}$ is the trace or the isotropic part of the stress tensor, which, when divided by 3, is the equivalent mean flow stress, and f is the void volume fraction, which is defined as the ratio of voids to the total volume of the material (Gurson, 1977). This dilatant plastic yield criterion is also known under the name of 'Gurson's criterion' (Needleman, 1994).

We start with $f=0$ at the beginning of ductile deformation and assume that the void evolution is described by two terms:

$$\frac{df}{dt} = \frac{df_{\text{nuc}}}{dt} + \frac{df_{\text{gr}}}{dt} \tag{7}$$

where the subscripts nuc and gr refer to void nucleation and growth rate. The void growth rate is governed by the isotropic stress tensor, and the viscous strain rate:

$$\frac{df_{\text{gr}}}{dt} = \frac{1}{3}(1-f)e_{ij}^{\text{visc}} \sigma_{kk}\delta_{ij} \tag{8}$$

voids grow in tension and shrink in compression. The nucleation of voids is controlled by the viscous strain energy density:

$$\frac{df_{\text{nuc}}}{dt} = A e_{ij}^{\text{visc}} \frac{3\sigma_{ij}^d}{2k_{(T,e)}} \tag{9}$$

with A describing a normal distribution of newly created voids.

This is the most simple approach to dilatant plasticity that neglects the mechanical interaction between microvoids and other factors (see Regenauer-Lieb, 1998 for a full approach). The interaction of microvoids becomes important at a critical f. In the solution presented here, we stay below such a critical value (no ductile fracture).

3. Numerical approach

We have obtained fully coupled finite-element solutions for both thermal–mechanical and thermal–mechanical–volatile feedback mechanisms. The mechanical equilibrium and energy conservation equations are solved simultaneously along with the rheology. Implicit and explicit techniques are employed prior to, and during, nucleation of shear zones (bifurcation), respectively. The codes ABAQUS STANDARD (ABAQUS/Standard, 1996) releases 5.6 and 5.7 are used (20 000 element plane strain elements) on the Cray C90 and the IBM SP2 at the Minnesota Supercomputer Institute. Since we are dealing with finite amplitude visco-elasto-plasticity, the need for defining an objective rate of the Jaumann stress tensor, τ_{ij}^A, which considers the advection of the stress, especially the rotational component, which is so essential along faults, must be discussed briefly. We note that the Jaumann stress is not considered much by geophysicists in finite-amplitude aspects of viscoelasticity.

The Jaumann stress, τ_{ij}, is defined by the true Cauchy stress, σ_{ij} with respect to the Jacobian of the elastic reference volume, V_0, and the current volume, V:

$$\tau_{ij} = \frac{dV}{dV_0} \sigma_{ij} \tag{10}$$

the elastic spin of such a reference volume is:

$$\Omega_{ij} = \frac{1}{2}\left(\frac{\partial \dot{x}_i}{\partial x_j} - \frac{\partial \dot{x}_j}{\partial x_i}\right) \tag{11}$$

where x_j is the position vector and $\dot{x}_j$ the velocity.

Now, the problem that has been overlooked in continuum mechanics until the late 1960s is that the constitutive equations cannot only rely on a fixed reference frame. In a body that experiences finite strain, the stress should also be invariant with respect to the time derivatives in a rotating coordinate system. Several valid approaches have emerged (see discussion by Maugin, 1992). Here, we only need to consider a deforming body that has an elastic spin. A consistent way of defining

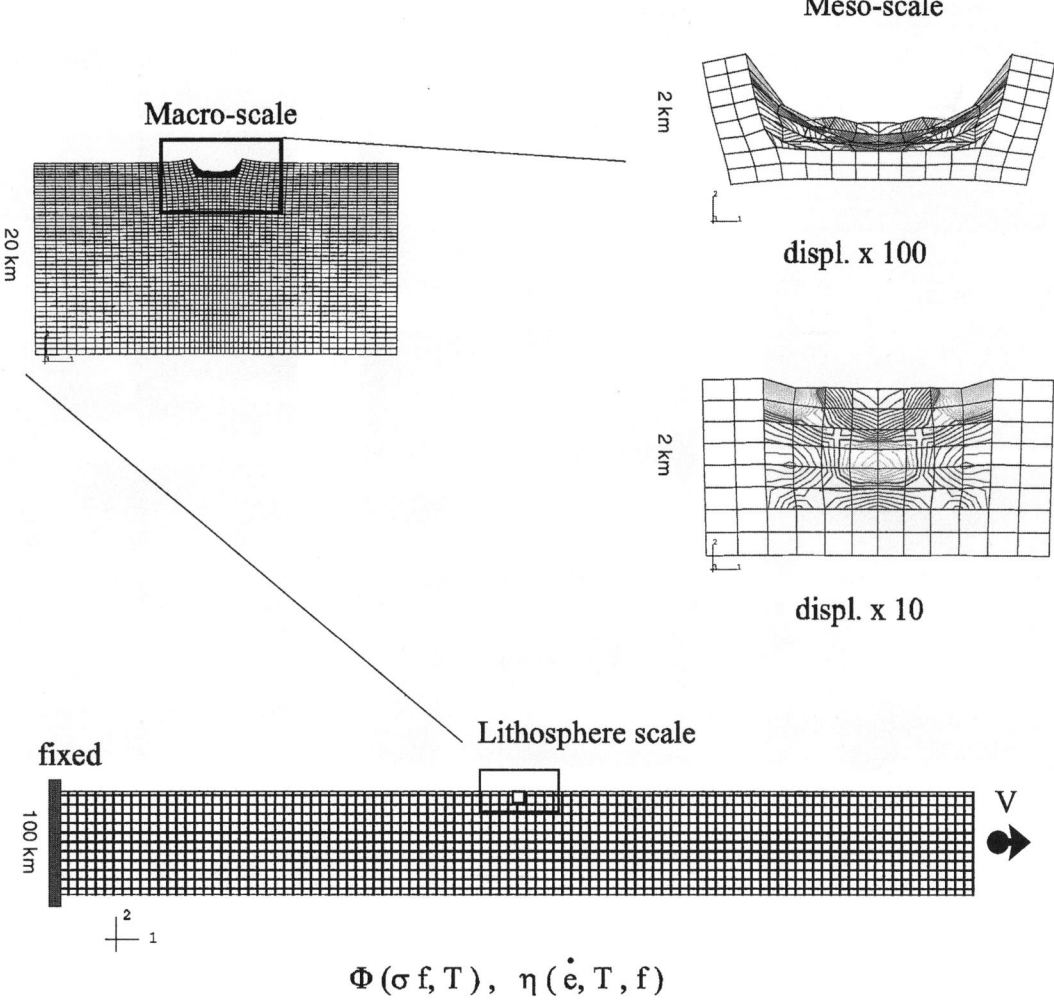

Fig. 2. Same model set-up shown at three different scales, here called the meso-, macro- and lithospheric scales. A small, soft elastoplastic inclusion is imbedded in a dunitic lithosphere (the crust is shaved off). The lithosphere is stretched with constant velocity boundary conditions (0.2–5 cm/yr). The meso- and macro-scales are shown with a displacement magnification (10, 100) to highlight deformation within the soft inclusion. The contour lines in the meso-scale show shear heating at 455 yrs referring to the third frame of panel 1 (Fig. 3).

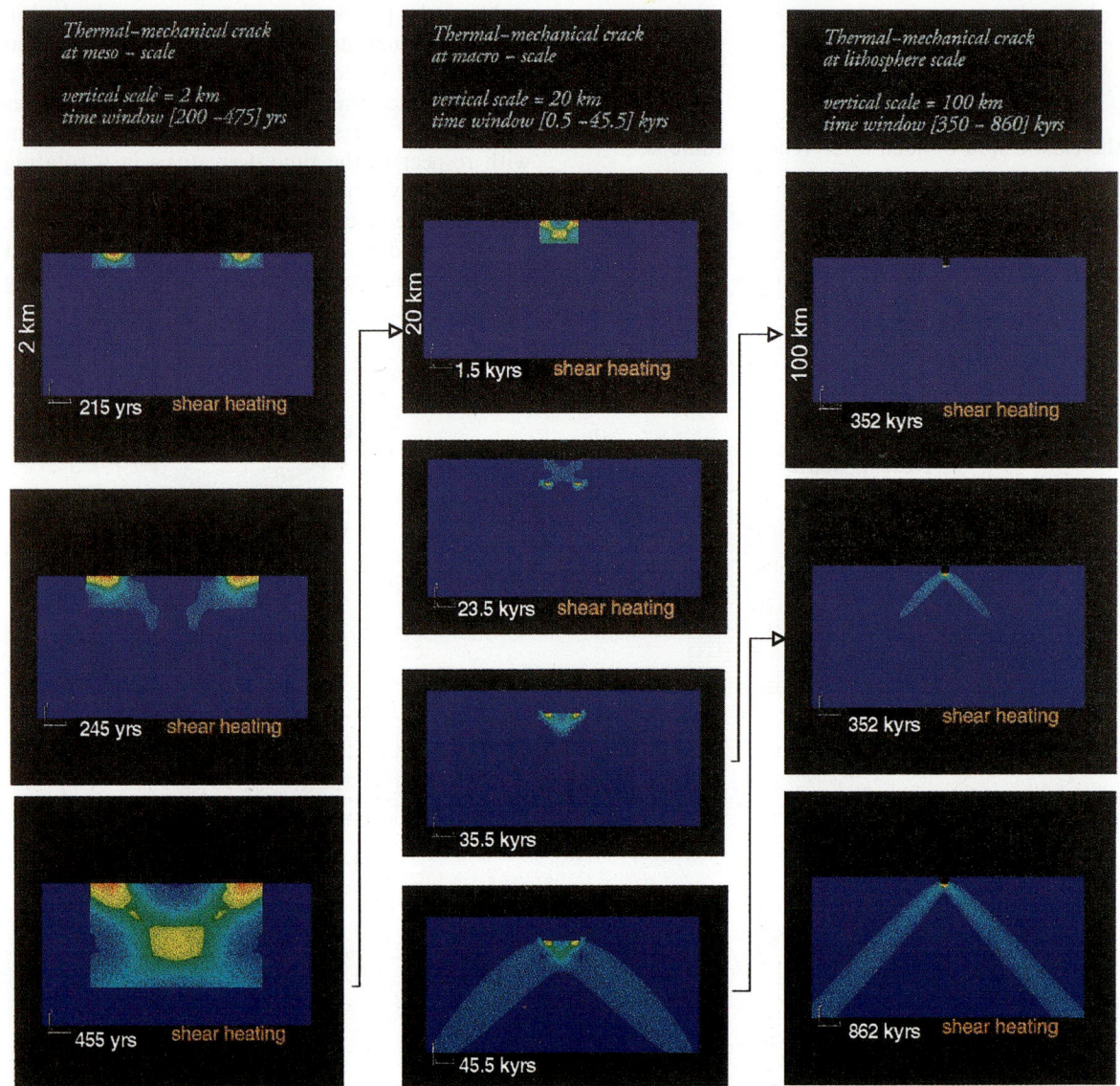

Fig. 3. Three panels of movie frames in three different space scales discussed in Fig. 2. The frames illustrate the formation and propagation of a thermal–mechanical crack (mode 2). The thermal crack migrates very rapdily on a meso-scale, stabilizes on the boundary of the soft inclusion, shoots out of the soft inclusion at 23.5 kyrs and then finally propagates through the entire lithosphere (@ 43.5 kyrs for the continental geotherm shown in panel 2, and @ 700 kyrs for the isothermal 100 km thick reference model shown in panel 3). The maximum amplitude (red) of heating jumps over four orders of magnitude when going from meso- to larger spatial scales. A maximum heating of 0.01 degrees at 455 yrs changes suddenly at 23.5 kyrs to 0.1 degree with another jump at 35 kyrs to several degrees at 45.5 kyrs, which corresponds to the 352 kyrs case in the reference model. In this and subsequent figures, a linear scaling is used where the blue color corresponds to zero magnitude. Note that due to the shear heating jumps, heating which is still present inside the soft inclusion after 23.5 kyrs, can no longer be resolved with a linear scaling.

the Jaumann stress rate, which is advected and rotates along with the elastic spin, is the corotational Jaumann stress rate:

$$\frac{d\tau_{ij}^{\Delta}}{dt} = \frac{d\sigma_{ij}}{dt} + \sigma_{ik}\Omega_{kj} + \sigma_{jk}\Omega_{ki}. \quad (12)$$

For a von Mises solid, only the elastic spin is important, while the plastic spin is zero. In the general case of an anisotropic medium, however, terms including the plastic spin must be introduced into Eq. (12) for considering plastic rotations in the advected stress rate (Dafalias, 1985; Drozdov, 1996).

4. Model description

We propose that the constitutive equations given above can portray accurately the mechanical behavior of the lithosphere over short to intermediate time-scales, of the order of 10^5 years. For investigating the mechanical changes owing to the transition from one feedback mechanism to another, an essential model set-up shall be discussed. This model set-up was therefore designed to be devoid of any kinematic or geometrical complexities, which are very common amongst general geophysical/geological applications. Additionally, our choice of free slip upper and lower boundaries is guided by the availability of semi-analytical rigid–plastic solutions to the uncoupled system (Dewhurst and Collins, 1973). For understanding the role of feedback processes, we have not included any such complexities because we assume the lid to be slipping freely between the asthenosphere and the lower crust.

In our model set-up (Fig. 2), we investigate a continental lithosphere under a one-sided constant extension (1 mm/yr–10 cm/yr). A local perturbation (soft inclusion) in the otherwise homogeneous lithospheric plate is allowed, which is given by a dimension of the order of 1.5% of the vertical direction (100 km) and 0.3% of the model dimension in the horizontal scale (1000 km). The imperfection is embedded in the top middle portion of the model. A plane strain vertical cross section through the lithosphere is modeled. The perturbation is chosen here to be a weaker substance with a Young modulus an order of magnitude lower and a yield strength two orders of magnitude lower than the remainder of the mantle. A 30 km thick crust has been stripped off in the model, and the mantle has a continental geotherm (Regenauer-Lieb et al., 1999). For the analysis of thermal–mechanical feedback, we use a slightly different reference model (Regenauer-Lieb and Yuen, 1998, 2000), which is only discussed on a lithosphere scale. On this scale, the effect of instability within the soft inclusion cannot be observed, and the strength of the inclusion is set to nil. Another simplification is only to observe the temperature effect due to plastic dissipation, and the plate is therefore set to isothermal (978 K) as the starting model with zero heat flow on the boundaries throughout the duration of the numerical experiment. This set-up resembles the classical notch tension test, as described in the mechanical fracture literature (Knott, 1973). We will examine this point in the discussion.

5. Results

In the thermal–mechanical model without volatiles, plastic instabilities were encountered on three different time and length scales (Fig. 3). Accordingly, we have subdivided the model into three different scales, the meso-scale (2×4 km), the macro-scale (20×40 km) and the lithospheric scale (100×200 km). The meso and macro-scales have been calculated using one and the same high-resolution model (total model dimension 1000 km $\times$ 100 km, 200 m local resolution), while for the lithospheric scale model of the same dimensions, a lower-resolution model was used (1 km local resolution), and the soft inclusion was replaced by a simple notch.

The first meso-scale instability occurs only about 200 years after initial loading and is confined to the inside of the soft inclusion. The instability nucleates on the interface between the soft inclusion and the top of the dunitic mantle, corresponding to Moho level. The second instability occurs after 20 kyrs have elapsed. At this stage, the shear zone that is constricted to the inside of soft inclusion propagates into the deeper mantle below the soft inclusion. Finally, the third instability rips through the entire lithosphere. The individual movie frames shown in Fig. 3 are available in an animated format (Regenauer-Lieb et al., 1999; or http://bobby.msi.umn.edu/klaus/paper.htm).

Multiple scale features develop from the three bifurcations: the smaller the spatial scale, the faster is the shear zone propagation. The first two panels (meso-macro) show the three instabilities for the case of a lithosphere, pulled with a velocity of 5 cm/yr, and the third panel shows the case of the thermal–mechanical reference model that is being pulled with 2 cm/yr. The reference model also explores Young modulus an order of magnitude lower that has the effect of scaling up the time for the onset of the instability by the same order of magnitude. Whole-scale lithosphere failure is observed in the second panel at 45 kyrs, whereas when using the reference model (shown in the third panel in Fig. 3), the same type of instability occurs at 800 kyrs. Note that in the middle panel at 35.5 kyrs, shear heating jumps in magnitude when the shear zone propagates from the soft into the strong material. Owing to this jump, the shear heating, still present inside the soft inclusion, can no longer be resolved with a linear scaling and is masked by the larger anomaly in the strong material. The reference model (Regenauer-Lieb and Yuen, 1998, 2000) only features the large-scale instability and is best suited to investigate the effects of thermal–mechanical feedback.

This is shown in Fig. 4. Initially, from 0 to 350 kyrs, the plate is charged up with elastic energy before localized plastic yielding starts around the imperfection. At this point, the magnitude of heat release is insignificant, as can be seen in the temperature–time evolution, and the total elastic energy of the plate continues to rise. One can compare this elastic loading phase with the analogy of pulling an elastic bow. After 700 kyrs, the temperature rises as the zone of plastic deformation leaves the vicinity of the notch and travels through the lithosphere in the shape of a thermal–mechanical crack. When the crack hits the bottom

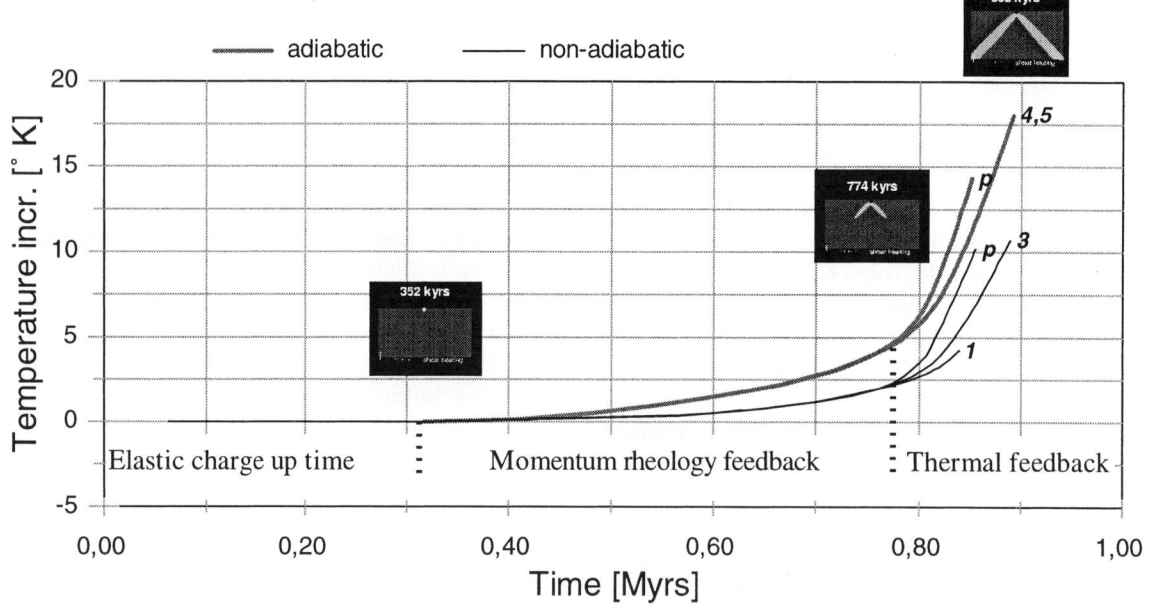

Fig. 4. Isothermal reference model corresponding to the third, lithosphere-scale panel of Fig. 3 (corresponding frames are shown) with thermal mechanical feedback (Fig. 1a) becoming important only after about 800 kyrs. Before this time, the system is chiefly controlled by rheology–moment feedback (Fig. 1b). For the time frame of initial elastic charging, we draw the analogy of an elastic bow. Thermal runaway ('the shooting arrow') is observed for a wide range of power law exponents (>1, p is the extreme case for ideal plastic) and for conductive (thin-) and adiabatic (thick line) cases.

of the plate, the elastic energy stored in the system releases itself in the shear zone leading to a thermal runaway, which, in the analogy, is the shooting of an arrow. This phenomenon was observed for a power-law flow law within a very wide range of power law exponents from $n=1$ (Newtonian) to $n=$ infinity (ideal plastic). This shows that the thermal runaway depends critically on the conversion of elastic into plastic energy (Fig. 1a). It goes without saying that for such short time-scale processes, heat conduction can be neglected. This is justified, when comparing the adiabatic solution with the solution in which heat conduction has been considered (Fig. 4).

The next step is to incorporate void–volatile interactions into the model calculations. Although volatiles have long been recognized as weakening agents in rheological flow laws of the lithosphere (Kirby, 1983), the aspect of a feedback mechanism between volatiles and shear deformation has only been investigated recently. Bercovici (Bercovici, 1998) has incorporated an ad-hoc fluid-dynamical model of void–volatiles in studying the problem of localization at plate margins. He found this feedback mechanism to be more efficient than the viscous heating mechanism. A different elasto-plastic model (Regenauer-Lieb, 1999) was used in an approach to model anomalous degassing along two deep seated shear zones in the Quaternary Eifel volcanic province in Europe (Downes, 1990). An estimated amount of 1 million tonnes of CO_2 per year (Puchelt, 1983) are released along these shear zones within an area of about 1400 km^2. The CO_2 springs carry along a distinct mantle helium signature (Griesshaber et al., 1992). In the elasto-plastic approach (Regenauer-Lieb, 1999), the matrix weakening mechanism by a self-organization of volatiles into void sheets was investigated. Volatile-rich shear zones were predicted, propagating in the direction of the observed mantle shear zones only if a ductile fracture mechanism by void–volatile interaction and void coalescence (Tvergaard and Needleman, 1992; Needleman, 1994) were permitted.

In this paper, we investigate the more general case where the presence of volatiles is well below the anomaly observed in the Eifel volcanic region. This means that volatiles can weaken the rock matrix, but local necking of the matrix in between the volatiles and the subsequent void coalescence is not allowed. The primary interest is whether this weakening mechanism acts earlier or later than the thermal–mechanical instability and whether it is a positive or a negative feedback to shear localization. Our calculations show that the elastic energy was the first wave that travelled ahead of the nascent shear zones. Then, the thermal crack mechanism with or without thermal mechanical feedback follows, and finally a wave of volatiles lags behind. Fig. 5 shows this for the case of the meso-scale instability, which provides a good time and space marker for distinguishing between the mechanisms.

The void–volatile mechanism definitely acts as a stabilizing mechanism that freezes in the shear zone once it has been formed by elasto-plastic and thermal feedback. Once the thermal wave has washed through and the shear zone has been formed, it enters the non-adiabatic realm, and conduction plays an increasing role. This weakens the country rock in the vicinity of the shear zone and finally must diffuse the shear zone. Comparing our reference model at 820 and 900 kyrs, it becomes evident that the void volatile wave finally catches up with the thermal wave at 900 kyrs and becomes the more important crack (Fig. 6).

6. Discussion

Our work has yielded the following conclusions:

(1) The major difference between a purely mechanical model and a thermal-void-mechanical model is that the shear zone stays localized in the fully coupled model, whereas it does not do so for the von Mises rheology without additional coupling. The pure von Mises model without thermal feedback is continually looking for and establishing better shear zones, as the geometry changes and the strongly sheared area broadens. With thermal–mechanical and mechanical–volatile feedback, the first shear zone is engraved in the material and weakened locally, so this can be used for a much longer time period. The same applies to a Coulomb model where some weakening mechanism is necessary to stabilize fault zones on a geological time-scale. In a fully coupled solution,

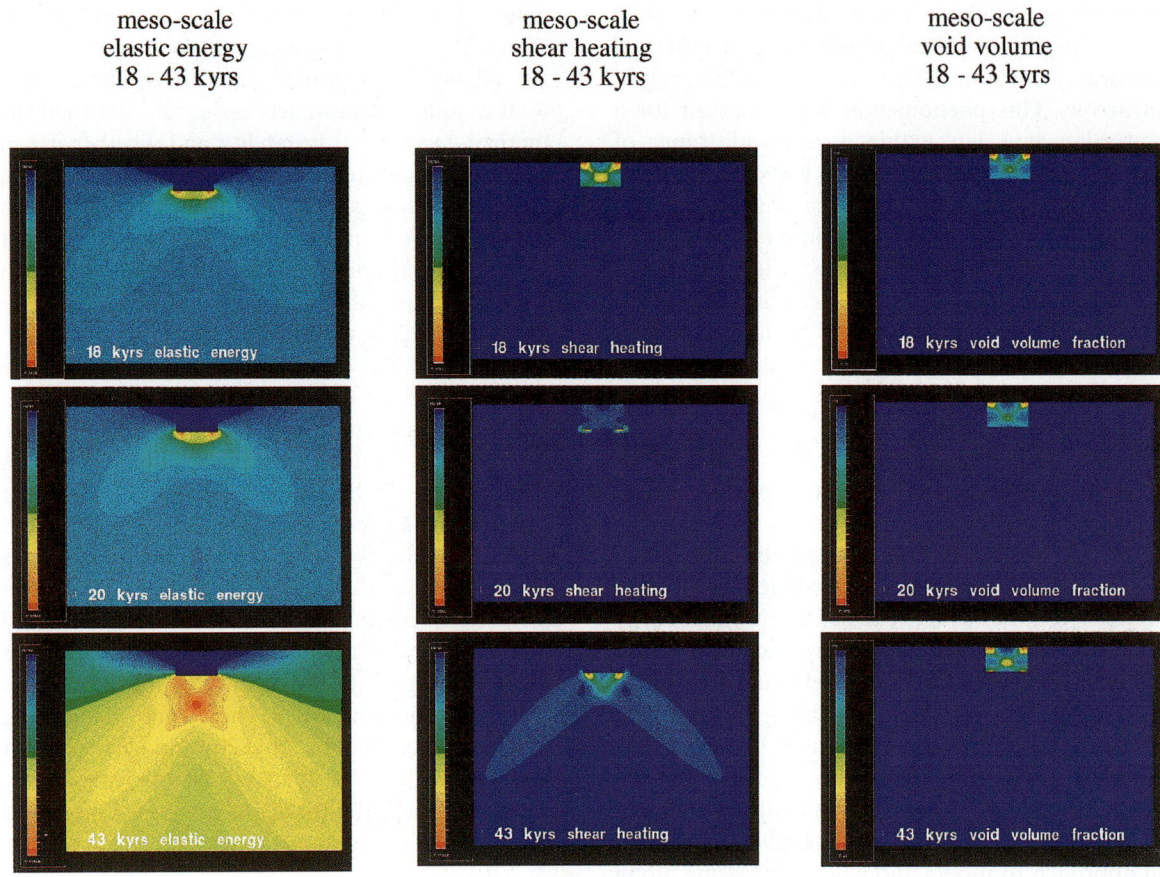

Fig. 5. Elastic energy, shear heating and void volatile for the meso-scale instability. We note here that the volatile wave tracks the thermal wave, which again follows behind the elastic energy wave. The discontinuous amplitude evolution of the shear heating wave has already been described in Fig. 3. It is several degrees before the onset of thermal runaway at 43 kyrs. The elastic energy is one order of magnitude smaller at this moment. The maximum void volume fraction (red) increases from 1×10^{-3} to 2×10^{-3} @ 20 kyrs to 1×10^{-2} @ 43 kyrs.

shear zones jump from one system to the next, only if the boundary conditions become severely incompatible with shearing along the old faults. We have not shown the case for ductile shear zones with a von Mises rheology and a feedback mechanism according to Fig. 1b, i.e. without thermal mechanical feedback. It suffices to note that the shear zone and the thermal crack initially also appear in such a model.

(2) A closer look at the experimental data of the brittle to ductile transition (BDT) reveals that there is a severe deficiency of knowledge of how the yield envelope develops with rising temperature. We need more experiments, microscopic observations and theoretical studies of the BDT phenomenon (Needleman and Tvergaard, 1995; Gumbsch et al., 1998). Some fracture mechanical data for experimentally deformed quartz aggregates exist (Hirth and Tullis, 1994). A transition of mode 1 fracture below (in the uppermost crust) to mode 2 fracture above (in the deep crust) the BDT has been inferred in our study (Fig. 7). This point is discussed below.

(3) In the parlance of solid mechanics, we are investigating problems that must be solved in the area where plasticity theory meets fracture mechanics. Our problem of a mechanical perturbation giving rise to local elastic stress amplifications

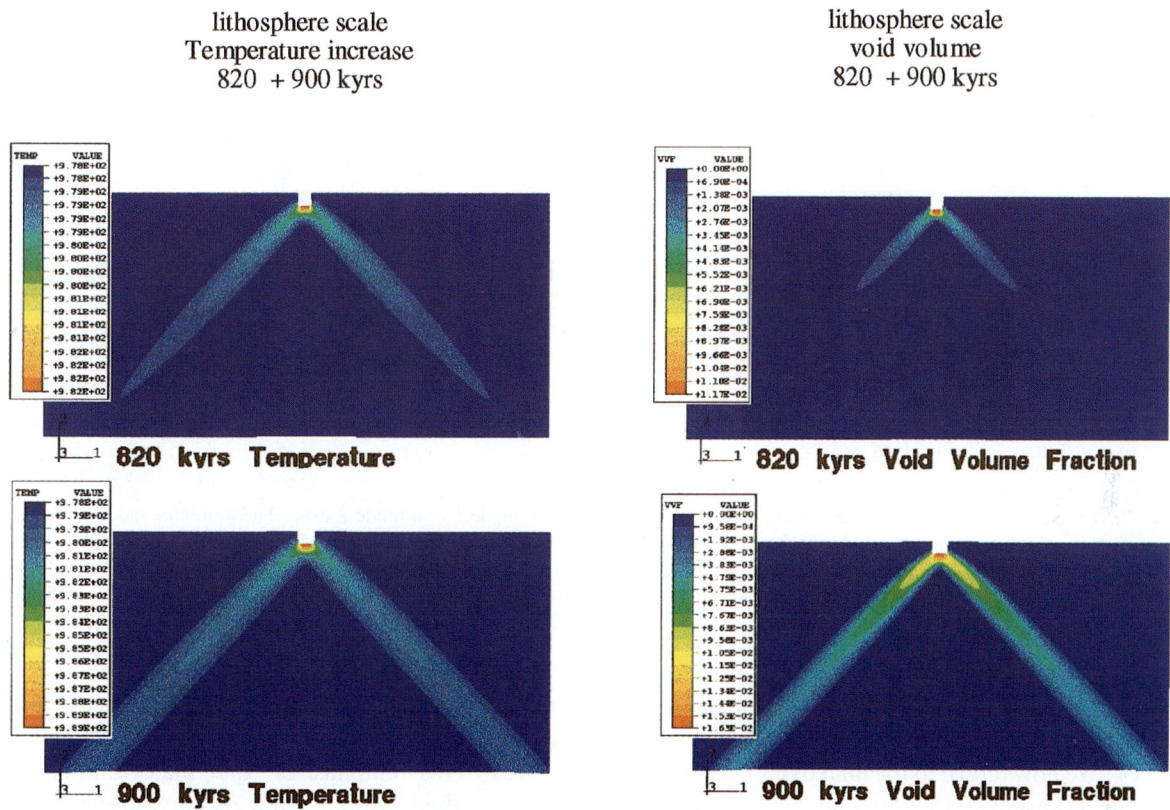

Fig. 6. Comparison of the thermal–mechanical wave with the volatile wave in the reference model at 820 and 900 kyrs. The void–volatile wave still lagging behind at 820 kyrs catches up with the thermal crack at 900 kyrs. The color scale is as follows: minimum temperature 978 K, maximum temperature (red) 982 and 989 K at 920 and 900 kyrs, respectively. The void volume fraction is 1.17×10^{-2} at 820 kyrs, increasing to 1.65×10^{-2} at 900 kyrs.

can be fully described in the inital elastic realm by linear elastic fracture mechanics. A brittle material would react to such a stress amplification by fracture in the direction of σ_1, which is a cleavage crack, also called a mode 1 fracture (Fig. 7). Our particular choice of a rheology, valid above the BDT (von Mises yield criterion) precludes such a mode 1 fracture (Fig. 7b) and only allows mode 2 instabilities. The results of our numerical experiments were anticipated more than 20 years ago (Knott, 1973b) from an extension of linear elastic fracture mechanics. The same type of plane strain perturbation was discussed, and two inverse dislocation pile-ups at 45° to the tensile axis (mode 2 fracture) were postulated (Fig. 7) for the case of a fully ductile material (no cleavage). Knott concluded that this type of fracture can only become unstable after a general yield. The same conclusion was drawn using plasticity theory. It was shown that this mechanism is the basis for shear band formation (Backofen, 1972). Backofen points out that a shear band, by the definition of shear, involves plane strain. Thus, our numerical experiment shows, at an elementary level, the interplay of fracture mechanics (mode 2 crack) evolving into a fully developed ductile shear band.

Volatiles, however, require an opening displacement and favor mode 1, which normally indicates the 'brittle' field. Our numerical result shows a clear control of mode 2 fracture in the ductile domain, even when volatiles are considered. Only for a tensile stress state with a very high volatile content can the volatile containing voids interact and coalesce so that mode 1 can be resumed in

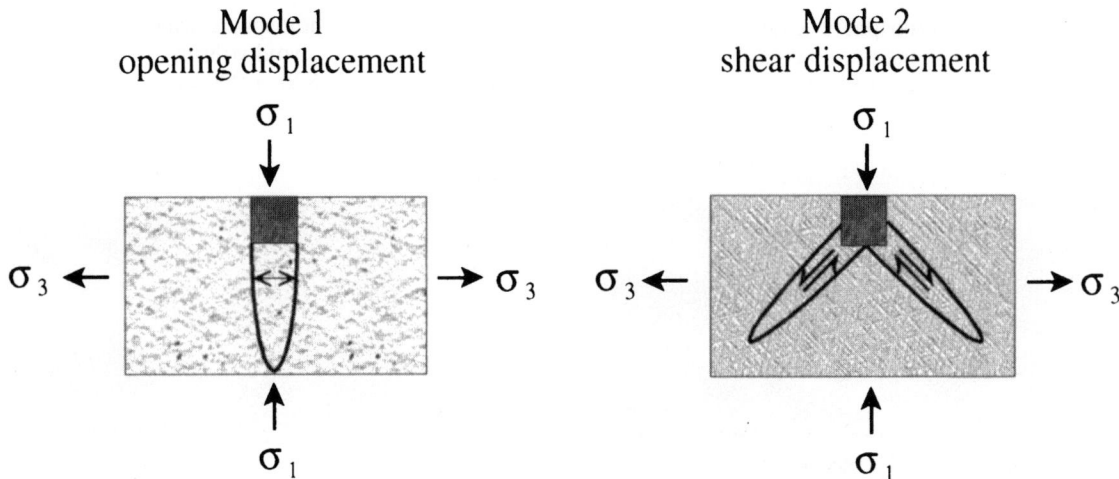

Fig. 7. Displacement modes of fracture mechanics. In plane strain, only mode 1 and mode 2 exist. The antiplane mode 3 requires strain in the third coordinate axis. Mode 1 creates an opening displacement in the direction of the principal compressive stress, nucleates on pre-existing imperfections and is a typical brittle fracture (cleavage). Mode 2 involves a pure shear displacement along the maximum shear stress trajectory and is the key to ductile fracture. This mode governs the mechanical instabilities described in this paper.

accord with the brittle field. This case is presented elsewhere (Regenauer-Lieb, 1999).

(4) We now return to the fluid dynamical models of the lithosphere (Schott and Schmeling 1998; Schott et al. 1999). To some degree, our work parallels the recent independent developments in this area of geophysics (Bercovici, 1996, 1998; Tackley, 1998). However, our model considers a different scale, which is the lithosphere instead of the entire lithosphere–mantle system. A persistent problem of the larger scale models of mantle convection is that they fail to produce transform faults. In this respect, it is useful to compare the vorticity in the solutions that, in theory, can produce transform faults in both scenarios. A fluid-dynamical vortex (Zhang and Yuen, 1996) is a 3-D feature requiring some vertical flushing. However, visco-elasto-plastic vortices (Regenauer-Lieb and Petit, 1997) can develop as 2-D long wavelength shear zones in plates (Fig. 8). We believe that the key to plate tectonics lies in the presence of such transform faults relying on the important role of elasticity, as discussed in the previous section. It follows that plates not only behave like plates, because they have a yield stress, but deform elastically prior to reaching the yield stress. This is a fundamental difference to the fluid dynamical approach.

(5) This emphasizes the major difference between fluid dynamics and elasto-visco-plasticity. Elasto-thermal-mechanical feedback gives rise to short-time-scale instabilities that are not possible in fluid-dynamical models or weakly plastic models with some plasticity added (Moresi and Solomatov, 1998; Trompert and Hansen, 1998). If we compare our solution with the recent models for mantle convection (Ogawa, Schott et al., this volume), we obtain, however, the same general style of self-organized criticality (Bak et al., 1987), which causes an avalanche style of deformation. Episodicity of tectonic phases can be expected even for large-scale constant velocity boundary conditions.

(6) Finally, we suggest that for the time-scales longer than that of adiabatic plastic deformation, the fourth feedback mechanism in Fig. 1 (grain size, volatiles) becomes prevalent. Volatiles (Regenauer-Lieb, 1999) and dynamic recrystallization (Kameyama et al., 1997) can exert an important effect on the healing of plate boundaries. This feedback mechanism can also be supported by other aspects that are self-induced and also caused

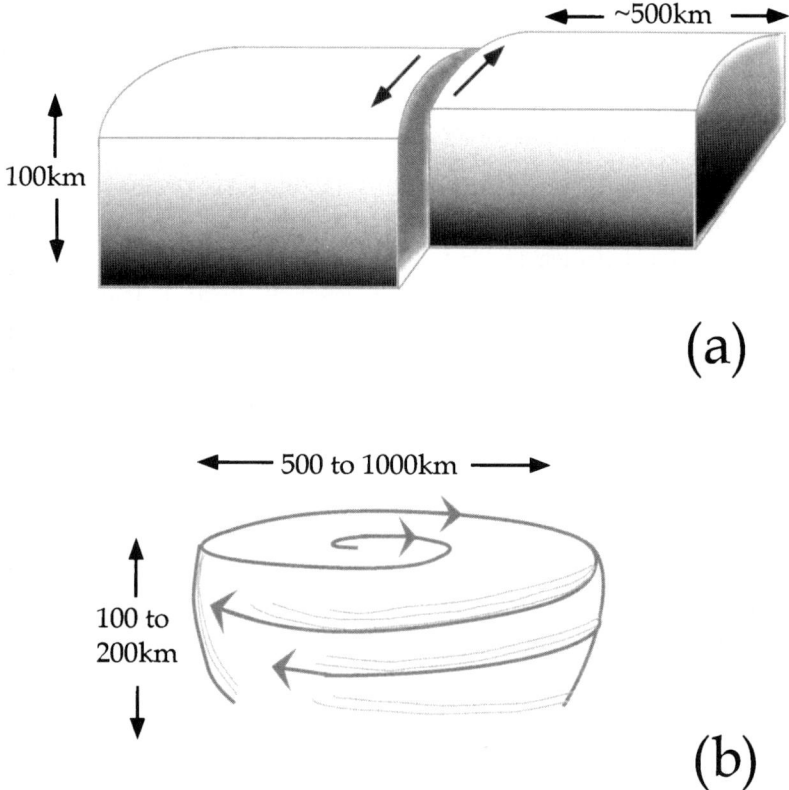

Fig. 8. Schematic drawing of the fluid dynamic vortex and elasto-plastic shear zone with curvature.

by the thermal feedback mechanism. Mechanical heating and influx of volatiles will cause compositional changes within the shear zone, while the evolving microstructure will cause anisotropy.

7. Concluding remarks

What can be learnt from this work for the problem of crustal growth in the Earth? Crustal growth would not be possible if there were a perfect linear viscous coupling between the convecting mantle and the lithosphere. Any heterogeneity would be recycled and mixed efficiently within the mantle (Ten et al., 1998, 2000). Crustal growth relies on two different non-linear processes. One is accretion and segregation on a long time-scale. This is controlled by fluid dynamics of the convecting mantle (see Ogawa, Schott et al., this volume). In this context, avalanche events must leave a positive balance between hot jets accumulating material to the lithosphere and cold downwellings recyling some of the material. This leads to the second process. A positive balance can be achieved by switching on a different kind of rheology inside the lithosphere, allowing decoupling between mantle flow and lithosphere dynamics. One option is to consider only plasticity inside the lithosphere. Such a solution would not alter the time-scale and still allow some avalanche events that swallow the entire lithosphere and episodically produce a new Earth surface, as argued, for instance, for Venus (Steinbach and Yuen, 1992; Trompert and Hansen, 1998). A more fundamental difference is obtained by considering the effect of increasing the Maxwell time-scale inside the lithosphere. This switches on totally different time-scales and vortices that can decouple vertical- from horizontal motion after the accretion and cooling.

Acknowledgements

We would like to thank Paul Sylvester and the GSA for stimulating this contribution, and we acknowledge Louis Moresi and Hans Mulhaus for very helpful comments. For assistance in graphical design, we would like to thank Christa Lieb, Melanee Lundgreen and Kerri Root. This research has been supported by the ETH Zurich (ETHZ contribution number 1078), the geophysics program of the National Science Foundation and the geosciences program of the Department of Energy.

References

ABAQUS/Standard, 1996. User's Manual Version 5.6 Vol. 1. Hibbit Karlsson and Sorenson.

Backofen, W.A., 1972. Deformation Processing. Addison-Wessley Series in Metallurgy and Materials. Addison-Wesley, Reading, MA. 326 pp.

Bak, P., Tang, C., Wiesenfeld, K., 1987. Self organized criticality: An explanation of 1/f noise. Phys. Rev. Lett. 59, 381.

Bercovici, D., 1996. Plate generation in a simple model of lithosphere-mantle flow with dynamic self-lubrication. Earth Planet. Sci. Lett. 144, 41–51.

Bercovici, D., 1998. Generation of plate tectonics from lithosphere-mantle flow and void–volatile self-lubrication. Earth Planet. Sci. Lett. 54, 139–151.

Brune, J.N., Ellis, M.A., 1997. Structural features in a brittle–ductile wax model of continental extension. Nature 387, 67–70.

Buck, W.R., Poliakov, A.N.B., 1998. Abyssal hills formed by stretching oceanic lithosphere. Nature 392, 272–275.

Chery, J., Vilotte, J.P., Daigniers, M., 1991. Thermomechanical evolution of a thinned continental lithosphere under compression: implication for the Pyrenees. J. Geophys. Res. 96 B3, 4385–4412.

Dafalias, Y.F., 1985. The plastic spin. J. Appl. Mech. 52, 865–871.

Davy, P., Hansen, A., Bonnet, E., Zhang, S.Z., 1995. Localization and fault growth in layered brittle–ductile systems: implications for deformations of the continental lithosphere. J. Geophys. Res. 100, 6281–6294.

Dewhurst, P., Collins, I.F., 1973. A matrix technique for constructing slip-line field solutions to a class of plane strain plasticity problems. Int. J. Num. Meth. Eng. 7, 357–378.

Dodd, B., Baiy, Y. (Eds.), 1987. Ductile Fracture and Ductility Academic Press, New York, 309 pp.

Downes, H., 1990. Shear zones in the upper mantle — relations between geochemical enrichment and deformation in mantle peridotites. Geology 18, 374–377.

Drozdov, A.D., 1996. Finite Elasticity and Viscoelasticity. World Scientific, Singapore.

Evans, B., 1984. The effect of temperature and impurity content on indentation hardness of quartz. J. Geophys. Res. 89, B6, 4213–4222.

Goetze, C., 1978. The mechanisms of creep in olivine. Phil. Trans. R. Soc. Lond. 288, 99–119.

Griesshaber, E.O, Nions, R.K., Oxburgh, E.R., 1992. Helium and carbon isotope systematics in crustal fluids from the Eifel, the Rhine Graben and Black Forest, FRG. Chem. Geol. 99, 213–235.

Gruntfest, I.J., 1963. Thermal feedback in liquid flow: plane shear at constant stress. Trans. Soc. Rheol. 7, 195–208.

Gumbsch, P., Riedle, J., Hartmaier, A., Fischmeister, H.F., 1998. Controlling factors for the brittle-to-ductile transition in Tungsten single crystals. Science 282, 1293–1295.

Gurson, A.L., 1977. Continuum theory of ductile rupture by void nucleation and growth: Part 1 — yield criteria and flow rules for porous ductile materials. J. Eng. Mater. Techn. 99, 2–15.

Hirth, G., Tullis, J., 1994. The brittle–plastic transition in experimentally deformed quartz aggregates. J. Geophys. Res. 99, B6, 11 731–11 747.

Hobbs, B.E., Ord, A., 1988. Plastic instabilities: implications for the origin of intermediate and deep focus earthquakes. J. Geophys. Res. 93, B9, 10 521–10 540.

Kagan, Y.Y., 1992. Seismicity: turbulence of solids. Nonlinear Sci. Today 2, 1–13.

Kameyama, M.C., Yuen, D.A., Fujimoto, H., 1997. The interaction of viscous heating with grain-size dependent rheology in the formation of localized slip zones. Geophys. Res. Lett. 24 (20), 2523–2526.

Kirby, S.H., 1983. Rheology of the lithosphere. Rev. Geophys. Space Phys. 21, 1458–1487.

Knott, J.F. (Ed.), Fracture Mechanics 1973. Butterworth, London, 273 pp.

Maugin, G.A., 1992. The Thermodynamics of Plasticity and Fracture. Cambridge University Press, Cambridge. 350 pp

Moresi, L., Solomatov, V., 1998. Mantle convection with a brittle lithosphere: thoughts on the global tectonic styles of the Earth and Venus. Geophys. J. Int. 133, 669–682.

Needleman, A., 1994. Computational modeling of material failure. Appl. Mech. Rev. 47, 34–42.

Needleman, A., Tvergaard, V., 1995. Analysis of a brittle–ductile transition under dynamic shear loading. Int. J. Solids Struc. 32, 17–18, 2571–2590.

Obata, M., Karato, S., 1995. Ultramafic pseudotachylite from Ivrea-Verbano Zone, northern Italy. Tectonophysics 242, 313–328.

Poirier, J.-P. (Ed.), Creep of Crystals: High Temperature Deformation Processes in Metals Ceramics and Minerals 1985. Cambridge University Press, Cambridge.

Puchelt, H., 1983. Carbon dioxide in the Rhenish Massif. In: Fuchs, K.von Gehlen, K., Mälzer, H., Murawski, H., Semmel, A. (Eds.), Plateau Uplift. Springer, Berlin, p., 152

Regenauer-Lieb, K., Petit, J.P., 1997. Cutting of the European continental lithosphere: Plasticity theory applied to the present Alpine collision. J. Geophys. Res. 102, 7731–7746.

Regenauer-Lieb, K., 1998. Energy estimates for large scale continental deformation: focussing of the strain energy density in the ductile part of the lithosphere due to kinematics and rheology (Energieabschätzungen für großräumige plattentektonische Deformation: Kinematisch- und rheologisch bedingte Fokussierung der Deformationsenergiedichte im duktilen Bereich der Lithosphäre). Habilitation, University of Mainz, 135 pp.

Regenauer-Lieb, K., Yuen, D., 1998. Rapid conversion of elastic energy into shear heating during incipient necking of the lithosphere. Geophys. Res. Lett. 25 (14), 2737–2740.

Regenauer-Lieb, K., 1999. Dilatant plasticity applied to Alpine Collision: ductile void growth in the intraplate area beneath the Eifel Volcanic Field. J. Geodynam. 27, 1–21.

Regenauer-Lieb, K., Petit, J.P., Yuen, D., 1999. Adiabatic shear bands in the lithopshere: Numerical and experimental approaches. Electron. Geosci. 4 (2).

Regenauer-Lieb, K., Yuen, D., 2000. Quasi-adiabatic thermal–mechanical instabilities associated with necking processes of an elasto-viscoplastic lithosphere. Phys. Earth Planet. Inter. 118 in press.

Schott, B., Schmeling, H., 1998. Delamination and detachment of a lithospheric root. Tectonophysics 296, 3–4, 225–247.

Schott, B., Yuen, D.A., Schmeling, H., 1999. Viscous heating in heterogeneous media as applied to the thermal interaction between the crust and mantle. Geophys. Res. Lett. 26 (4), 513–516.

Steinbach, V., Yuen, D.A., 1992. The effects of multiple phase transitions on Venusian mantle convection. Geophys. Res. Lett. 19, 2243–2246.

Tackley, P., 1998. Self-consistent generation of tectonic plates in three-dimensional mantle convection. Earth Planet. Sci. Lett. 157, 9–22.

Ten, A.A., Podladchikov, Y.Y., Yuen, D.A., Larsen, T.B., Malevsky, A.B., 1998. Comparison of mixing properties in convection with the particle-line method. Geophys. Res. Lett. 25 (16), 3205–3208.

Ten, A.A., Podladchikov, Y.Y., Yuen, D.A., 2000. Visualization and analysis of mixing dynamical properties in convecting systems with different rheologies. Electron. Geosci. 4 (3).

Trompert, R., Hansen, U., 1998. Mantle convection simulations with rheologies that generate plate-like behaviour. Nature 395, 686–689.

Tvergaard, V., Needleman, A., 1992. Elastic–viscoplastic analysis of ductile failure. In: Finite Inelastic Deformation Theory and Applications. Springer, Berlin, pp. 3–14.

Zhang, S., Yuen, D.A., 1996. Intense local toroidal motion generated by variable viscosity compressible convection in 3_D spherical shell. Geophys. Res. Lett. 23, 3135–3138.

Zhong, S., Gurnis, M., Moresi, L., 1998. Role of faults, nonlinear rheology, and viscosity structure in generating plates from instantaneous mantle flow models. J. Geophys. Res. 103, B7, 15 255–15 268.

TECTONOPHYSICS

www.elsevier.com/locate/tecto

Growth and recycling of early Archaean continental crust: geochemical evidence from the Coonterunah and Warrawoona Groups, Pilbara Craton, Australia

Michael G. Green [a,*], Paul J. Sylvester [b], Roger Buick [a]

[a] *School of Geosciences, FO5, The University of Sydney, Sydney, NSW 2006, Australia*
[b] *Department of Earth Science, Memorial University of Newfoundland, St. John's, NF A1B 3X5, Canada*

Received 25 February 1999; accepted for publication 24 September 1999

Abstract

In the Pilgangoora Belt of the Pilbara Craton, Australia, the Coonterunah Group and Carlindi granodiorite underlie the Warrawoona Group beneath an erosional unconformity, providing evidence for emergent continental crust at ~3.47 Ga. The basalts either side of the unconformity are remarkably similar, with N-MORB-normalised enrichment factors for LILE, Th, U and LREE greater than those for Ta, Nb, P, Zr, Y and M-HREE. Geological and geochemical evidence shows that both the Coonterunah and Warrawoona Groups were erupted onto continental basement, and that the basalts assimilated up to 25% crustal material. Geochemical modelling suggests that the mantle source of the basalts was as depleted in incompatible elements as the present-day mantle. Therefore, the early development of the Pilbara Craton records the interaction of depleted mantle with pre-3.5 Ga continental basement. Since only very small relics of pre-3.52 Ga rocks have been discovered in the Pilbara, it is likely that the pre-3.5 Ga continental basement has been substantially recycled. This suggests that the recycling of continental material was an important process in the Archaean, and consequently that many estimates of continental growth rates in the early Archaean are too low. The evidence is most consistent with steady-state crustal evolution models in which large volumes of continental crust were extracted from the mantle early in Earth's history, and then efficiently recycled. © 2000 Elsevier Science B.V. All rights reserved.

Keywords: basalt geochemistry; crustal growth; early Archaean; Pilbara Craton; recycling

1. Introduction

Unravelling early Earth processes is hampered by the limited outcrop of pre-3.0 Ga terrains, with the oldest-known rocks, the 4.03 Ga Acasta gneisses (Bowring and Williams, 1999), leaving over 500 million years with no known terrestrial rocks after planetary accretion. The absence of early rocks is partly filled by <4.28 Ga detrital zircons (Froude et al., 1983; Compston and Pidgeon, 1986; Maas et al., 1992), but global conclusions are difficult to draw from these small relics. Moreover, all known pre-3.5 Ga terrains are strongly deformed and highly metamorphosed, making interpretation very difficult. Therefore, the processes controlling crustal evolution during the first billion years of Earth's history are poorly constrained.

* Corresponding author. Tel.: +61-2-9351-3682; Fax: +61-2-9351-0184.
E-mail address: migreen@mail.usyd.edu.au (M.G. Green)

0040-1951/00/$ - see front matter © 2000 Elsevier Science B.V. All rights reserved.
PII: S0040-1951(00)00058-5

Given this, there are two general models to explain the timing of global differentiation of continental crust from the mantle. According to these, the present-day volume of continental crust formed either very early in Earth's history, with subsequent steady-state recycling (Armstrong, 1968, 1981, 1991; Bowring and Houch, 1995), or through gradual growth punctuated by periods of rapid growth, with less recycling (Hurley and Rand, 1969; Reymer and Schubert, 1984; Taylor and McLennan, 1985, 1995; McCulloch and Bennett, 1993, 1994; Collerson and Kamber, 1999). In other words: does the present age distribution of continental crust record merely its preservation, and hence represent only the portions of continental lithosphere large enough and thick enough to resist the Earth's dynamic processing, or does it document the gradual extraction of continents from a progressively depleting mantle, and hence reflect the age spectrum of continental growth? Resolving these issues is critical to understanding planetary differentiation.

Two types of early crust have been proposed for planets: a primary crust formed by accretionary heating, for example the lunar ferroan anorthosites, and a secondary crust formed by partial melting of the mantle, such as lunar mare and terrestrial oceanic crust (Taylor, 1989). The earliest terrestrial crust would not have been composed of ferroan anorthosites because the Earth is relatively poor in Ca and Al, plagioclase converts to garnet at shallow terrestrial depths and plagioclase will not float on wet terrestrial basaltic magmas (Taylor, 1989). Therefore, it seems probable that the earliest terrestrial crust was basaltic, and most likely oceanic depending on the timing, volume and source of aqueous degassing (de Wit and Hynes, 1995). Although this basaltic carapace would have been chemically distinct from the underlying mantle due to partial melting and crystal fractionation processes, the formation of more evolved magma is required to make continental crust. The most plausible process for producing continental crust is by modification of basaltic crust during subduction-like petrogenesis (Ellam and Hawkesworth, 1988). It is envisaged that the first continental-type magmas would have erupted onto or intruded into the basaltic crust. In the steady-state models, the production of evolved magma and consequent depletion of the mantle would have been rapid and early, but direct geological evidence of this differentiation episode would be scarce in the rock record because of effective recycling. In the gradual growth models, the extraction of continental crust from primitive or underdepleted (composition between primitive and depleted, where depleted is defined as elemental abundances equivalent to modern N-MORB source) mantle would have been delayed, so that after 1.5 billion years only a small fraction of the present-day crust had formed. Again, relics of early differentiation would be rare in pre-3.0 Ga terrains, but in this case through minimal creation rather than through vigorous recycling.

Around 3.5 billion years ago, the Pilbara and Kaapvaal Cratons began to assemble, preserving the oldest-known low-grade supracrustal rocks. In the gradual growth models, the appearance of these two crustal blocks in the geological record represents one of the first periods of continental growth. However, the roughly coeval appearance of two ancient continental blocks which were physically able to survive for 3.5 billion years does not necessarily prove that they grew from primitive or underdepleted mantle. On the contrary, the establishment of relatively stable continental crust may require the interaction and recycling of older continental material to form lithospheric domains which were large, cool and rigid enough to survive 3.5 billion years of Earth dynamics. Therefore, providing evidence of early Archaean crustal recycling would indicate that the volume of continental crust extracted from the mantle was greater during the Earth's early history than suggested by the present geological record.

Buick et al. (1995) imply that crustal evolution in the Pilbara Craton began with the basalt-dominated Coonterunah Group, into which granodiorite subsequently intruded, forming a buoyant, emergent continental basement, that was then eroded and unconformably covered by further basaltic eruptions. This paper discusses the composition and petrogenesis of these basalts, which preserve strong evidence of contamination by even older continental crust. Since very little of this older continental basement has been preserved, it must have been efficiently recycled. This shows that crustal growth models which exclude significant recycling probably underestimate the volume of continental crust in the early Archaean.

2. Geological setting

The Pilbara Craton in Western Australia (Fig. 1) comprises volcanosedimentary greenstone belts which surround large ovoid granitoid complexes. Prior to 1995, the base of the supracrustal volcanosedimentary succession was thought to be the ~3.45 Ga Warrawoona Group (Hickman, 1983; Barley, 1993). However, the discovery of an erosional unconformity in the Pilgangoora Belt (Buick et al., 1995), in an area previously mapped as Warrawoona Group, has provided a new basal unit to the stratigraphy, the ~3.52 Ga Coonterunah Group.

The Coonterunah Group is composed of tholeiitic basalt and gabbro with minor felsic volcanic rocks, magnesium basalt, chert and clastic sedimentary units. An eruption age of 3515.2 ± 2.7 Myr has been determined by SHRIMP U–Pb zircon dating of a hyaloclastic rhyolite high in the stratigraphy (Buick et al., 1995). The metamorphic grade varies from mid-greenschist (actinolite, chlorite, epidote) to lowermost amphibolite facies (actinolite to low-Al hornblende transition; Spear, 1995). The spatial distribution of the metamorphic facies has yet to be determined, but may be influenced by contact metamorphism from the large intrusive Carlindi granodiorite. This granodiorite was in turn intruded by quartz-phyric microgranite with a SHRIMP U–Pb zircon age of 3467.6 ± 3.7 Myr (Buick et al., 1995). The terrain was then uplifted, with associated broad open folding and large-scale block rotation, allowing oblique to semi-conformable erosion of the Coonterunah Group and Carlindi Batholith. The upper units of the basalt-dominated ~3.46 Ga Warrawoona Group were then deposited upon the resulting unconformity.

In the Pilgangoora Belt, the units now uniformly dip subvertically with younging away from the Carlindi Batholith. The unconformity crops out over a strike length of more than 50 km (Fig. 2). Locally, the Coonterunah Group is up to 6 km thick, the Warrawoona Group is up to 4 km thick and the Carlindi granodiorite extends for over 30 km from where it is cut by the unconformity. There is little evidence for polyphase intrusion in the Carlindi granodiorite, with recent geological and geophysical mapping failing to detect any major tectonic breaks or compositional variations. The western portion of the Pilgangoora Belt was intensely deformed with a well-developed bedding-parallel schistosity in the greenstones and marginal granodiorite undergoing folding during two local tectonic events. One of these events is probably dated by galena Pb–Pb ages of ~2640 to 2760 Ma for vein-related mineralisation in the Carlindi granodiorite and within the Coonterunah–Warrawoona basalts (Richards et al., 1981). The other may be related to a Pb–Pb isotopic age of ~2900 Ma for peak metamorphic fold-related gold mineralisation in the extreme western Pilgangoora Belt (Neumayr et al., 1998), although these rocks may not correlate with the Coonterunah–Warrawoona basalts. Unravelling stratigraphic relationships in this deformed western domain is difficult, and the unconformity cannot be traced

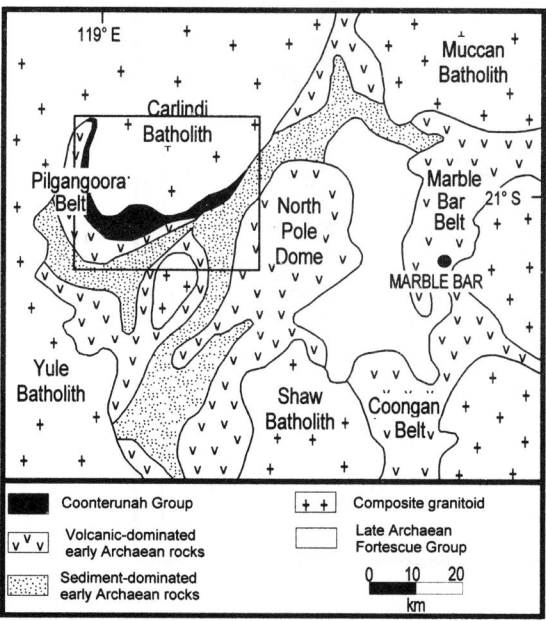

Fig. 1. Geological map of part of the eastern Pilbara Craton, showing the location of the Pilgangoora Belt and general distribution of rock-types. Box represents outline of Fig. 2.

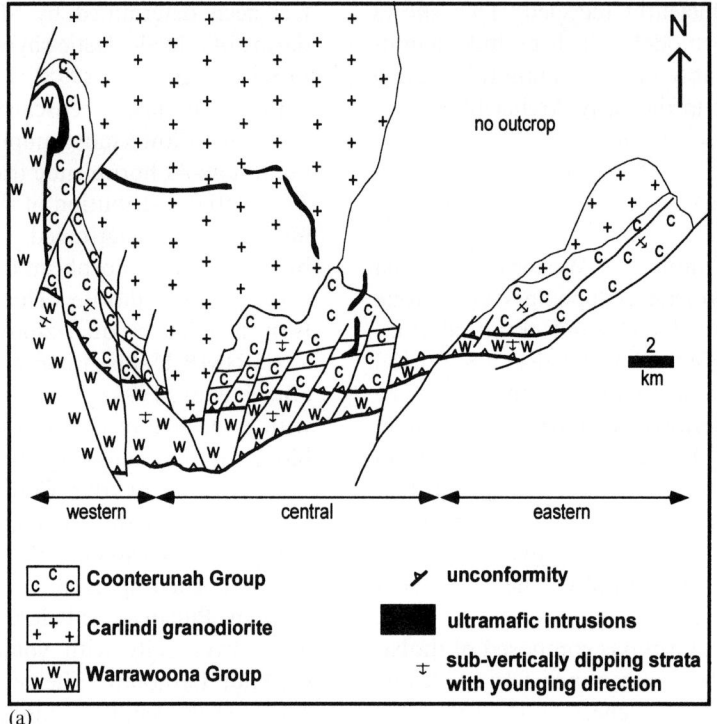

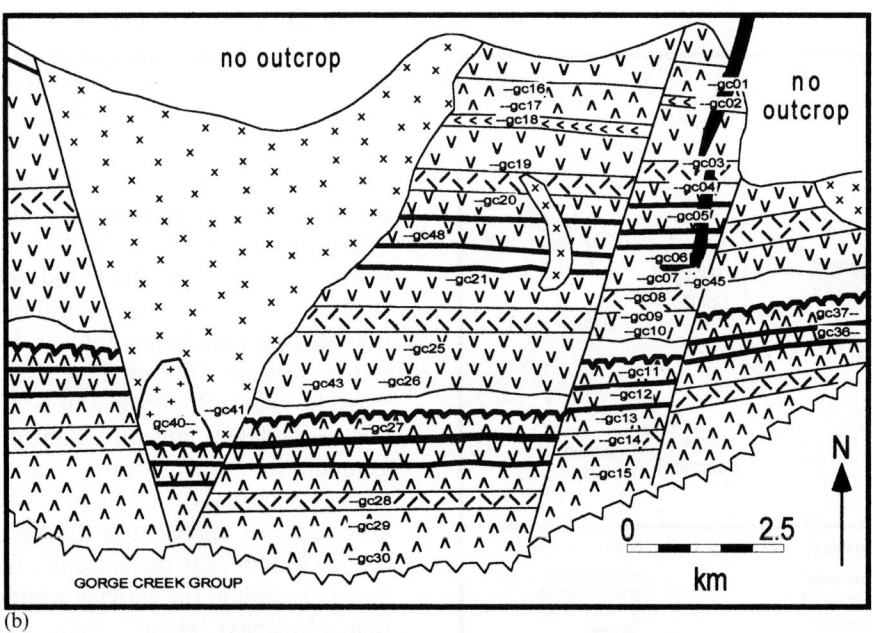

Fig. 2. (A) Geological map of Pilgangoora Belt showing the Coonterunah and Warrawoona Groups, the Carlindi granodiorite and the structural domains. (B) Schematic map of the central domain of the Pilgangoora Belt with sample sites marked (minus 0697 postscript). Rock-type symbols are described in Fig. 3.

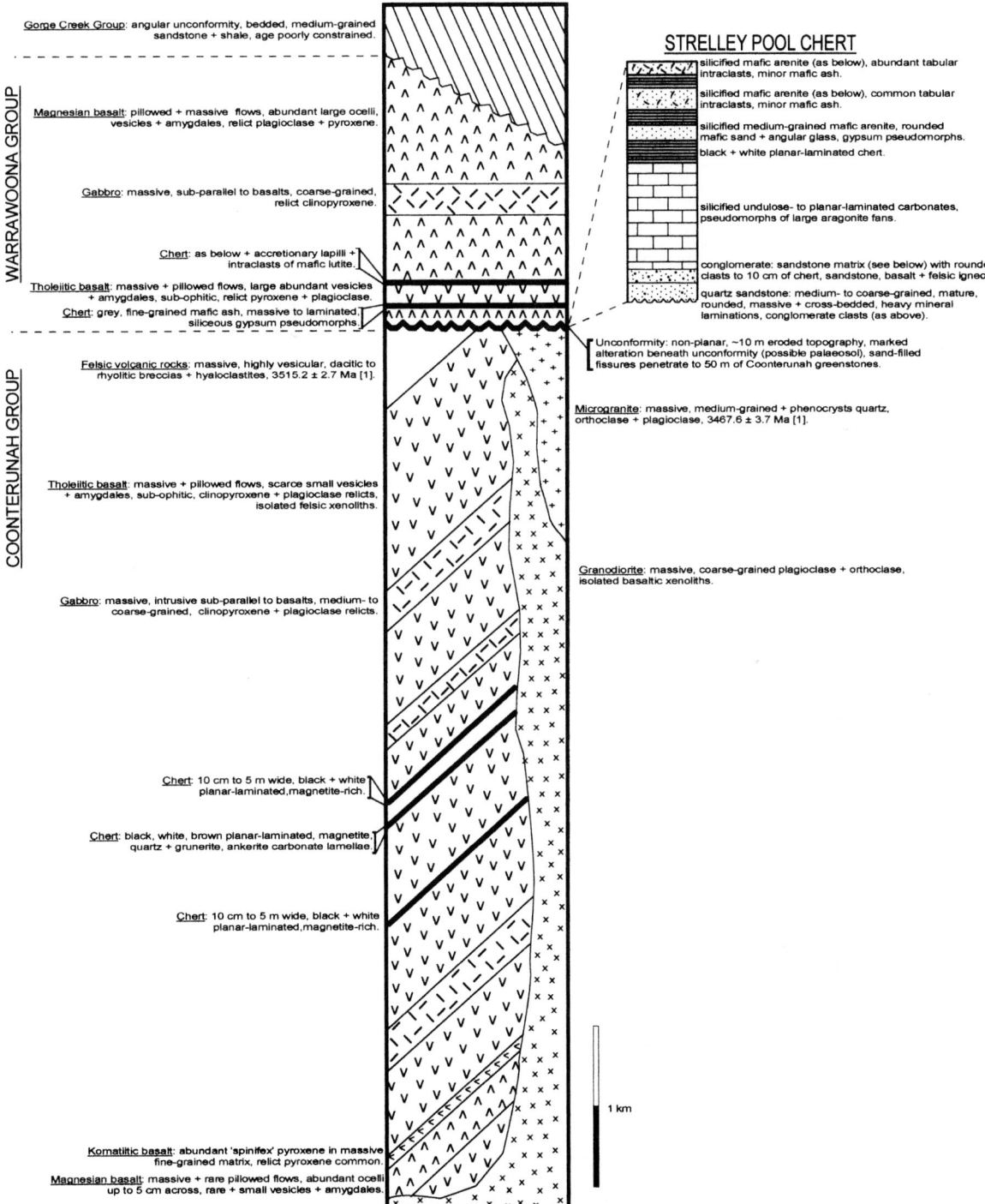

Fig. 3. Generalised stratigraphic section of the Coonterunah–Warrawoona unconformity succession with lithological descriptions and zircon SHRIMP ages from Buick et al. (1995).

Table 1
XRF–ICPMS analyses of Coonterunah–Warrawoona basalts and gabbros, and Carlindi granitoids. Major elements (wt%), trace elements and REE (ppm). Kom = komatiitic basalt, Mg = magnesian basalt, Thol = tholeiitic basalt, Gabb = gabbro, Gdio = granodiorite, Mgr = microgranite

SNo.		SiO_2	TiO_2	Al_2O_3	Fe_2O_3	MnO	MgO	CaO	Na_2O	K_2O	P_2O_5	Cr_2O_3	Sr	Rb	Ba	Th	U	Ta	Nb	Zr	Y	La	Ce	Nd	Sm	Eu	Gd	Dy	Er	Yb	Lu	Nb/U
Coonterunah																																	
gc020697	Kom	53.25	0.63	14.67	8.54	0.15	7.71	10.80	2.98	0.29	0.05	0.068	98.7	12.7	47.94	0.363	0.085	0.118	1.911	42.52	16.32	2.79	6.83	4.66	1.43	0.50	2.00	2.47	1.63	1.70	0.250	22.5
gc180697	Kom	53.13	0.60	14.27	8.89	0.16	7.83	11.43	2.95	0.14	0.05	0.079	132.5	4.5	27.79	0.335	0.076	0.111	1.757	40.39	15.53	2.73	6.51	4.43	1.34	0.58	1.87	2.33	1.54	1.58	0.229	23.0
gc010697	Mg	50.50	0.62	11.83	10.12	0.21	11.62	11.63	1.82	0.41	0.04	0.109	100.9	20.9	77.48	0.343	0.071	0.113	1.804	41.40	14.48	2.15	5.85	4.08	1.27	0.33	1.75	2.26	1.48	1.56	0.230	25.5
gc170697	Mg	52.77	0.57	13.02	9.30	0.19	9.90	10.95	2.79	0.13	0.05	0.101	122.3	2.0	39.17	0.328	0.070	0.104	1.681	37.80	12.89	2.14	5.71	3.97	1.23	0.40	1.65	2.08	1.32	1.35	0.199	23.9
gc160697	Mg	51.84	0.56	12.87	9.32	0.21	11.21	9.77	2.94	0.12	0.05	0.107	91.2	2.4	18.69	0.354	0.074	0.110	1.634	38.85	13.75	2.46	6.07	4.25	1.33	0.38	1.77	2.21	1.42	1.52	0.223	22.0
gc040697	Thol	49.75	1.00	14.54	13.27	0.22	5.85	11.82	2.26	0.40	0.07	0.029	173.9	18.1	61.81	0.818	0.071	0.170	2.806	60.51	22.65	3.29	8.83	6.90	2.23	0.78	2.98	3.50	2.20	2.23	0.327	39.3
gc050697	Thol	50.77	1.81	13.42	16.69	0.21	7.76	6.55	2.25	0.12	0.14	0.018	117.8	6.2	24.28	0.682	0.141	0.332	5.272	118.13	44.94	7.32	18.41	14.40	4.52	1.28	5.96	6.88	4.13	4.23	0.622	37.4
gc060697	Thol	50.05	1.56	15.10	13.44	0.21	6.08	9.26	2.90	0.77	0.24	0.024	272.3	20.1	68.98	1.191	0.331	0.483	8.188	123.99	27.49	10.31	25.23	15.91	4.06	1.33	4.46	4.43	2.53	2.46	0.365	24.8
gc070697	Thol	49.45	1.32	14.41	16.28	0.29	5.64	8.83	3.11	0.62	0.11	0.014	194.2	14.4	130.54	1.023	0.247	0.323	5.112	84.83	30.36	6.50	15.54	10.32	2.93	0.87	3.84	4.58	2.86	2.99	0.448	20.7
gc090697	Thol	56.26	1.03	15.77	10.28	0.15	4.17	6.46	3.99	0.47	0.19	0.004	254.2	10.4	60.53	2.960	0.649	0.704	10.582	230.20	37.66	21.29	46.74	23.90	5.45	1.38	5.77	5.80	3.42	3.48	0.506	16.3
gc190697	Thol	49.78	0.95	13.70	14.67	0.23	6.83	10.12	2.22	0.44	0.05	0.024	91.2	13.6	40.46	0.321	0.073	0.144	2.285	46.47	18.44	3.03	7.61	5.69	1.81	0.72	2.44	2.90	1.82	1.82	0.274	31.2
gc200697	Thol	49.18	1.02	14.21	13.95	0.26	7.08	10.69	2.30	0.69	0.08	0.027	112.0	19.7	59.10	0.321	0.070	0.175	2.829	64.20	25.98	3.39	9.05	7.27	2.35	0.84	3.21	3.92	2.53	2.54	0.362	40.6
gc210697	Thol	49.97	1.64	17.25	11.91	0.17	4.20	9.57	3.23	1.19	0.23	0.025	200.3	54.3	148.50	1.466	0.350	0.585	10.642	132.01	30.09	10.46	24.72	15.30	3.73	1.35	4.51	4.77	2.75	2.73	0.391	30.4
gc250697	Thol	50.58	1.14	13.31	10.11	0.21	2.55	9.35	0.57	2.30	0.17	0.004	66.4	73.0	291.58	3.599	0.828	0.587	8.364	202.12	24.31	18.85	39.56	20.26	4.55	1.13	4.26	3.87	2.44	2.74	0.417	10.1
gc430697	Thol	48.71	1.25	15.21	10.97	0.20	4.32	7.00	3.84	0.46	0.27	0.003	86.0	15.7	90.60	3.413	0.759	0.705	11.420	229.02	29.73	22.59	48.58	25.27	5.66	1.56	5.33	5.21	2.88	3.03	0.453	15.0
gc450697	Thol	54.89	0.91	15.48	9.22	0.10	4.95	3.19	3.31	1.48	0.14	0.003	92.0	39.4	114.89	3.850	0.853	0.555	7.951	166.08	20.01	16.76	34.72	17.44	3.94	1.10	3.52	3.29	1.88	2.06	0.313	9.3
gc480697	Thol	55.81	1.33	15.38	10.97	0.28	4.03	7.78	3.16	0.70	0.23	0.014	165.6	24.7	78.23	2.919	0.633	0.657	10.300	206.35	34.05	21.10	45.16	23.24	4.89	1.37	5.39	5.22	3.05	2.98	0.447	16.3
gc030697	Gabb	49.92	0.99	13.26	15.10	0.22	6.81	9.97	2.53	0.22	0.06	0.021	89.9	4.5	19.91	0.362	0.083	0.168	2.490	50.64	20.56	3.38	8.63	6.42	2.05	0.77	2.75	3.21	2.01	2.07	0.308	30.0
gc080697	Gabb	49.43	0.96	15.70	12.47	0.17	8.02	8.38	2.95	0.82	0.09	0.043	175.5	19.2	58.08	0.594	0.149	0.184	2.632	52.54	14.75	5.36	11.48	7.23	1.94	0.84	2.27	2.36	1.43	1.47	0.219	17.6
gc100697	Gabb	50.19	1.12	11.42	15.18	0.24	8.16	9.58	1.81	1.20	0.10	0.011	83.9	29.1	27.47	0.380	0.083	0.187	2.912	159.84	26.27	3.97	10.59	8.45	2.71	0.91	3.54	4.09	2.51	2.60	0.386	34.9
gc260697	Gabb	48.50	0.99	13.43	16.43	0.23	6.85	7.84	2.62	0.63	0.07	0.027	114.4	14.6	70.70	0.298	0.069	0.155	2.523	49.78	21.07	4.22	10.07	7.00	2.14	0.86	2.78	3.26	1.99	1.97	0.287	36.4
Carlindi																																	
gc410697	Gdio	70.62	0.24	15.19	2.23	0.02	0.90	2.11	5.20	1.89	0.06	0.009	352.4	39.8	547.18	4.454	1.007	0.367	4.311	105.75	5.14	19.61	32.49	11.78	1.82	0.92	1.33	0.88	0.44	0.39	0.063	4.3
gc400697	Mgr	72.63	0.15	14.95	1.31	0.02	0.68	1.19	4.52	3.77	0.05	0.012	181.6	86.1	437.57	5.499	2.052	0.604	5.214	92.44	5.61	16.35	29.53	9.58	1.57	0.69	1.20	0.90	0.45	0.47	0.071	2.5
Warrawoona																																	
gc120697	Mg	44.90	1.33	10.79	15.92	0.17	9.62	7.27	0.62	0.02	0.10	0.007	43.6	0.3	7.24	0.719	0.171	0.293	4.688	89.98	24.93	5.54	15.33	11.98	3.60	1.23	4.19	4.18	2.24	2.11	0.296	27.4
gc270697	Mg	49.64	0.64	11.67	10.62	0.20	8.26	10.26	3.22	0.10	0.05	0.109	93.3	2.8	74.99	0.386	0.093	0.143	2.352	42.19	15.28	3.14	7.75	5.54	1.69	0.61	2.06	2.35	1.49	1.50	0.213	25.2
gc370697	Mg	53.37	0.62	11.79	7.16	0.14	5.83	11.05	4.56	0.15	0.06	0.061	50.1	2.4	19.34	0.369	0.088	0.136	2.355	42.39	14.25	3.17	7.64	5.29	1.63	0.60	1.98	2.25	1.36	1.39	0.193	26.9
gc110697	Thol	53.18	2.03	13.82	18.86	0.09	6.42	0.27	0.01	0.03	0.14	0.011	4.2	1.0	17.08	0.842	0.264	0.380	6.430	135.32	16.87	6.49	16.34	12.38	3.67	1.27	3.70	3.04	1.67	1.32	0.240	24.4
gc130697	Thol	48.49	1.46	13.24	13.74	0.25	5.20	7.92	3.66	0.64	0.13	0.045	128.3	13.1	60.60	0.675	0.163	0.286	4.492	92.89	35.50	6.49	16.35	11.72	3.44	1.19	4.42	5.12	3.27	3.36	0.501	27.6
gc150697	Thol	47.27	0.98	15.47	10.24	0.17	6.27	12.25	1.67	1.30	0.09	0.052	191.0	27.1	274.64	1.004	0.225	0.234	3.728	66.19	18.68	5.30	12.49	8.42	2.40	0.80	2.95	3.21	1.84	1.89	0.275	16.6
gc290697	Thol	47.69	2.74	12.43	16.54	0.23	4.53	9.15	2.42	0.58	0.30	0.009	104.2	11.5	310.29	3.023	0.685	0.597	9.602	159.84	43.83	15.54	33.87	20.43	5.43	1.80	6.24	6.75	3.98	3.86	0.575	14.0
gc300697	Thol	52.45	2.09	17.60	11.71	0.14	3.99	2.58	6.70	0.28	0.19	0.005	75.1	5.8	81.37	2.798	0.598	0.556	9.030	92.89	38.65	7.97	22.00	15.88	4.76	1.26	5.70	6.53	4.01	4.02	0.590	15.1
gc360697	Thol	50.70	1.83	13.16	16.88	0.24	4.84	9.68	1.62	0.01	0.14	0.010	398.8	0.1	17.87	0.811	0.196	0.331	5.455	166.05	29.50	7.24	18.00	13.35	4.04	1.37	4.95	5.08	2.69	2.41	0.329	27.8
gc140697	Gabb	46.25	1.15	15.21	10.05	0.14	6.23	12.51	2.69	0.88	0.11	0.048	158.0	20.4	344.86	1.534	0.352	0.350	5.355	85.71	26.12	7.79	17.95	11.33	3.20	1.09	3.71	4.19	2.54	2.59	0.379	15.2
gc280697	Gabb	51.75	1.78	12.56	15.10	0.21	5.74	5.60	2.84	2.03	0.18	0.010	102.7	28.9	201.37	0.909	0.215	0.353	6.070	117.31	43.44	6.67	17.98	13.86	4.17	1.33	5.31	6.32	3.98	4.00	0.565	28.3

across a fault on the western limb of the regional Pilgangoora fold. The eastern margin of the Pilgangoora Belt is juxtaposed against the ~2950 Ma Lalla Rookh sediments along a major shear zone (Krapez, 1993). Galena associated with the shear zone has a Pb–Pb age of ~3100 to 3200 Ma (Richards et al., 1981), and hence reactivation of the shear zone is necessary to account for the translation of younger rocks. The rocks in the central and eastern portion of the Pilgangoora Belt are well preserved with prehnite–pumpellyite to lowermost amphibolite-facies metamorphism, but little development of structural fabrics. The detailed geology and geochemistry of igneous rocks in the Coonterunah Group, Carlindi granodiorite and Warrawoona Group are described below.

3. Analytical methods

Surface samples of 31 basalts and gabbros, one granodiorite and one microgranite (Fig. 2B) were collected from the central portion of the Pilgangoora Belt where preservation is best. Most samples were not visibly altered and were located away from large-scale faults and fractures. Three altered basalts were collected to determine the alteration characteristics. Samples were cleaned to minimise contamination from oxidised surfaces and veins. Twenty-seven samples were collected along two traverses to detect any systematic stratigraphic variations. Rock names are based on field criteria, where tholeiitic basalt is fine-grained, sub-ophitic and has abundant leucoxene, indicating a high TiO_2 content; gabbro is coarser grained; magnesium basalt has abundant large ocelli in a fine-grained relatively pale matrix; and komatiitic basalt has elongate pyroxene pseudomorphs in a dark matrix. Chemical criteria are not used to reclassify the rocks after analysis, due mainly to post-magmatic mobility of major elements. Generally the field criteria were a good proxy for the original basalt composition.

Samples were analysed at the Australian National University; in the Department of Geology for major elements using glass-fusion XRF techniques and in the Research School of Earth Sciences for trace elements using the acid dissolution, solution nebulisation ICP-MS techniques described by Eggins et al. (1997). Previous studies at ANU found that acid dissolution of low-grade basalts may produce a Zr–Hf-bearing, but otherwise trace-element-free refractory residue, so samples were re-analysed for Zr using pressed-powder XRF methods at Memorial University of Newfoundland. The data are presented in Table 1.

4. Coonterunah Group

The Coonterunah Group is dominantly composed of basalt and gabbro with significant rhyolitic to andesitic volcanic and volcaniclastic deposits and minor sedimentary units. The sedimentary units are typically grey to brown carbonates and plane-laminated ferruginous cherts. The absence of shallow-water sedimentary structures and the general paucity of vesicles in basalts led Buick et al. (1995) to interpret a depositional environment well beneath wave base with high confining pressures.

Tholeiitic basalts are typically fine-grained, massive and contain actinolite, low-Al hornblende, albite, chlorite, epidote, margarite and quartz with minor leucoxene, magnetite and ilmenite. In places, very fine-grained relics of igneous clinopyroxene, plagioclase and hornblende are preserved. Although metamorphic recrystallisation is relatively complete, igneous textures have been preserved by pseudomorphic replacement. For instance, a sub-ophitic felted texture of intergrown plagioclase and pyroxene is preserved in many tholeiitic samples (Fig. 4A). Preserved pillows are rare and are generally less than 2 m in diameter, with pipe and spherical vesicles, chilled rinds and interpillow triple junctions (Fig. 4B). Where pillows are not developed, thin semi-planar flows may be defined by truncated vesicles (Fig. 4C). The relict primary structures indicate that the younging of the rocks is uniformly away from the granodiorite batholith. A single small outcrop of massive tholeiitic basalt in the middle of the stratigraphic succession contains medium- to coarse-grained granitic (quartz, K-feldspar, plagioclase) xenoliths (Fig. 4D). These unfoliated xenoliths are less than 10 cm across, with nebulous margins and

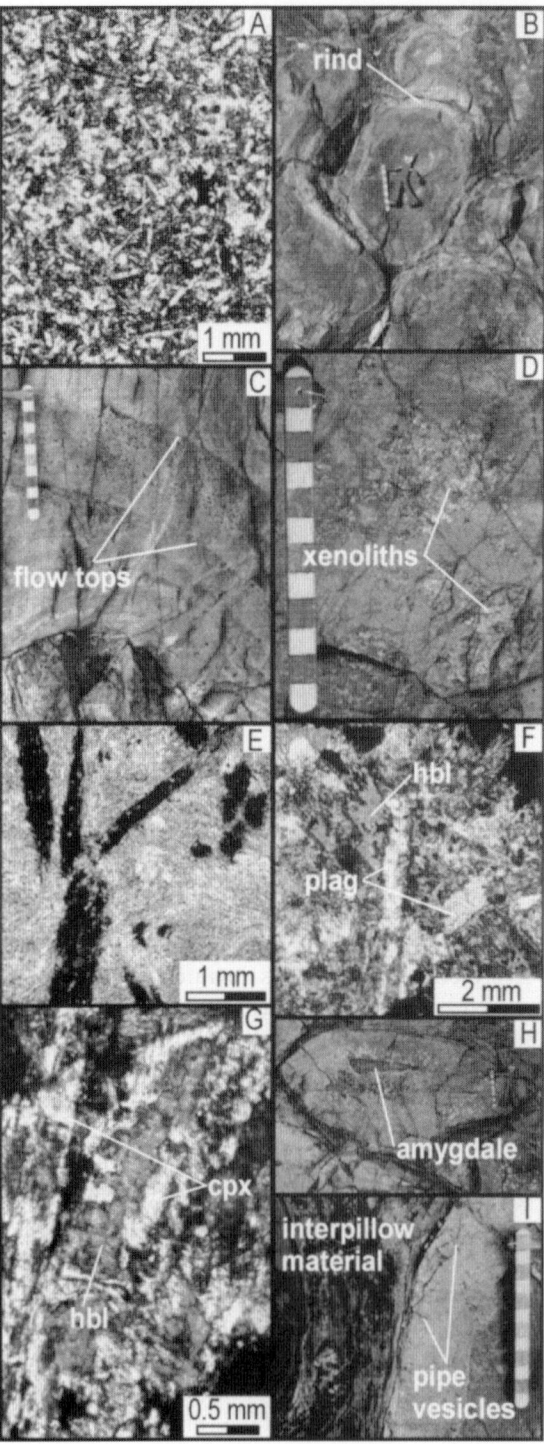

irregular shapes. Although an absolute age has yet to be determined for the xenoliths, they may be approximately coeval with the basalts and related to felsic volcanics lower in the sequence or significantly older and derived from earlier crust.

Gabbros have a similar mineral composition to the tholeiitic basalts, but with a remnant medium-grained igneous texture. Since the metamorphic grade of the gabbros is the same as the basalts, and both basalt and gabbro are truncated by the granodiorite, the gabbros must have crystallised before the granodiorite intrusion and, therefore, are deemed to be part of the Coonterunah Group. Although some of the gabbros have intrusive relationships with the volcanic units, others may represent the base of thick basaltic flows.

Magnesian basalts are restricted to the lower part of the stratigraphy and contain abundant spherical ocelli of very fine-grained albite and actinolite in a fine-grained actinolite–albite matrix. Accessory minerals include chlorite, clinozoisite and Fe–Ti-oxide minerals. Relict igneous minerals have not been found. The flows are commonly pillowed. A single komatiitic basalt unit, less than 20 m thick, immediately overlies the uppermost

Fig. 4. *Coonterunah Group*. (A) Polarised light photomicrograph of tholeiitic basalt gc430697 with remnant ophitic texture; pale laths are altered plagioclase crystals and intercrystalline material is fine-grained actinolite, chlorite, chlorite and albite, probably after pyroxene and glass. (B) Well-preserved pillow basalts with bleached, vesicle-poor rinds and scarce amygdales, cm-scale. (C) Small-scale flow features in tholeiitic basalt, where flow tops are semi-planar and marked by the truncation of vesicle-rich layers, cm-scale. (D) Massive granitic xenoliths in tholeiitic basalt, cm-scale. (E) Plane light photomicrograph of komatiitic basalt gc180697 with coarse-grained pyroxene laths altered to dominantly FeTi-oxides in a fine-grained massive, fibrous groundmass of actinolite, chlorite and albite. *Warrawoona Group*. (F) Polarised light photomicrograph of gabbro gc280697 with subophitic texture, where plagioclase is intergrown with hornblende which has completely replaced a pyroxene crystal. (G) Polarised light micrograph of tholeiitic basalt gc290697 where relict pyroxene survives hornblende replacement of a large magmatic pyroxene crystal. (H) Extremely well-preserved pillow in magnesian basalt domain where a large amygdale is surrounded by small, white amygdales, cm-scale. (I) Margin of an extremely well-preserved tholeiitic pillow basalt with radially oriented pipe vesicles, cm-scale.

magnesian basalt and consists of fine-grained Fe–Ti-oxides and low-Al hornblende after elongate clinopyroxene needles (up to 10 cm long) in a fine-grained actinolite–hornblende–albite groundmass (Fig. 4E).

All basalts have been moderately affected by post-magmatic alteration, which produced variability of some major and trace element abundances (Fig. 5). For major elements, the main effects of alteration were to increase K_2O and reduce Na_2O, Al_2O_3 and MgO, most obvious in sample gc250697 which was mapped as a moderately altered and bleached tholeiitic basalt. Thus, classifying the rocks using the major elements is

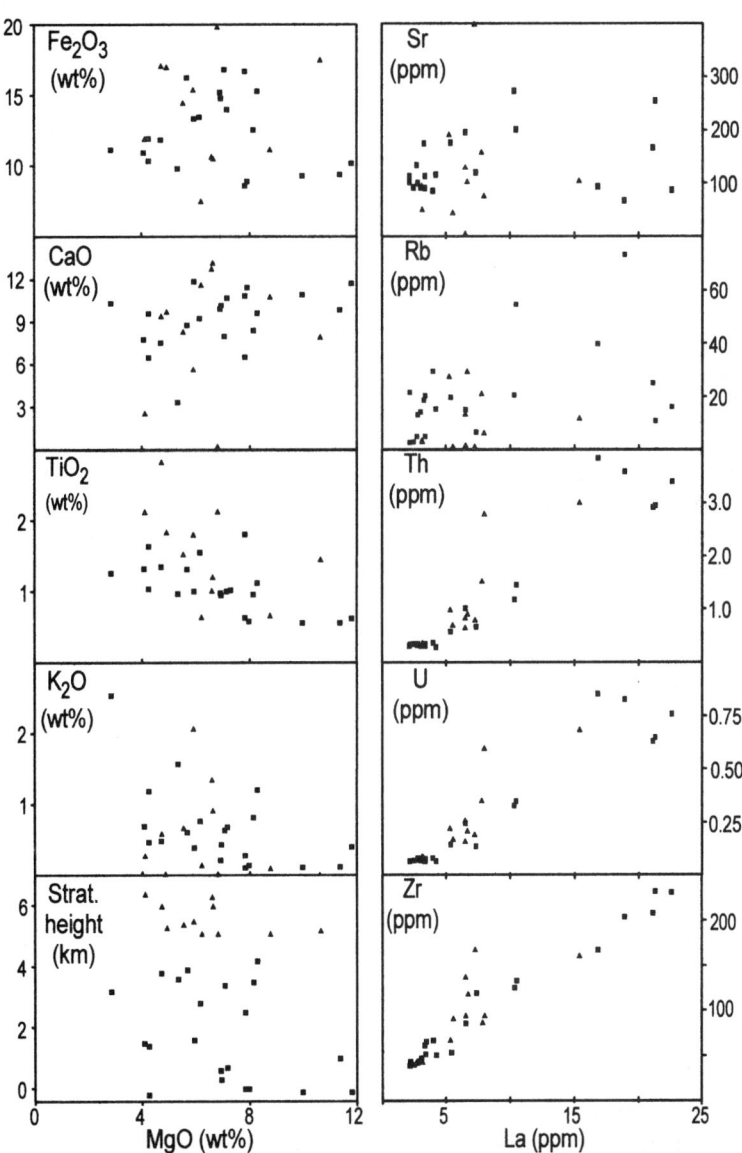

Fig. 5. Elemental variation diagrams show both magmatic variation and post-magmatic alteration. Squares represent Coonterunah and triangles represent Warrawoona samples. Stratigraphic height is measured from a komatiitic basalt reference near the base of the Coonterunah Group (see Fig. 3).

problematic. For example, only four of the six magnesian basalts have greater than 9% MgO, the usual geochemical signature for such rocks (Sylvester et al., 1997b). Further problems are highlighted by sample gc120697, which has the highest MgO concentration in the Warrawoona Group, but for most other elements has a more evolved chemistry than other magnesian basalts.

For trace elements, abundances of large ion lithophile elements (LILE) such as Sr, K and Rb, which are known to be mobile during sea-floor alteration (Shiano et al., 1993), are the most variable. In contrast, the high field strength elements (HFSE), such as Ti, Th, Zr and LREE, are essentially immobile and thus preserve their original magmatic concentrations. This is shown in Fig. 5, where Sr, Rb, Th, Zr and U are plotted against La. There is a random scatter of Sr and Rb, but more linear (magmatic) trends for Th, Zr and U. Due to the post-magmatic alteration of the LILE, the interpretations of this study are based mainly on the immobile HFSE.

Tholeiitic basalts are generally characterised by higher TiO_2 and Fe_2O_3 and lower MgO and CaO relative to the gabbros, with more extreme differences for komatiitic and magnesian basalts. MgO concentrations are all less than 12%, indicating the absence of ultramafic magmas. The basalts lie along the tholeiitic–subalkalic trend, and three tholeiitic and the komatiitic basalts have SiO_2 contents within the basaltic andesite field (Le Maitre et al., 1989). For least altered samples, normative calculations show that plagioclase, clinopyroxene and orthopyroxene can account for the bulk composition with minor amounts of quartz or olivine. This is generally consistent with the relict igneous minerals plagioclase, clinopyroxene and hornblende, and the interpretation of remnant mineral textures. Importantly, there are no systematic variations of basalt composition with stratigraphic height (Fig. 5), except that the Coonterunah magnesian and komatiitic basalts are restricted to the base of the sequence. Hence, there was no profound systematic temporal change of the source composition or petrogenetic processes.

N-MORB-normalised (Sun and McDonough, 1989) trace element patterns (Fig. 6) are all characterised by the same shape; high LILE, Th, U and LREE and lower Ta, Nb, P, Zr, Ti, Y and M-HREE. Generally, tholeiitic basalts are the most enriched and variable followed by gabbros and then magnesian basalts, although there are some discrepancies resulting from post-magmatic alteration. There are distinct negative Ta, Nb and P anomalies. Relative to N-MORB, the background signature after removal of anomalies has an overall negative slope, produced by greater concentrations of incompatible elements (Fig. 6).

Chondrite-normalised (Sun and McDonough, 1989) REE patterns (Fig. 7) for magnesian and komatiitic basalts and gabbros are generally flat at ratios approximately 10 times chondritic, with slight LREE enrichment in a few samples. Gabbros have a more enriched and less uniform REE content than the magnesian basalts, possibly reflecting their greater stratigraphic extent and multiple emplacement events. Tholeiitic basalts have extremely variable REE compositions with strong LREE enrichment, and with the greatest enrichment of LREE found at the highest REE contents. There are small negative and positive Eu anomalies, which are probably controlled by plagioclase removal and accumulation, respectively. There are no Ce anomalies.

Significantly, there are no major changes in the shapes of the trace element and REE patterns of the komatiitic, magnesian and tholeiitic basalts and gabbros, and thus the differences in enrichment between the samples probably reflect straightforward differentiation or partial melting processes. Simply stated, the data are consistent with the komatiitic and magnesian basalts being more primitive than the gabbros, which are more primitive than the tholeiitic basalts.

5. Carlindi granodiorite

The Carlindi granodiorite is medium- to coarse-grained and mostly massive, but is foliated near (<1 km) the northernmost part of the Pilgangoora Belt. The foliation is folded subparallel to the greenstone contact. The granodiorite intrudes the Coonterunah Group and is itself intruded by a quartz-phyric microgranite with a zircon U–Pb SHRIMP age of 3467.6±3.7 Ma (Figs. 2 and 3;

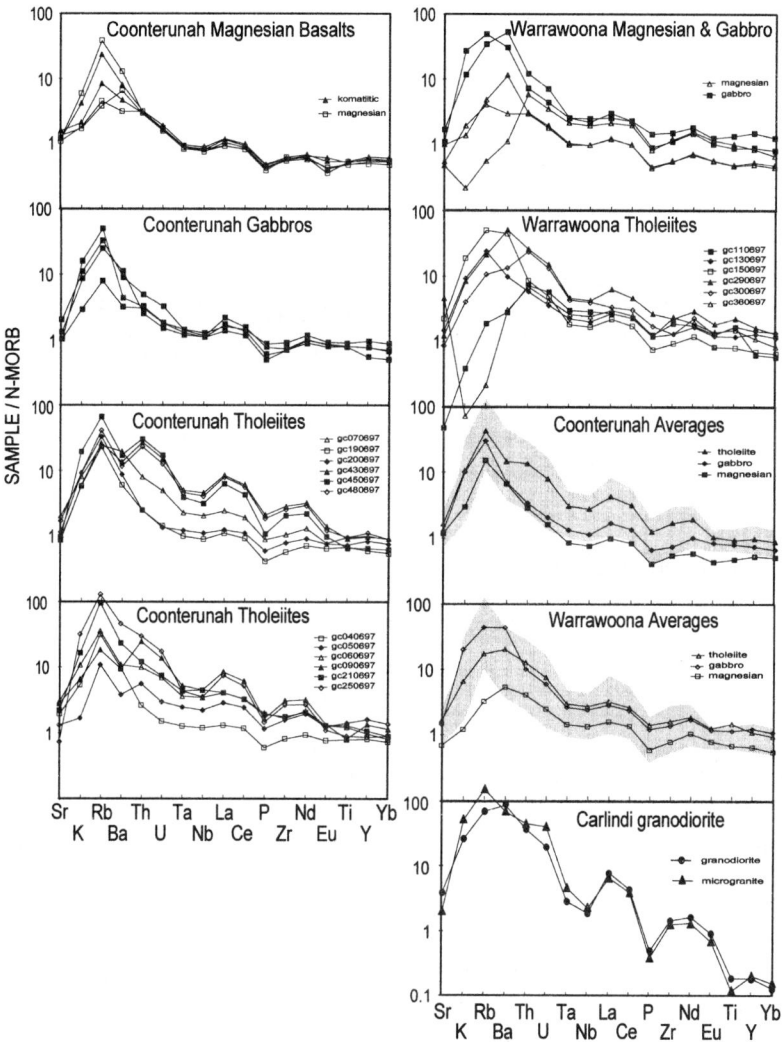

Fig. 6. N-MORB-normalised (Sun and McDonough, 1989) trace element patterns for Coonterunah and Warrawoona basalts and gabbros and Carlindi granodiorite and microgranite. Shaded area in average plots represents the range of Coonterunah tholeiitic basalts.

Buick et al., 1995). This constrains the age of the granodiorite to between 3515 and 3468 Ma. The granodiorite and microgranite have N-MORB-normalised trace element patterns similar to the Coonterunah basalts, with enrichment of LILE, Th, U and LREE relative to Ta, Nb, P, Zr, Ti, Y and M-HREE (Fig. 6). However, the overall negative slope and the magnitudes of the anomalies are enhanced. The chondrite-normalised REE plots show significant enrichment relative to the HREE. The granodiorite and microgranite are more depleted in HREE than the basalts (Fig. 7). A pronounced positive Eu anomaly and a slight positive Ce anomaly are developed. The geochemical characteristics of the granodiorite and quartz-phyric microgranite are very similar, providing evidence for a similar source and petrogenetic history.

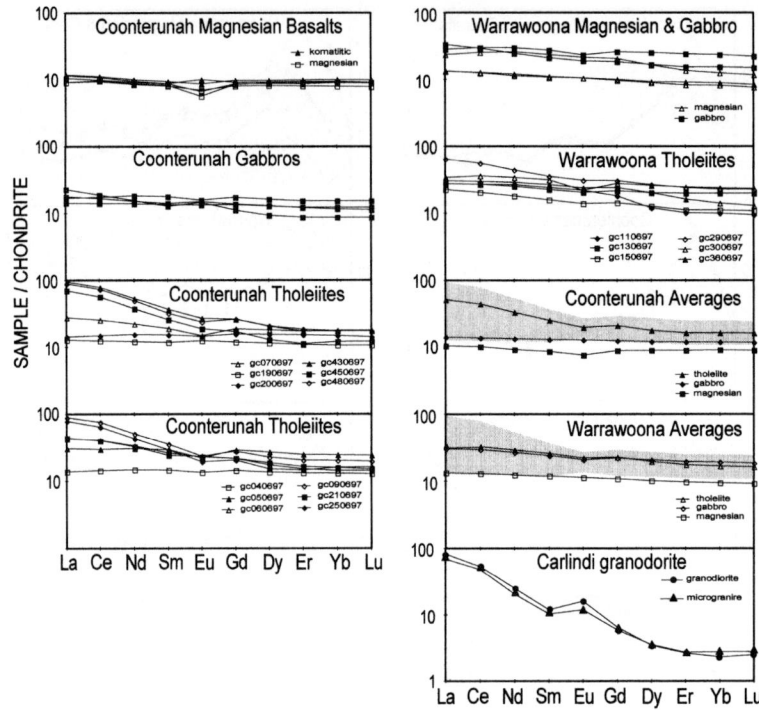

Fig. 7. Chondrite-normalised (Sun and McDonough, 1989) REE patterns for Coonterunah and Warrawoona basalts and gabbros and Carlindi granodiorite and microgranite. Shaded area in average plots represents the range of Coonterunah tholeiitic basalts.

6. Warrawoona Group

The Coonterunah Group and Carlindi granitoids were uplifted, exposed and eroded to form an unconformity which is generally semi-conformable. However, in the eastern part of the Pilgangoora Belt the unconformity is angular, indicating some pre-erosional deformation. Immediately beneath the unconformity, there is an increase in alteration and decrease in grain size which may represent a palaeosol. Sediment-filled fractures cutting down into the Coonterunah and Carlindi units also provide evidence of exposure and weathering.

Overlying the unconformity is the basal unit of the Warrawoona Group in the Pilgangoora Belt, an ~20 m thick silicified sedimentary package known as the Strelley Pool Chert. This comprises a basal sandstone with minor conglomerate, silicified laminated carbonates and mafic volcaniclastics (Lowe, 1983; Buick and Barnes, 1984; DiMarco and Lowe, 1989). Three important features of this unit in the Pilgangoora Belt are: (1) the basal sandstone is dominantly composed of rounded to well-rounded, sub-spherical quartz grains; (2) the sandstone contains abundant detrital zircon; and (3) the sedimentary structures and pseudomorphically replaced minerals indicate shallow-water deposition.

This sedimentary horizon is overlain by up to 5 km of basalt and gabbro. Generally, the stratigraphic succession proceeds from basal magnesian basalt to tholeiitic basalt and then more magnesian basalt (Figs. 2B and 3), but there are tholeiitic domains within the magnesian basalt areas and vice versa. Between the different basalt units are silicified mafic volcaniclastic units, which have wave ripples and pseudomorphs after evaporite minerals indicating continued shallow-water deposition. The stratigraphy is believed to correlate with the uppermost part of the Warrawoona Group in the North Pole Dome, where similar rocks have been dated at ~3.46 Ga (Thorpe et al., 1992; Buick et al., 1995). There is no development of structural fabrics, so post-depositional deformation has been minimal. The Warrawoona succession is unconformably overlain by sedimentary units of the Gorge Creek Group and possibly the Sulphur Springs Group, both of which were deposited after 3.24 Ga (Brauhart et al., 1998).

Warrawoona tholeiitic basalts are fine-grained, massive and composed of prehnite, actinolite, chlorite, clinozoisite, quartz, carbonate and albite with abundant relict clinopyroxene, plagioclase, hornblende and FeTi-oxides (Fig. 4F and G). The assemblage indicates prehnite–pumpellyite to lowermost greenschist-facies metamorphism (Spear, 1995). Magnesian basalts have abundant actinolite–albite–clinozoisite ocelli in a fine-grained pale actinolite–chlorite–albite matrix. Gabbros of both tholeiitic and magnesian composition have larger grain sizes and better preservation of igneous plagioclase, clinopyroxene and Fe–Ti-oxides. Some intrusive relationships have been observed, but other gabbros, as in the Coonterunah Group, may represent the bases of basaltic flows. Pillowed outcrops are common in both tholeiitic and magnesian basalts with exceptional preservation of pipe vesicles, carbonate- and silica-filled amygdales, altered and chilled margins, interpillow hyaloclastite and spalled pillow-margin fragments (Fig. 4H and I).

Post-magmatic alteration is more variable in the Warrawoona basalts than the Coonterunah Group, possibly reflecting LILE homogenisation during the higher temperature metamorphism experienced by the Coonterunah Group. Overall, the geochemistry of the Warrawoona basalts is similar to the Coonterunah basalts with enrichment in LILE, Th, U and LREE relative to Ta, Nb, P, Zr, Ti, Y and M-HREE (Figs. 5–7). The Warrawoona basalts are enriched in Ba relative to the Coonterunah basalts, which may be related to Ba mobility during hydrothermal alteration. In the North Pole Dome, there are thick beds and veins of barite in the Warrawoona Group (Buick and Dunlop, 1990), which may indicate widespread Ba alteration.

7. Interpretation

The most important observation is that the Coonterunah and Warrawoona basalts have very similar geochemical characteristics. Therefore, the mantle sources and petrogenetic processes responsible for forming the two basalt suites must have been similar, even though the successions are separated by an erosional unconformity and ∼55 million years. It follows that constraining the source and petrogenesis of the Warrawoona basalts should also constrain those of the Coonterunah basalts. All of the basalts have a crustal signature compared with N-MORB: that is, significant Ta–Nb depletion associated with LILE, Th, U and LREE enrichment (Fig. 6). These characteristics are most readily explained by addition of a crustal component to a mantle-derived magma, either by modification of the mantle source region, for example during subduction-related, metasomatic processes (Brenan et al., 1994; Pearce et al., 1995), or by assimilation of crustal material during passage of the magma through continental crust (Barley, 1986). There are no general geochemical criteria to discriminate between these two mechanisms. In the Pilgangoora Belt, however, geological considerations favour the latter alternative.

The Warrawoona Group was deposited onto the subaerially eroded Coonterunah and Carlindi basement. The basal unit of the Warrawoona Group above the unconformity is a mature quartz-rich sandstone with detrital zircon, providing evidence that the basement was chemically evolved. The sandstone was deposited in shallow water, as shown by symmetrical wave ripples and intercalated evaporitic carbonates. Furthermore, primary features from other sedimentary and basaltic units in the Warrawoona Group clearly show that deposition and eruption continued in a shallow-water environment. The rocks have been subjected to only very low-grade metamorphism with minimal deformation, indicating that since the deposition of the Warrawoona Group, the Pilgangoora Belt has escaped major uplift and erosion and also deep burial and high-grade metamorphism, and thus has been relatively isostatically stable. Therefore, the Warrawoona basalts were evidently erupted onto a chemically evolved silicic basement that has remained relatively rigid, cool and buoyant since ∼3.47 Ga, by all definitions continental crust (Buick et al., 1995). Since the geochemical likeness between the Warrawoona and Coonterunah basalts indicates similar petrogenetic processes, it is likely that the Coonterunah basalts were also erupted onto continental basement. The granitic xenoliths in the Coonterunah basalts may provide direct evidence of assimilation of this basement.

REE modelling provides further evidence for crustal contamination of the Warrawoona and Coonterunah basalts. The models suggest that the REE contents of the most evolved basalts are reproduced more precisely by combined assimilation–fractional crystallisation of the least evolved basalts than by fractional crystallisation alone. In the models, plagioclase and clinopyroxene are assumed to be the main crystallising phases because they are the dominant minerals in the basalts, not only as relics and pseudomorphs, but also in normative calculations. Orthopyroxene is included as it accounts for a large mode in the normative calculation and can significantly fractionate the REE. Although no relict olivine has been found, it is included in the models because it features in many of the normative calculations and probably crystallised from the more primitive melt compositions. Conversely, relict igneous hornblende is found in places but has not been included in the models due to its low modal abundance (<5%) and the complexity of assigning partition coefficients to the hornblende series. The models assume a 40% plagioclase, 30% clinopyroxene, 20% orthopyroxene and 10% olivine assemblage. Partition coefficients are taken from Fujimaki et al. (1984) and Schwandt and McKay (1998). Varying the modal abundance of the minerals by up to 20% has no profound influence on the REE models, but does control whether the final magma has a basaltic composition. This is especially noticeable for Al_2O_3, which is controlled solely by plagioclase fractionation. At greater than 60% fractional crystallisation the Al_2O_3 concentration increases unrealistically to over 20% (assuming An_{80}), although lower anorthite/albite ratios limit this increase. More complex models, where the modal mineralogy varies during fractional crystallisation, have not been considered.

The models have been assigned a starting composition based on local igneous compositions, for example, a melt with the composition of the least evolved magnesian basalt (gc170697) modified by addition of 10% granodiorite (gc410697), and then numerically evolved using Rayleigh fractional crystallisation with the aforementioned partition coefficients. The aim is to produce REE patterns similar to the most evolved Coonterunah-Warrawoona tholeiitic basalts within a plausible percentage of mass fractionation. The local granodiorite is used as the additive in the models, although it did not crystallise until after the eruption of the Coonterunah Group. It is the most plausible contaminant for the Warrawoona Group, and its composition probably reflects that of the now-lost basement as it is a typical Archaean TTG (cf. Martin, 1994). Three models are represented in Fig. 8.

The first model shows straightforward fractional crystallisation of a melt with a magnesian basalt (gc170697) composition and no assimilation. Two problems arise from this model: the patterns produced are flatter than those of the Pilbara basalts, and the amount of fractionation required to explain the most evolved basalts is unrealistically high ($\sim 90\%$). Therefore, given the assumptions used here, the most evolved tholeiitic basalts cannot be simply explained by varying amounts of fractional crystallisation of the least evolved magnesian basalts. The other models show 10% and 20% assimilation, respectively, of the local granodiorite (gc410697) into a melt with a magnesian basalt composition (gc170697) and then fractional crystallisation. The assimilation models produce patterns with a shape consistent with the evolved Pilbara basalts, except for some MREE depletion. Adding hornblende to the fractionating assemblage would only enhance this depletion as, for basaltic melts, the MREE are more compatible in hornblende than the LREE and HREE (Green and Pearson, 1985). The amount of mass fractionation required to attain the LREE and HREE concentrations of the most evolved basalt is much lower than for the assimilation-free model, although still high (up to 75%). In addition, the basalts have higher concentrations of many incompatible elements than the granodiorite, which suggests that the actual contaminants were more enriched in incompatible elements than the granodiorite, or that there was selective assimilation of certain elements. However, given the number of assumptions required for the models,

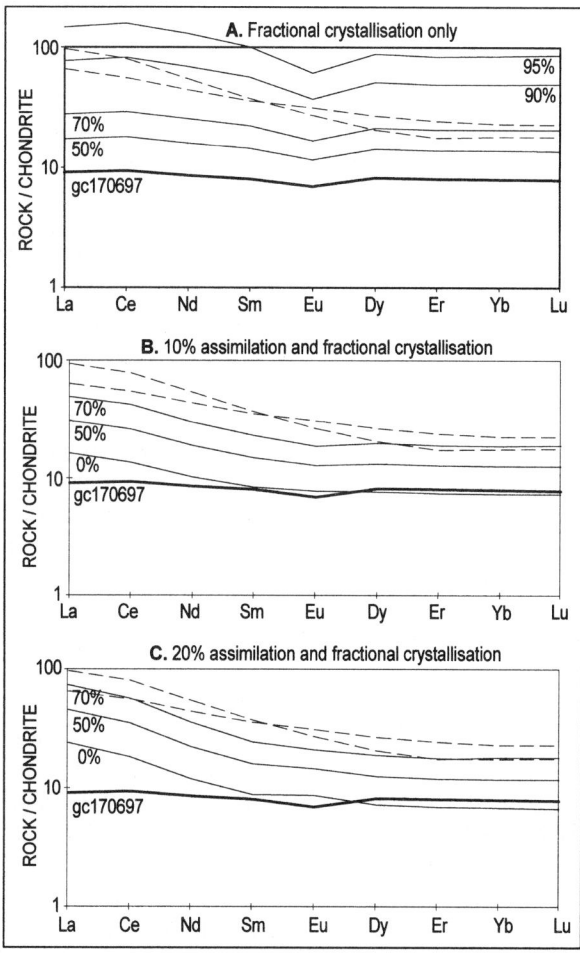

Fig. 8. Fractional crystallisation models with a starting composition of the least evolved magnesian basalt (gc170697) and assimilation of the local granodiorite (gc410697). The two most evolved tholeiitic basalts (gc290697, gc430697) are plotted (dashed lines) to assess the ability of the models. (A) No crustal assimilation with 50, 70, 90 and 95% fractional crystallisation. (B) 10% assimilation and 0, 50 with 70% fractional crystallisation. (C) 20% assimilation with 0, 50 and 70% fractional crystallisation.

particularly contaminant composition, mineral modes and partition coefficients, it is apparent that the crustal assimilation–fractional crystallisation models can match the observed compositions of the Coonterunah and Warrawoona basalts far better than do the simple fractionation models.

Assuming that the Coonterunah and Warrawoona basalts were produced by crustal assimilation and fractional crystallisation, the amount of contamination can be estimated from mixing curves of HFSE, given that the elements have similar partition coefficients during typical basaltic petrogenesis and that the elemental ratio in the contaminant differs markedly from that in the mantle source. This approach works because the elements should not fractionate during source melting and basalt crystallisation (Sylvester et al., 1997a), and hence, any variations should be caused by either crustal addition or source heterogeneity, provided that they are immobile during post-magmatic metamorphism and alteration. As shown previously, post-magmatic mobility of HFSE was minor in the Coonterunah and Warrawoona basalts.

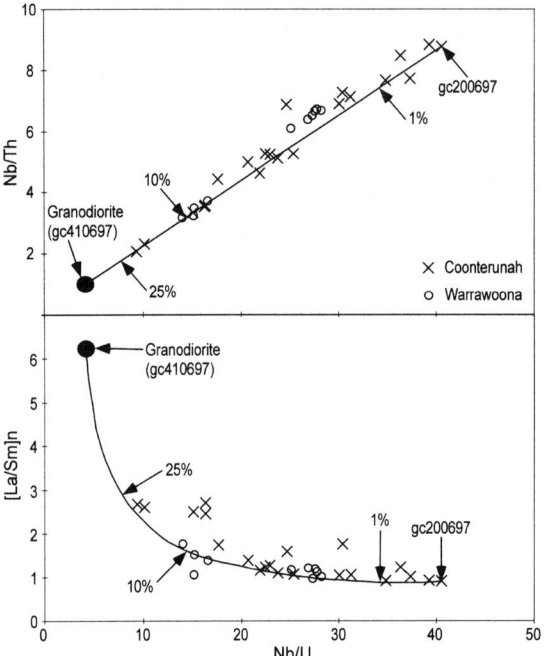

Fig. 9. Mixing models of least contaminated sample defined by highest Nb/U and Nb/Th (gc200697) and local granodiorite (gc410697) show estimates of the percentage contamination.

Two elements particularly useful for this approach are Nb and U; they have similar partition coefficients and much greater Nb/U ratios in the mantle than the crust (Hofmann et al., 1986; Sims and DePaolo, 1997). As explained above, the Pilbara basalts have generally retained their magmatic Nb and U compositions (Fig. 5), indicating that there was negligible post-magmatic mobility of these elements, a feature of other Archaean basalt suites (Sylvester et al., 1997a; Collerson and Kamber, 1999).

In Fig. 9, Nb/U is plotted against Nb/Th and La/Sm as both Th and the LREE are enriched in continental crust relative to the mantle (Rudnick and Fountain, 1995), and thus mixing produces a wide range of values. The basalt with the highest Nb/U (gc200697=40.6) is interpreted to be the least contaminated, which is borne out by its high Nb/Th and low LREE values. Assuming that the mantle source also had a Nb/U ratio of 40.6, and the local granodiorite is the contaminant (Nb/U = 4.3), it can be shown that the maximum level of contamination was <25%. The two most contaminated samples (gc250697, gc450697) have very evolved trace element patterns, as expected for crustal contamination (Figs. 6 and 7). Thermal considerations suggest that crustal assimilation would have been greatest in primitive magmas, with up to 30% and 10% assimilation possible for komatiites and tholeiites, respectively (Huppert and Sparks, 1985). Contamination would also have been enhanced at shallower crustal levels and for contaminants with low fusion temperatures (Reiners et al., 1995). Therefore, the estimates of up to 25% crustal assimilation for the Archaean Pilbara basalts seem reasonable.

The least contaminated basalts (for instance, gc200697) can provide constraints on the composition of the mantle source. They have flat, approximately 10 times chondritic REE patterns very similar to modern ocean-floor basalts (Sun and McDonough, 1989), but are enriched in K, Rb, Ba, Th and U and depleted in Ta, Nb, P, Nd, Eu, Ti, Y and Yb relative to average N-MORB (Fig. 6). However, as these are the characteristics expected from contamination, these differences are probably due to the small degrees of contamination in even these Pilbara basalts. Without this contamination, the N-MORB normalised patterns of the Coonterunah and Warrawoona basalts would be similar to modern ocean-floor basalts derived from the depleted mantle. Therefore, the Coonterunah and Warrawoona basalts are interpreted to have formed by eruption from a depleted upper mantle source through continental crust, and not by subduction-related mantle enrichment.

8. Discussion

Between 3.52 and 3.45 Ga, the proto-Pilbara was an active volcanic domain with generally bimodal volcanism in most greenstone belts and synchronous, voluminous granitoid emplacement. In the Phanerozoic, this would be most consistent with a convergent margin setting, but it is not yet clear that plate tectonics in the modern mode operated during the first billion years of Earth's history (Davies, 1992, 1995). Assuming, however, that actualistic geotectonics can be applied to

~3.5 Ga rocks, some possible tectonic settings can be eliminated. In the Pilgangoora Belt, the basalts erupted onto continental basement, thus eliminating juvenile volcanic arc systems, oceanic plateaux or ocean-floor settings, all invoked in tectonic models of other Archaean greenstone belts (de Wit et al., 1987; Abouchami et al., 1990; Desrochers et al., 1993). In addition, the lack of regional deformation and tectonic contacts in the Pilgangoora Belt is inconsistent with the accretion of allochthonous blocks, ruling out ophiolitic obduction. Elsewhere in the Pilbara Craton, tectonic interpretations of the Warrawoona Group have favoured an initial oceanic plateau or marginal basin setting which then evolved into a convergent volcanic-arc environment (Barley, 1993; Krapez, 1993; Barley and Pickard, 1999). Clearly, such interpretations are inadequate for the Warrawoona Group in the Pilgangoora Belt, though a marginal basin floored by continental is a plausible setting.

Gradual growth models of crustal evolution (Hurley and Rand, 1969; Reymer and Schubert, 1984; Taylor and McLennan, 1985; McCulloch and Bennett, 1993, 1994; Collerson and Kamber, 1999) predict that by 3.5 Ga only a small fraction of the present-day continental volume had been extracted from the mantle. According to such models, early Archaean continental crust must have been either thinner or smaller in total area than its modern counterpart. The preserved rock record seems to support this view, as only 0.5% of the present area of the continents is older than 3 Ga (de Wit, 1998). Thus, there should be evidence at 3.5 Ga of continental growth onto oceanic crust and magma extraction from primitive to underdepleted mantle. Neither of these is evident in the oldest-known supracrustal successions in the Pilbara Craton with the Coonterunah and Warrawoona basalts derived from depleted mantle and erupted onto continental basement. Therefore, the Pilbara basalts do not record the formation of juvenile continental crust.

The alternative steady-state models for crustal evolution (Armstrong, 1968, 1981, 1991; Bowring and Housh, 1995) predict that large volumes of continental crust did indeed form before 3.5 Ga and that the paucity of pre-3.5 Ga rocks reflects recycling. The Pilgangoora basalts preserve strong geochemical evidence of contamination by pre-3.5 Ga continental crust, even though tangible relics of this basement are restricted to ~3.58 Ga anorthosite énclaves in the South Daltons pluton (McNaughton et al., 1988), a xenocrystic zircon in the North Pole Dome (Thorpe et al., 1992) and ~3.58 to 3.66 Ga detrital zircon grains within middle to late Archaean sedimentary packages (Smithies et al., 1999; Nelson, 1998). It is conceivable that the pre-3.5 Ga rocks are now buried deeply in the Pilbara Craton; but this is unlikely in the absence of evidence for a general age increase with the present-day depth in this or any other craton. It is much more probable that the now-lost basement block was efficiently recycled into the mantle or incorporated into other Pilbara rocks or both.

The original size and thickness of this pre-3.5 Ga basement are difficult to constrain. However, the crustal signature evident in the Pilgangoora basalts, that is, significant Ta–Nb depletion associated with LILE, Th, U and LREE enrichment, is also found in the Warrawoona basalts in the Marble Bar Belt (Gruau et al., 1987) and the North Pole Dome (Green et al., in preparation). Furthermore, both belts display additional evidence for early crustal assimilation. In the Marble Bar Belt, Nd isotopic data can best be explained by assimilation of older REE-enriched crustal material producing an anomalously old isochron age (Hamilton et al., 1981; Gruau et al., 1987; McNaughton et al., 1988, 1993). In the North Pole Dome, a ~3.72 Ga xenocrystic zircon provides direct evidence of crustal assimilation (Thorpe et al., 1992). Including the Pilgangoora Belt, these three belts represent substantial volumes of basalt (each ~10 km thick, >50 km strike length), providing a minimum estimate of ~150 km for the extent of the pre-3.5 Ga basement. Further evidence of this early basement includes small, isolated énclaves and detrital zircon grains. The relevant constraints upon the lithospheric thickness are that the block was emergent at ~3.47 Ga and that thermal gradients inferred from early Archaean metamorphic assemblages from elsewhere in the Pilbara Craton (Shaw Batholith; Bickle et al., 1985) and other ancient terrains (England and Bickle, 1984) suggest minimum crustal thicknesses similar to the modern day. If so, then the pre-3.5 Ga basement was probably at least 30 km thick, assuming that the volume of ocean water was not significantly less than today (de Wit and Hynes, 1995).

Clearly, it would be premature to claim that the pre-3.5 Ga basement represented a significant fraction of the present continental volume. However, the evidence that such continental basement existed in the early Archaean and is now gone suggests that there was efficient early crustal recycling and that the present age distribution of continental crust is a product of preservation and recycling. This lends support to models for the early formation of substantial volumes of continental crust.

Acknowledgements

Funded by Australian Research Council grants to R. Buick. Constructive comments were provided by J.R. Ridley. Reviews by M.E. Barley and K.C. Condie greatly improved this paper.

References

Abouchami, W., Boher, M., Michard, A., Albarede, F., 1990. A major 2.1 Ga event of mafic magmatism in West Africa: an early stage of crustal accretion. J. Geophys. Res. 95, 17 605–17 629.

Armstrong, R.L., 1968. A model for the evolution of strontium and lead isotopes in a dynamic Earth. Rev. Geophys. 6, 175–199.

Armstrong, R.L., 1981. Comment on 'Crustal growth and mantle evolution: inferences from models of element transport and Nd and Sr isotopes'. Geochim. Cosmochim. Acta 45, 1251

Armstrong, R.L., 1991. The persistent myth of crustal growth. Aust. J. Earth Sci. 38, 613–640.

Barley, M.E., 1986. Incompatible-element enrichment in Archean basalts: a consequence of contamination by older sialic crust rather than mantle heterogeneity. Geology 14, 947–950.

Barley, M.E., 1993. Volcanic, sedimentary and tectonostratigraphic environments of the ∼3.46 Ga Warrawoona Megasequence: a review. Precambrian Res. 60, 47–67.

Barley, M.E., Pickard, A.L., 1999. An extensive, crustally-derived, 3325 to 3310 Ma silicic volcanoplutonic suite in the eastern Pilbara Craton: evidence from the Kelly Belt, McPhee Dome and Corunna Downs Batholith. Precambrian Res. 96, 41–62.

Bickle, M.J., Morant, P., Bettenay, L.F., Boulter, C.A., Blake, T.S., Groves, D.I., 1985. Archean tectonics of the Shaw Batholith, Pilbara Block, Western Australia: structural and metamorphic tests of the batholith concept. In: Ayres, L.D., Thurston, P.C., Card, K.D., Weber, W. (Eds.), Evolution of Archean Supracrustal Sequences. Geol. Assoc. Can. Spec. Pap. 28, 325–341.

Bowring, S.A., Housh, T., 1995. The Earth's early evolution. Science 269, 1535–1540.

Bowring, S.A., Williams, I.S., 1999. Priscoan (4.00–4.03 Ga) orthogenesis from northwestern Canada. Contrib. Miner. Petrol. 134, 3–16.

Brauhart, C.W., Groves, D.I., Morant, P., 1998. Regional alteration systems associated with volcanogenic massive sulfide mineralization at Panorama, Pilbara, Western Australia. Econ. Geol. 93, 292–302.

Brenan, J.M., Shaw, H.F., Phinney, D.L., Ryerson, F.J., 1994. Rutile–aqueous fluid partitioning of Nb, Ta, Hf, Zr, U and Th: implications for high field strength element depletions in island-arc basalts. Earth Planet. Sci. Lett. 128, 327–339.

Buick, R., Barnes, K.R., 1984. Cherts in the Warrawoona Group: early Archaean silicified sediments deposited in shallow-water environments. In: Muhling, J.R., Groves, D.I., Blake, T.S. (Eds.), Archaean and Proterozoic Basins of the Pilbara, Western Australia: Evolution and Mineralization Potential. Univ. West. Aust. Spec. Publ. 9, 37–53.

Buick, R., Dunlop, J.S.R., 1990. Evaporitic sediments of Early Archaean age from the Warrawoona Group, North Pole, Western Australia. Sedimentology 37, 247–277.

Buick, R., Thornett, J.R., McNaughton, N.J., Smith, J.B., Barley, M.E., Savage, M., 1995. Record of emergent continental crust ∼3.5 billion years ago in the Pilbara craton of Australia. Nature 375, 574–577.

Collerson, K.D., Kamber, B.S., 1999. Evolution of the continents and the atmosphere inferred from Th–U–Nb systematics of the depleted mantle. Science 283, 1519–1522.

Compston, W., Pidgeon, R.T., 1986. Jack Hills, evidence of more very old detrital zircons in Western Australia. Nature 321, 766–769.

Davies, G.F., 1992. On the emergence of plate tectonics. Geology 20, 963–966.

Davies, G.F., 1995. Punctuated tectonic evolution of the earth. Earth Planet. Sci. Lett. 136, 363–379.

Desrochers, J.P., Hubert, C., Ludden, J.N., Pilote, P., 1993. Accretion of Archean oceanic plateau fragments in the Abitibi greenstone belt, Canada. Geology 21, 451–454.

de Wit, M.J., 1998. On Archean granites, greenstones, cratons and tectonics: does the evidence demand a verdict? Precambrian Geol. 91, 181–226.

de Wit, M.J., Hynes, A., 1995. The onset of interaction between the hydrosphere and oceanic crust and the origin of the first continental lithosphere. In: Coward, M.P., Ries, A.C. (Eds.), Early Precambrian Processes, Geol. Soc. London, Spec. Publ. 95, 1–9.

de Wit, M.J., Hart, R.A., Hart, R.J., 1987. The Jamestown Ophiolite Complex, Barberton mountain belt: a section through 3.5 Ga oceanic crust. J. Afr. Earth Sci. 6, 681–730.

DiMarco, M.J., Lowe, D.R., 1989. Petrography and provenance of silicified early Archaean volcaniclastic sandstones, eastern Pilbara Block, Western Australia. Sedimentology 36, 821–836.

Eggins, S.M., Woodhead, J.D., Kinsley, L.P.J., Mortimer, G.E., Sylvester, P.J., McCulloch, M.T., Hergt, J.M., Handler, M.R., 1997. A simple method for the precise determination of 40 trace elements in a geologic sample by ICP-MS using enriched isotope internal standardisation. Chem. Geol. 134, 311–323.

Ellam, R.M., Hawkesworth, C.J., 1988. Is average continental crust generated at subduction zones? Geology 16, 314–317.

England, P.C., Bickle, M.J., 1984. Continental thermal and tectonic regimes during the Archaean. J. Geol. 92, 353–367.

Froude, D.O., Ireland, T.R., Kinny, P.D., Williams, I.S., Compston, W., Williams, I.R., Myers, J.S., 1983. Ion microprobe identification of 4100–4200 Myr-old terrestrial zircons. Nature 304, 616–618.

Fujimaki, H., Tatsumoto, M., Aoki, K., 1984. Partition coefficients of Hf, Zr and REE between phenocrysts and groundmasses. Proc. 14th Lunar and Planetary Science Conf., Part 2. J. Geophys. Res. 89, B662–B672.

Green, T.H., Pearson, N.J., 1985. Experimental determination of REE partition coefficients between amphibole and basaltic to andesitic liquids at high pressures. Geochim. Cosmochim. Acta 49, 1465–1468.

Gruau, G., Jahn, B.M., Glikson, A.Y., Davy, R., Hickman, A.H., Chauvel, C., 1987. Age of the Archean Talga–Talga Subgroup, Pilbara Block, Western Australia, and early evolution of the mantle: new Sm–Nd isotopic evidence. Earth Planet. Sci. Lett. 85, 105–116.

Hamilton, P.J., Evensen, N.M., O'Nions, R.K., Glikson, A.Y., Hickman, A.H., 1981. Sm–Nd dating of the North Star Basalt, Warrawoona Group, Pilbara Block, Western Australia. Geol. Soc. Aust. Spec. Publ. 7, 187–192.

Hickman, A.H., 1983. The geology of the Pilbara Block and its environs. Geol. Surv. West. Aust. Bull. 127

Hofmann, A.W., Jochum, K.P., Seufert, M., White, W.M., 1986. Nb and Pb in oceanic basalts: new constraints on mantle evolution. Earth Planet. Sci. Lett. 79, 33–45.

Huppert, H.E., Sparks, R.S.J., 1985. Cooling and contamination of mafic and ultramafic magmas during ascent through continental crust. Earth Planet. Sci. Lett. 74, 371–386.

Hurley, P.M., Rand, J.R., 1969. Pre-drift continental nuclei. Science 164, 1229–1242.

Krapez, B., 1993. Sequence stratigraphy of the Archaean supracrustal belts of the Pilbara Block, Western Australia. Precambrian Res. 60, 1–45.

Le Maitre, R.W., Bateman, P., Dudek, A., Keller, J., Lameyre, J., Le Bas, M.J., Sabine, P.A., Schmid, R., Sørensen, H., Streckeisen, A., Woolley, A.R., Zanettin, B., 1989. A Classification of Igneous Rocks and Glossary of Terms. Blackwell, Oxford.

Lowe, D.R., 1983. Restricted shallow-water sedimentation of early Archean stromatolitic and evaporitic strata of the Strelley Pool Chert, Pilbara Block, Western Australia. Precambrian Res. 19, 239–283.

Maas, R., Kinny, P.D., Williams, I.S., Froude, D.O., Compston, W., 1992. The Earth's oldest known crust: a geochronological and geochemical study of 3900–4200 Ma old detrital zircons from Mt. Narryer and Jack Hills, Western Australia. Geochim. Cosmochim. Acta 56, 1281–1300.

Martin, H., 1994. The Archean grey gneisses and the genesis of continental crust. In: Condie, K.C. (Ed.), Archaean Crustal Evolution. Elsevier, Amsterdam, pp. 205–259.

McCulloch, M.T., Bennett, V.C., 1993. Evolution of the early Earth: constraints from ^{143}Nd–^{142}Nd isotopic systematics. Lithos 30, 237–255.

McCulloch, M.T., Bennett, V.C., 1994. Progressive growth of the Earth's continental crust and depleted mantle: geochemical constraints. Geochim. Cosmochim. Acta 58, 197–214.

McNaughton, N.J., Green, M.D., Compston, W., Williams, I.S., 1988. Are anorthositic rocks basement to the Pilbara Craton? Geol. Soc. Aust. Abstr. 21, 272–273.

McNaughton, N.J., Compston, W., Barley, M.E., 1993. Constraints on the age of the Warrawoona Group, eastern Pilbara Block, Western Australia. Precambrian Res. 60, 69–98.

Nelson, D.R., 1998. Compilation of SHRIMP U–Pb zircon geochronological data, 1997. Western Australia Geological Survey, Record 1998/2.

Neumayr, P., Ridley, J.R., McNaughton, N.J., Kinny, P.D., Barley, M.E., Groves, D.I., 1998. Timing of gold mineralization in the Mt York district, Pilgangoora greenstone belt, and implications for the tectonic and metamorphic evolution of an area linking the western and eastern Pilbara Craton. Precambrian Res. 88, 249–265.

Pearce, J.A., Baker, P.E., Harvey, P.K., Luff, I.W., 1995. Geochemical evidence for subduction fluxes, mantle melting and fractional crystallization beneath the South Sandwich Island Arc. J. Petrol. 36, 1073–1109.

Reiners, P.W., Nelson, B.K., Ghiorso, M.S., 1995. Assimilation of felsic crust by basaltic magma: thermal limits and extents of crustal contamination of mantle-derived magmas. Geology 23, 563–566.

Reymer, A., Schubert, G., 1984. Phanerozoic addition rates to the continental crust and crust growth. Tectonics 3, 63–77.

Richards, J.R., Fletcher, I.R., Blockley, J.G., 1981. Pilbara galenas: precise isotopic assay of the oldest Australian leads, model ages and growth-curve implications. Miner. Deposita 16, 7–30.

Rudnick, R.L., Fountain, D.M., 1995. Nature and composition of the continental crust: a lower mantle perspective. Rev. Geophys. 33, 267–309.

Schwandt, C.S., McKay, G.A., 1998. Rare earth element partition coefficients from enstatite/melt synthesis experiments. Geochim. Cosmochim. Acta 62, 2845–2848.

Shiano, P., Dupré, B., Lewin, E., 1993. Application of element concentration variability to the study of basalt alteration (Fangataufa atoll, French Polynesia). Chem. Geol. 194, 99–124.

Sims, K.W.W., DePaolo, D.J., 1997. Inferences about mantle magma sources from incompatible element concentration ratios in oceanic basalts. Geochim. Cosmochim. Acta 4, 765–784.

Smithies, R.H., Hickman, A.H., Nelson, D.R., 1999. New constraints on the evolution of the Mallina Basin, and their bearing on relationships between the contrasting eastern and western granite–greenstone terranes of the Archaean Pilbara Craton, Western Australia. Precambrian Res. 94, 11–28.

Spear, F.S., 1995. Metamorphic Phase Equilibria and Pressure-–Temperature–Time Paths. Mineralogical Society of America, Washington.

Sun, S.-S., McDonough, W.F., 1989. Chemical and isotopic systematics of oceanic basalts: implications for mantle composition and processes. In: Saunders, A.D., Norry, M.J. (Eds.), Magmatism in the Ocean Basins. Geol. Soc. London, Spec. Publ. 42, 313–345.

Sylvester, P.J., Campbell, I.H., Bowyer, D.A., 1997a. Niobium/uranium evidence for early formation of the continental crust. Science 275, 521–523.

Sylvester, P.J., Harper, G.D., Byerly, G.R., Thurston, P.C., 1997b. Volcanic aspects. In: de Wit, M.J., Ashwal, L.D. (Eds.), Greenstone Belts. Clarendon Press, Oxford, pp. 55–90.

Taylor, S.R., 1989. Growth of planetary crusts. Tectonophysics 161, 147–156.

Taylor, S.R., McLennan, S.M., 1985. The Continental Crust: Its Composition and its Evolution. Blackwell, Oxford.

Taylor, S.R., McLennan, S.M., 1995. The geochemical evolution of the continental crust. Rev. Geophys. 33, 241–265.

Thorpe, R.I., Hickman, A.H., Davis, D.W., Mortensen, J.K., Trendall, A.F., 1992. U–Pb zircon geochronology of Archaean felsic units in the Marble Bar region Pilbara Craton, Western Australia. Precambrian Res. 56, 169–189.

TECTONOPHYSICS

www.elsevier.com/locate/tecto

Age constraints on recycled crustal and supracrustal sources of Archaean metasedimentary sequences, Eastern Goldfields Province, Western Australia: evidence from SHRIMP zircon dating

B. Krapez [a], S.J.A. Brown [a,*], J. Hand [b], M.E. Barley [a], R.A.F. Cas [b]

[a] *Centre for Research in Strategic Mineral Deposits, Department of Geology and Geophysics, University of Western Australia, Nedlands WA 6907, Australia*
[b] *Department of Earth Sciences, Monash University, Clayton VIC 3168, Australia*

Received 3 March 1999; accepted for publication 20 September 1999

Abstract

SHRIMP U–Pb zircon dating of sedimentary, volcanic and subvolcanic-intrusive rocks from unconformity bounded sequences in the Eastern Goldfields Province defines maximum depositional ages for those sequences. The Kambalda Sequence, which records a backarc basin, was deposited prior to 2700 Ma and from at least 2715 Ma. The Spargoville Sequence, which records an arc-adjacent, volcano-bound basin, was deposited between 2700 and 2683 Ma. The Kalgoorlie Sequence, which records an intra-arc rift basin, comprises two tectonic stages with depositional ages of 2681–2670 and 2661–2655 Ma. Deposition of the Spargoville and Kalgoorlie Sequences overlapped with the intrusion of granitoids, and subvolcanic porphyries, thereby indicating that those granitoids were emplaced during crustal extension. Younger, submarine-fan (Kurrawang Sequence) and braided-fluvial (Merougil and Jones Creek Sequences) deposits of a remnant-ocean basin were deposited at <2655 Ma, prior to regional compression and metamorphism, which apparently occurred between 2650 and 2630 Ma.

When coupled to palaeocurrent data, provenance dating of detrital zircon suites fingerprints southeast-oriented amalgamation of felsic detritus from multiple sources within longitudinal depositional systems, and longitudinal reworking of felsic detritus from single-sourced transverse systems. Most felsic detritus was recycled from coeval volcanoplutonic arcs and the orogen to the remnant-ocean basin sequences, with zircon ages ranging from ∼2730 to ∼2660 Ma. Older felsic sources, with age ranges of 3570–3130, 3030–2900, 2870–2770 and ∼2760 Ma, could have been structural enclaves within the volcanoplutonic arcs or orogen, but were more likely within a distant cratonic basement. When coupled with stratigraphic data, it is concluded that detrital-zircon age distributions preserve evidence for the sedimentary recycling of coeval volcanoplutonic arc or orogenic sources, uplifted basin-floor and dissected-arc successions, and cratonic basement. Virtually all source terrains have apparently vanished, presumably through the combination of tectonic and sedimentary recycling, and tectonic severance. Increased use of detrital zircon suites from Archaean sedimentary sequences is advocated in order to quantify the balance between sedimentary recycling of felsic rocks from contemporaneous volcanoplutonic sources, which themselves document crustal growth in arc environments, and from relict volcanoplutonic and crustal-basement sources, which document the degree of crustal recycling. © 2000 Elsevier Science B.V. All rights reserved.

Keywords: Archaean; Eastern Goldfields Province; provenance; SHRIMP; zircon

---

* Corresponding author. Tel.: +61-8-9380-3465; fax: +61-8-9380-1178.
E-mail address: sbrown@geol.uwa.edu.au (S.J.A. Brown)

0040-1951/00/$ - see front matter © 2000 Elsevier Science B.V. All rights reserved.
PII: S0040-1951(00)00059-7

1. Introduction

A genetic relationship between provenance tectonic settings and detrital compositions of sandstones was advocated long ago by Krynine (1942), and further emphasized by Folk (1968) and Pettijohn et al. (1972). Since then, the correlation of sandstone compositions to global tectonic settings has become a standard basin-analysis technique (Dickinson and Suczek, 1979; Valloni and Maynard, 1981; Ingersoll et al., 1984). Heavy-mineral suites are also provenance indicators. In particular, Folk (1968) emphasized the provenance value of morphological varieties of detrital zircon. With the advent of techniques that allow rapid and precise isotopic analyses of single zircon

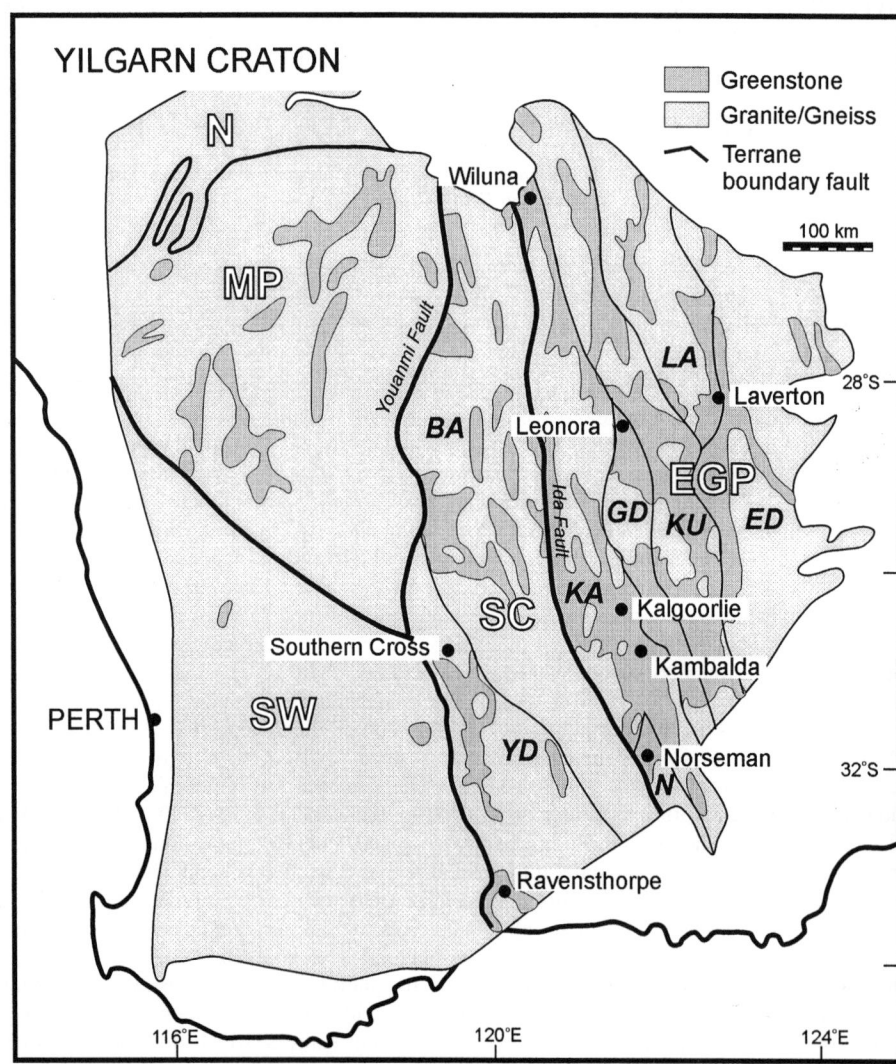

Fig. 1. Geographic distribution of greenstone belts and granitoid-gneiss in the Archaean Yilgarn Craton, showing provinces and terranes with terrane-boundary faults (after Myers, 1997; cf. tectonic provinces of Krapez, 1997, fig. 7). N=Narryer Terrane, MP=Murchison Province, SW=Southwestern Province, SC=Southern Cross Province, EGP=Eastern Goldfields Province, BA=Barlee Terrane, YD=Yellowdine Terrane, KA=Kalgoorlie Terrane, N=Norseman Terrane, GD=Gindalbie Terrane, KU=Kurnalpi Terrane, LA=Laverton Terrane, ED=Edjudina Terrane.

grains, for example by thermal ionization mass spectrometry (TIMS) and sensitive high-resolution ion microprobe (SHRIMP), provenance analysis can also determine the ages of those grains. If supported by isotopic dating of igneous and metamorphic rocks, isotopic dating of detrital zircon grains in sandstones can identify contributions from intrabasinal and extrabasinal sources, and thereby quantify recycling of felsic supracrustal and crustal rocks, the principal sources of detrital zircon. This is significant for studies of Precambrian sandstones, because provenance dating can identify contributions from vanished sources (Froude et al., 1983).

We report here the results of a detrital-zircon provenance study of metamorphosed sandstones from the Late Archaean Eastern Goldfields Province of Western Australia. The study is part of a project that covers all the Province, but here presents data from the Kalgoorlie, Gindalbie and Norseman Terranes, and the Kathleen Valley–Lawlers Domain (Figs. 1 and 2). The paper reports SHRIMP U–Pb detrital-zircon ages from a set of unconformity bounded sequences. The data indicate that some sediment in those sequences was derived from unknown felsic volcanoplutonic-arc and basement sources. The study also highlights the difficulty of estimating maximum depositional ages due to Pb loss in, and metamorphic recrystallization of, zircon.

2. Tectonostratigraphic overview of the Eastern Goldfields Province

Archaean supracrustal rocks of the Yilgarn Craton (Fig. 1) crop out in fault-bounded domains, terranes and superterranes (i.e. provinces) (Myers, 1997). The structural geometry of the Eastern Goldfields Province has been interpreted to record Cordilleran-style terrane accretion (Myers, 1997), although Krapez (1997) interprets the geometry as a fold-thrust belt that developed during transcurrent closure of a marginal sea. There is no doubt that the original geometry of supracrustal and crustal elements has been disrupted, but the domains and terranes are not exotic to the original geotectonic setting (Swager, 1997).

More than one ~45 myr tectonic supercycle and ~180 myr tectonic megacycle (Krapez, 1997) are recorded in the Yilgarn Craton (Barley et al., 1998a; cf. Busby et al., 1998). The youngest megacycle, from ~2772 to ~2590 Ma (Barley et al., 1998a), comprises arc-adjacent, rift, marginal and remnant-ocean basin successions preserved mainly within the Eastern Goldfields and Murchison Provinces (Fig. 1). Therefore, for provenance studies, provenance ages <2772 Ma record recycling from the youngest megacycle, whereas provenance ages >2772 Ma record recycling from older tectonic megacycles.

The youngest megacycle in the Eastern Goldfields Province comprises, in part, northeastern volcanic-arc and southwestern backarc-basin lithotectonic associations (e.g. Barley et al., 1989), with an original geometry analogous to the Indonesian Tectonic Zone (Krapez, 1997). Recent stratigraphic interpretations (Krapez, 1997; Krapez et al., 1997; Barley et al., 1998a,b) indicate that sequences of tholeiitic and high-MgO basalt are backarc spreading-related, calc-alkalic volcanic-volcaniclastic sequences are arc-related, and combined volcaniclastic-epiclastic sedimentary and basalt-rhyolite volcanic sequences are rift-related. The recent work also identified a set of unconformity bounded sequences in the Kalgoorlie, Gindalbie and Norseman Terranes (Fig. 2), and in the Kathleen Valley–Lawlers Domain of the Kurnalpi Terrane (Figs. 1 and 2).

The sequences are of second-order rank (see Krapez, 1996, 1997 for terminology) but are erosional and structurally bound remnants, such that their present distributions do not define the original extent of depositional basins. Indeed, stratigraphic interpretations suggest that those basins extended beyond the present structural confines of the Eastern Goldfields Province. In the southwestern Eastern Goldfields Province, and from oldest to youngest, the sequences are named the Kambalda Sequence, the Spargoville Sequence, the Kalgoorlie Sequence, the Kurrawang Sequence and the Merougil Sequence, whereas the Jones Creek Sequence is named from the Kathleen Valley–Lawlers Domain.

Based on the basin analysis reported in Krapez (1997), Krapez et al. (1997), Barley et al. (1998b)

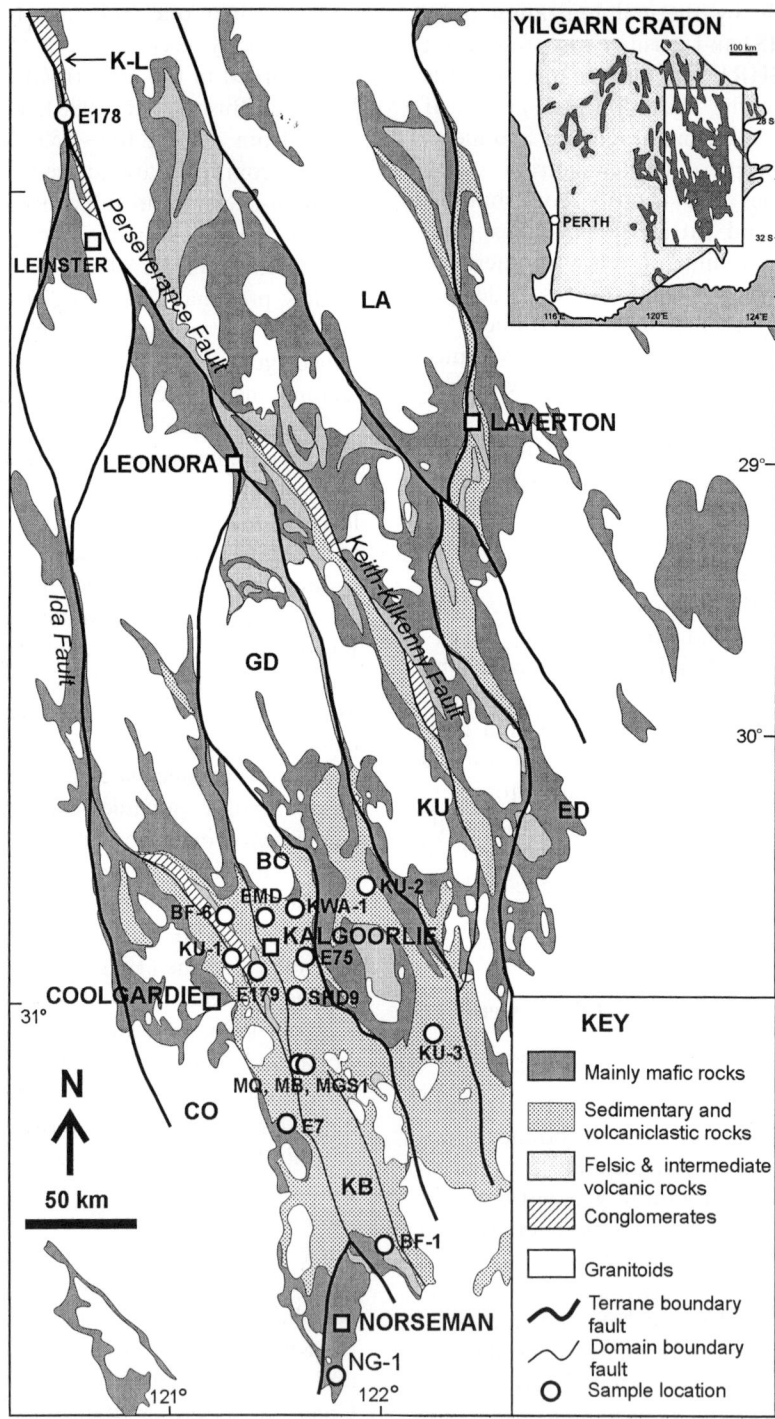

Fig. 2. Simplified geological map of the Eastern Goldfields Province, showing the broad distribution of greenstone-belt lithologies, and terranes and domains discussed in the text. Terrane and domain boundaries are taken from Myers (1997). In addition to the terrane names defined in Fig. 1, CO=Coolgardie Domain, KB=Kambalda Domain, BO=Boorara Domain, K–L=Kathleen Valley–Lawlers Domain.

and Hand (1998), the sequences are interpreted to record opening and closing phases of a marginal sea. The Kambalda Sequence is interpreted to record a sediment-starved backarc basin. The Spargoville Sequence is interpreted to represent the preserved deposits of an arc-adjacent, volcano-bound basin, and the Kalgoorlie Sequence the preserved fill of an intra-arc rift basin, probably associated with strike-slip tectonics. The Kurrawang and Merougil Sequences are interpreted to record closure of the marginal sea, and to represent respectively flysch and molasse stages of a remnant-ocean basin, probably with a transcurrent western margin (Krapez, 1997, figs. 13 and 14).

3. Stratigraphic framework

3.1. Kambalda Sequence

The Kambalda Sequence is named after the Kambalda Group (Swager et al., 1992), which is best preserved in the northwest of the Kambalda Domain of the Kalgoorlie Terrane (Fig. 2), although a basal stratigraphic contact is not exposed. There, the sequence comprises, from oldest to youngest, the Lunnon Basalt, the Kambalda Komatiite, the Devon Consols Basalt and the Paringa Basalt (Fig. 3). Foster et al. (1996) defined a Re–Os isochron age of 2706 ± 36 Ma from the Kambalda Komatiite. In the southeast of the Kambalda Domain, the Buldania Formation (Fig. 3) comprises equivalents of the Lunnon Basalt and the Kambalda Komatiite.

Packages of 'interflow' sedimentary rocks in the Kambalda Sequence comprise siliciclastic or feldspathochloritic turbidites and carbonaceous shales (Bavington, 1981). The packages represent distal lowstand-fan sediments and are the basal units to volcanosedimentary depositional sequences (see Krapez, 1996, 1997 for terminology). The best-known package is the Kapai Slate (Fig. 3), which comprises two facies assemblages, namely carbonaceous shales at the top of the Devon Consols Basalt, and incised turbidites and carbonaceous shales at the base of the Paringa Basalt. Claoué-Long et al. (1988) assign a SHRIMP U–Pb zircon age of 2692 ± 4 Ma to the Kapai Slate, but with zircon grains as old as 3441 ± 18 Ma. Nelson (1996) dated a similar unit in the Boorara Domain (Fig. 2) at 2708 ± 7 Ma, with zircon grains as old as 2890 ± 12 Ma.

In the Norseman Terrane (Fig. 2), the Kambalda Sequence comprises, from oldest to youngest, sedimentary rocks of the Upper Penneshaw Formation and the Noganyer Formation, and basalt of the Woolyeenyer Formation (Fig. 3). A synvolcanic dolerite sill in the Woolyeenyer Formation has a SHRIMP U–Pb zircon age of 2714 ± 5 Ma (Hill et al., 1992), which may correlate the formation to the Lunnon Basalt. Basalts of the Lower Penneshaw Formation (Fig. 3) are stratigraphic components of an older, unnamed sequence and depositional basin.

Upper greenschist to amphibolite facies, metasedimentary rocks of the Upper Penneshaw Formation (Fig. 3) are described only from diamond drill-core sections (Hallberg, 1970). Hill et al. (1992) interpret the formation as comprising internally differentiated felsic volcanic rocks, with a SHRIMP U–Pb zircon age of 2938 ± 10 Ma, whereas Nelson (1995) interprets the metasedimentary rocks as biotite rhyolites, with a SHRIMP U–Pb zircon age of 2930 ± 4 Ma. Both SHRIMP studies record zircon xenocrysts, some as old as 3129 ± 6 Ma. There is significant conflict as to the meaning of all the zircon ages because Hallberg (1996, pers. commun.) maintains that there are only metasedimentary rocks in the Upper Penneshaw Formation.

Depositional sequences in the Noganyer Formation (Fig. 3) comprise epiclastic turbidites and carbonaceous shales. Thinly bedded turbidites are composites of silicified sandstone–siltstone (cf. chert) and magnetic shale that resemble banded iron formation (Krapez, 1997). In confirmation of an epiclastic origin for the 'cherts', samples analysed by Campbell and Hill (1988) produced SHRIMP U–Pb ages of ~ 3670–3650 Ma from detrital zircon grains, as well as an age of 2706 ± 5 Ma from hydrothermal zircon rims and grains. Sandstones in the Noganyer Formation are of two types: one is quartzose to quartzofeldspathic with few lithic fragments, and the other is feldspathochloritic with clasts of felsic porphyry and

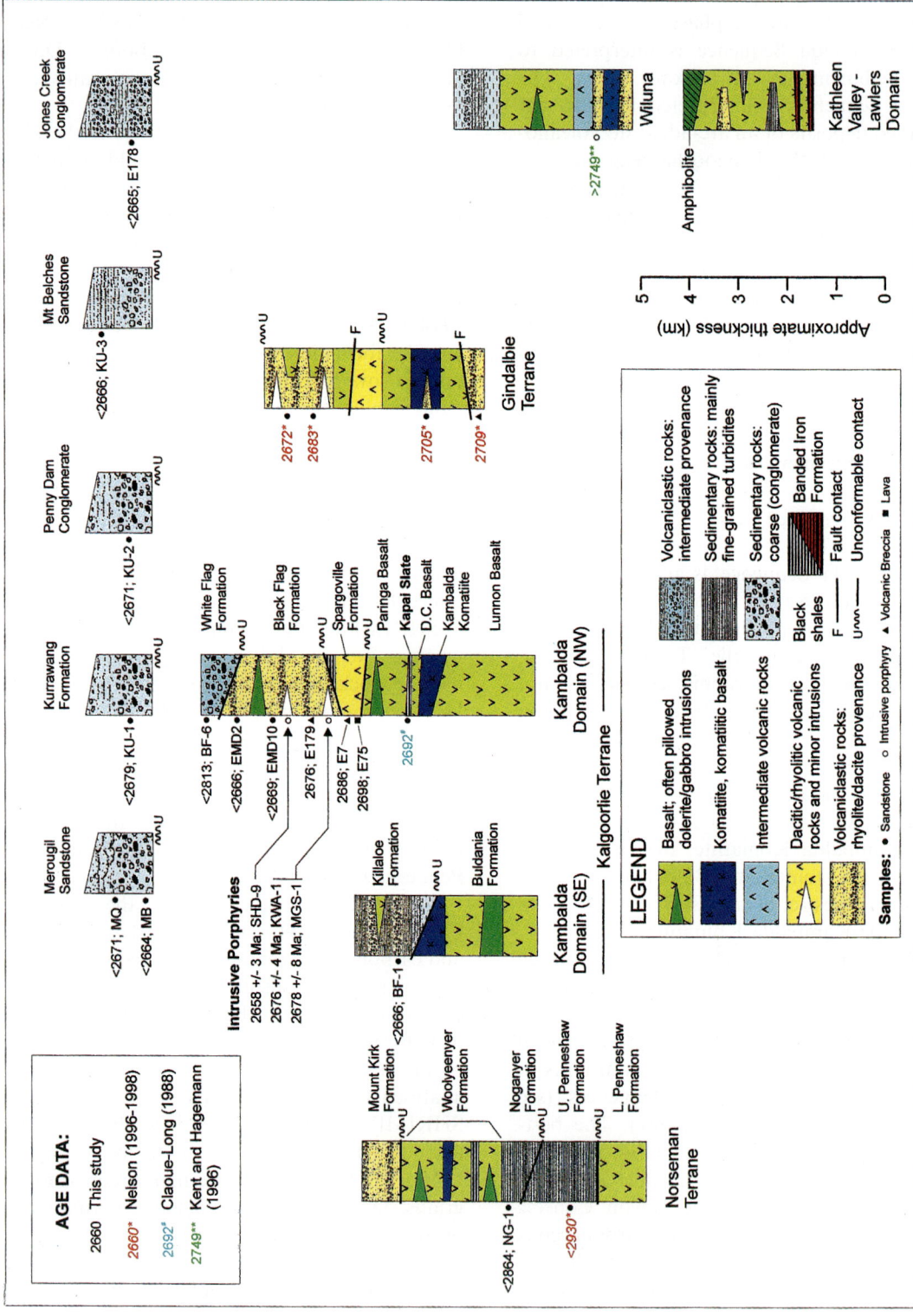

Fig. 3. Generalized stratigraphic relationships in the Eastern Goldfields Province; modified from Hallberg (1970, 1985), Swager et al. (1992), Krapez (1997), Nelson (1997), and Barley et al. (1998b).

dacite. The former is dominant, whereas the latter is present in only the basal part of the basal depositional sequence.

At the western margin of the Norseman Terrane, the uppermost section of the Kambalda Sequence, above the Woolyeenyer Formation, comprises a depositional sequence of lowstand-fan sandstone turbidites and carbonaceous shale overlain by tholeiitic amphibolite. Hill et al. (1992) obtained a SHRIMP U–Pb zircon age of 2688 ± 8 Ma from a felsic–volcanic unit, presumably from that depositional sequence. Recent mapping (Barley et al., 1998b) shows that the dated 'felsic–volcanic unit' is a quartz–albite porphyry that intruded the shear zone between the Norseman Terrane and the Coolgardie Domain.

Time equivalents of the Kambalda Sequence are known from the Gindalbie Terrane (Fig. 3), but lithostratigraphic units are not formally designated (Ahmat, 1993). A dacitic breccia from a felsic volcanic–volcaniclastic complex in the north of the Gindalbie Terrane has a SHRIMP U–Pb zircon age of 2709 ± 4 Ma (Nelson, 1997). The breccia also contains zircon xenocrysts with mean ages of 2737 ± 6 and 2761 ± 4 Ma. A felsic volcaniclastic unit interlayered with ultramafic lava flows, which may correlate to the Kambalda Komatiite, has an age of 2705 ± 4 Ma (Nelson, 1997).

3.2. Spargoville Sequence

The Spargoville and Kalgoorlie Sequences are components of the Black Flag Group (see, for example, Swager et al., 1992). The group derives from the Black Flag Beds (of original formation status) by incorporation with the Spargoville and White Flag Formations (Fig. 3). Volcanic and sedimentary rocks from the Black Flag Group have geochemical signatures that indicate derivation from a source with Tonalite–Trondjhemite–Granodiorite (TTG) characteristics (Barley et al., 1998b).

The Spargoville Sequence is named after the Spargoville Formation, which is best preserved in the Kambalda (Fig. 3) and Coolgardie Domains of the Kalgoorlie Terrane (Fig. 2). The sequence comprises rhyolitic and dacitic lava interlayered with resedimented volcaniclastic breccias and sandstones, and overlain by carbonaceous shale (Fig. 3). The base of the Spargoville Sequence is best exposed in the Coolgardie Domain (Fig. 2), where the basal surface is an incised, angular unconformity on equivalents of the Kambalda Komatiite and Devon Consols Basalt (Fig. 3). The Spargoville Sequence may also be present in the Gindalbie Terrane (Fig. 2), but sequence division there has not yet been accomplished (Fig. 3).

3.3. Kalgoorlie Sequence

The Kalgoorlie Sequence comprises the Black Flag and White Flag Formations, which are best preserved in the northwest of the Kambalda and Coolgardie Domains of the Kalgoorlie Terrane (Fig. 3). The Black Flag Formation is referred to as the Killaloe Formation in the southeast of the Kambalda Domain, and the Mount Kirk Formation in the Norseman Terrane (Fig. 3). A sequence of basalt and komatiite previously correlated to the Mount Kirk Formation in the southeastern Coolgardie Domain of the Kalgoorlie Terrane (Hill et al., 1992) is part of the Kambalda Sequence (Barley et al., 1998b). The basal surface of the Kalgoorlie Sequence in the Coolgardie Domain is an angular unconformity on the Spargoville and Kambalda Sequences, and in the southeast of the Kambalda Domain, an incised angular unconformity on the Kambalda Sequence (Fig. 3).

Depositional sequences in the Black Flag Formation (Fig. 3), some of which are also unconformity-bound, are composed of volcaniclastic and epiclastic turbidites, with intervening condensed sections of carbonaceous shale (Krapez, 1997; Krapez et al., 1997; Hand, 1998). Those sequences contain minor rhyolite–dacite pyroclastic and lava lithofacies, and are intruded by felsic, mafic and ultramafic sills. In the Boorara Domain (Fig. 2), equivalent depositional sequences comprise interlayered dacite and trachyte lava flows, heterolithic volcaniclastic turbidites and an olistostrome. The White Flag Formation comprises coarse-grained epiclastic turbidites, derived from an andesitic source, with olistoliths of pillowed and massive andesite. Depositional systems in the Kalgoorlie Sequence were longitudinal, with sediment transport to the southeast (Hand, 1998).

3.4. Kurrawang Sequence

The Kurrawang Sequence is named after the Kurrawang Formation of the Kalgoorlie Terrane (Figs. 2 and 3). Based on the basin analysis reported in Barley et al. (1998b), the Kurrawang Sequence is also represented by the Penny Dam Conglomerate and the Mt Belches Sandstone in the Gindalbie Terrane (Figs. 2 and 3). The sequence comprises an upwards fining succession of quartzofeldspatholithic conglomerate, sandstone and mudrock that represent high-density, coarse-grained to low-density fine-grained turbidites. Palaeocurrent data establish deposition on a southeast-oriented submarine-fan. The basal surface is an angular unconformity on the Kalgoorlie Sequence.

3.5. Merougil Sequence

The Merougil Sequence is named after the Merougil Sandstone of the Kalgoorlie Terrane (Figs. 2 and 3). The sequence comprises an upwards fining succession of quartzofeldspatholithic conglomerate and sandstone (Fig. 3). Sedimentological analyses (Krapez, 1997; Krapez et al., 1997; Barley et al., 1998b) establish that the Merougil Sequence records the deposits of fluvial bar and channel depositional systems. Palaeocurrent data indicate deposition on a southeast-oriented braid-plain. The basal surface is an angular unconformity on the Kalgoorlie Sequence.

3.6. Jones Creek Sequence

The Jones Creek Sequence is named after the Jones Creek Conglomerate of the Wiluna and Kathleen Valley–Lawlers Domains (Figs. 1–3). The sequence unconformably overlies unnamed volcanosedimentary strata (Fig. 3) and nonconformably overlies granitoid. The Jones Creek Sequence has two provenance facies, namely granitic and mafic–volcanic, although the mafic–volcanic facies contains some granitic detritus (Marston and Travis, 1976). Krapez et al. (1997) and Barley et al. (1998b) interpret the Jones Creek Sequence to be the deposit of a southeast-oriented braid-plain in a fault-bounded basin, and a time-stratigraphic equivalent of the Merougil Sequence.

4. Structural history of the Eastern Goldfields Province

Structural studies of the Eastern Goldfields Province have recognized two phases of compressive deformation. The first (D_1) was apparently south-over-north thrust stacking and recumbent folding (e.g. Swager et al., 1992), but those structures are not recognized in every domain. The relationship between D_1 and early (possibly pre-D_1) low-angle extensional structures, which are recorded in northeastern domains of the Province (Hammond and Nisbet, 1992), is not established. The extensional structures may be listric detachments of the rift basin defined by the Kalgoorlie Sequence, whereas D_1 structures, which involve strata of the Spargoville Sequence, may record core-complex uplifts and granitoid intrusions >2675 Ma (Swager, 1997). Post-D_1, pre-D_2 extension is apparently recorded by ∼2675 Ma porphyry intrusions (Swager, 1997), and probably also by layered mafic–ultramafic sills that intrude the Kalgoorlie Sequence (e.g. Golden Mile Dolerite).

D_2 is recorded by northnorthwest-trending upright folds. All strata, including the Kurrawang and Merougil Sequences, were affected by D_2. Voluminous 2675–2657 Ma granitoid intrusions are interpreted to have been synchronous with D_2 (Nelson, 1997), with post-D_2 extension linked to the uplift of granitoid–gneiss complexes at ∼2660 Ma (Swager, 1997). D_3 transcurrent shear zones and en échelon folds are constrained to the interval 2663–2645 Ma (Nelson, 1997; Swager, 1997).

Although some domain- and terrane-boundary faults may approximate original compressive structures, it is likely that many (e.g. the Ida Fault, Fig. 2) were reactivated during post-D_3 extensional collapse of the entire fold-thrust belt, with accompanying high heat-flow and lode-gold emplacement at <2640 Ma (Swager, 1997).

Most studies (e.g. Swager et al., 1992) have argued that regional metamorphism, characterized by low to intermediate pressures (<4.5 kbar), reached peak temperatures late during D_2–D_3 and was contemporaneous with syn-D_3 granite emplacement. Binns et al. (1976) recognized static and dynamic styles of deformation, with the static style varying from prehnite–pumpellyite to upper greenschist facies, and the dynamic style varying from upper greenschist to upper amphibolite facies. Post-D_2 and syn-D_3 granitoids resulted in thermal aureoles that were apparently superposed on elevated regional metamorphic temperatures. Barley and Groves (1987) also recognized seafloor hydrothermal metamorphism of mafic and ultramafic rocks of the Kambalda Sequence, the products of which were not everywhere destroyed by regional metamorphism.

5. Overview of detrital zircon dating

5.1. General considerations

For several reasons, detrital zircon suites in sedimentary rocks can be biased provenance indicators. Firstly, detrital zircon suites are biased to felsic volcanic and plutonic rocks. Ultramafic, mafic and most intermediate igneous rocks are under-represented by zircon in heavy-mineral suites because those rocks are inherently poor in zircon. Zircon grains of inherited (xenocrystic) origin in igneous source rocks can also be present in the detrital population, despite the possibility that rocks corresponding to the ages of those grains were not exposed during sediment deposition. In some cases, a lack of recognition of the inherited provenance of zircon in sedimentary rocks of andesitic volcaniclastic sequences has led to the misinterpretation of syneruptive ages from detrital-zircon populations (e.g. Smithies et al., 1999, p. 21 and fig. 4G). Secondly, because of inherent structural damage, high-U zircon grains are often preferentially destroyed during source-rock weathering and sediment transport, such that preserved detrital populations may be biased towards low-U grains. When present, high-U grains are difficult, or impossible, to date by U–Pb methods because of radiation damage, mobilization (loss) of radiogenic Pb, and high common-Pb contents. Thirdly, detrital zircon suites may contain grains that record metamorphic events in source terrains, and those events may also be multicyclic. Fourthly, zircon grains that record past episodes of felsic magmatism or metamorphism may have been recycled through older sedimentary rocks.

Other considerations are radiation or structural damage and recrystallization of zircon grains during metamorphism or hydrothermal alteration of the host rock (e.g. Pidgeon, 1992). Zircon grains that have sustained damage to their crystal lattice (i.e. metamict grains) are typically brown and translucent or black and opaque, with enhancement of growth zones high in U. The damage, and subsequent hydration, typically causes the formation of radial expansion cracks, which leads to Pb loss. Recrystallization or new growth of zircon may occur in response to high temperatures and interaction with metamorphic fluids (Wayne et al., 1992). Recrystallization appears as irregular patches that overprint and obscure growth zonation. The patches are typically clear in transmitted light and marginally brighter in backscatter electron (BSE) SEM images (Pidgeon, 1992). Hartmann et al. (1997) propose that recrystallization of zircon may take place by 'fracture sealing'. Recrystallized zones are often lower in U than the host grain, and therefore tend to be resistant to subsequent radiation damage. However, recrystallization is typically accompanied by radiogenic Pb loss, which is indicated by large age variations in spot analyses from the same recrystallized zone.

Although analyses from recrystallized zones are often concordant or near-concordant, the analyses could represent either the age of magmatic crystallization, or metamorphic or magmatic recrystallization, or, indeed, any age between those events due to partial Pb loss (Pidgeon, 1992). Therefore, ages obtained from recrystallized zones should be regarded as minimum ages. In some cases, in-situ recrystallized ages can be incorrectly assigned to detrital ages, with a resulting error in the interpreted maximum age of deposition. This is particularly a problem where there are few other constraints on the age of a depositional basin. Recrystallized zircon grains also survive erosion and transport as detrital grains, and can therefore yield information on recrystallization events (typically metamorphism) in provenance areas. Those events are indicated by multiple age estimations from the same grains (e.g. Martin et al., 1998, table 2), all of which predate the depositional age of the host sedimentary rock.

Because of the ambiguity associated with analyses of recrystallized zones, it is essential that each zircon grain on a SHRIMP mount be recorded on SEM BSE and catholuminescence (CL) images prior to SHRIMP analysis. Identification of cores, rims, resorption textures, structural damage (e.g. small fractures not visible in transmitted and reflected light), and recrystallized zones from SEM images can help selection of SHRIMP sampling sites and subsequent interpretations, unless the study is interested in the ages of recrystallization events.

5.2. Criteria for rejecting data

Because the exact mechanisms for Pb loss are not well-understood, criteria have been used to exclude data from zircon grains suspected of having lost radiogenic Pb.

Analyses that are discordant, or plot away from the concordia line, must have undergone some Pb (or U) loss or gain. Because it is often unclear at what stage and by what mechanism Pb became mobile, discordant analyses are regarded as less reliable than those that lie on, or close to, the concordia. Analyses that are discordant can be regarded only as minimum ages. In this study, analyses are considered acceptable if they lie within the statistically set limits of 5% of concordia (i.e. between 95 and 105% concordant). Nevertheless, it is accepted that there may be analyses within those limits that record post-depositional Pb loss, but these may prove impossible to eliminate if they are within operational error of the dating technique.

Excess common- or initial-Pb in a zircon grain will affect its isotopic composition and therefore also affect the age indicated by the zircon analysis. Although small amounts of common Pb can be corrected for, the corrections introduce additional uncertainty to the age determination because of assumptions about the isotopic composition of the common-Pb contaminant. Correspondingly, analyses with more than 2% ^{206}Pb attributable to decay of common Pb (as detected by ^{204}Pb) are excluded from mean-age calculations.

Nelson (1997) noted that analyses of zircon grains from the Eastern Goldfields Province that have values of U > 350 ppm yield spuriously young radiogenic ages, suggesting that the lattice in those grains has undergone significant radiation damage and subsequent Pb loss by diffusion. The same feature was noted during this study (in particular, samples MQ, E7, E179 and EMD10; refer to Tables 1 and 1A). Analyses with U concentrations > 350 ppm were therefore discarded from mean-age calculations.

5.3. Reporting of SHRIMP data

Individual zircon analyses (ages) are traditionally quoted as the mean plus or minus one standard error (mean $\pm 1\sigma$). A pooled age, or a sample estimate of a population mean, is traditionally calculated from a group of individual ages, and quoted as mean plus or minus two standard errors (mean $\pm 2\sigma$). The statistic 2σ is quoted as an estimate of the 95% confidence limits on the estimated mean age (the statistically correct values are $\pm 1.96\sigma$). The 95% confidence limits are decision statistics about which individual ages are either rejected or accepted as probable estimators of a sample mean. Individual ages have no significance other than as members of a statistical group used to define a mean (i.e. they are part of a sample age distribution), and are not independently quoted unless they fall outside all statistical groups.

Statistical validity of a sample mean age, and therefore of a statistical grouping of individual zircon ages, is established by Student's t-test on the differences between the mean and its individuals. The Null Hypothesis, that the differences between the individual ages and the sample mean age are insignificant, is tested by the χ^2 statistic. Calculated values of $\chi^2 \leq 1$ show that there is very little statistical variation between individual ages, i.e. $\geq 95\%$ of the individual ages are valid estimators of the sample mean. Calculated values of $\chi^2 > 1$ imply that there is a > 5% probability that some individual ages used to calculate the sample

Table 1
Rock descriptions and locations of sample collection sites

Sample	Unit	Description	Location	mE[a]	mN[a]	Mount No. (UWA)
Metasedimentary and volcaniclastic rocks						
E178	Jones Creek Conglomerate	Sandstone	Jones Creek	259009	6962544	97-11A
MQ	Merougil Sandstone	Sandstone	Speedway Quarry, Kambalda	368718	6543510	97-12B
MB	Merougil Sandstone	Sandstone	Merougil base — Morgans Island	370478	6542050	97-11B
KU-1	Kurrawang Formation	Sandstone	Kurrawang Syncline	340200	6589500	97-12C
KU-2	Penny Dam Conglomerate	Sandstone	Penny Dam	394815	6621836	97-25B
KU-3	Mt Belches Sandstone	Sandstone	Mt Belches	427970	6556400	97-20C
BF-1	Killaloe Formation	Sandstone	Buldania	403495	6463142	97-20A
EMD-10	Black Flag Formation	Dacite breccia	Eight Mile Dam, Hole EMD2	352503	6608636	97-23D
EMD-2	Black Flag Formation	Sandstone	Eight Mile Dam, Hole EMD2	352436	6608409	96-73A
E179	Black Flag Formation	Dacite conglomerate (clasts)	Gibson Honman Rock, blast hole	349750	6586400	97-17A
E7	Spargoville Formation	Dacite breccia	Widgiemooltha	362505	6518136	96-08A
BF-6	White Flag Formation	Sandy conglomerate matrix	White Flag Lake	333250	6609750	97-49C
NG1	Noganyer Formation	Sandstone	South Scotia	385050	6410000	97-25A
Intrusive porphyries and lava lithofacies						
E75	Black Flag Group	Dacite lava	Golden Ridge (hole HD2, 269.5 m)	369500	6591500	96-37
KWA-1	Kanowna Porphyry	Porphyritic rhyolite	Kanowna Belle	363857	6612165	97-02C
MGS-1	Morgans Island porphyry	Porphyritic rhyolite	Lake Lefroy	370500	6543200	97-02A
SHD-9	Mt Shea porphyry	Porphyritic rhyolite	Mt Shea	364972	6578916	97-02B

[a] Grid references are for the Australian Map Grid (AMG), Zone 51.

mean are invalid estimators of that mean. Values of $\chi^2 \leq 1$ can incorrectly imply statistical significance of a sample mean age if individual ages have large standard errors, or if there are too few individuals in a grouping. The statistical minimum number of individuals for a group is 15, but this is often not attained.

Detrital zircon ages are used to identify the ages of source rocks, which can be represented by many statistical groups or modes (i.e. sample age distributions) per sample age population, and by individual ages that fall outside those groups. Provenance dating does not give a depositional age for the host strata, but it can provide an estimate of the maximum depositional age. It is logical to consider that a maximum depositional age must be younger than the statistical limits placed on the age of the youngest source (i.e. the minimum age of the youngest age distribution). Therefore, maximum depositional ages must be the youngest 95 percentiles of the youngest detrital zircon age distributions, unless facies analysis defines a sample sedimentary rock as volcaniclastic and syneruptive. This convention is followed here, using the mean -1.96σ to define maximum depositional ages, although population estimates are quoted as mean $\pm 2\sigma$. For a one-tailed 5% test of significance, the decision statistic is mean -1.645σ. These statistics provide valid comparisons of maximum depositional ages between samples and stratigraphic units.

6. SHRIMP provenance dating programme

6.1. Sample locations and aims

We report here SHRIMP U–Pb zircon age data from 13 samples of sedimentary rocks, one sample of dacite lava, and three samples of intrusive porphyry from the Eastern Goldfields Province (Table 1). The samples, which cover the Kambalda, Spargoville, Kalgoorlie, Kurrawang, Merougil and Jones Creek Sequences (Table 2),

Table 2
Summary of SHRIMP age data

Sample	Genetic unit or time-equivalent	Standard CZ3		Grains analysed[a]	Youngest Grain[b] (Ma)	Youngest population[c]	n (χ^2)	Lower 95% CI (Young population)	Older populations/ranges[d] (Ma)
		n	Unc.(%)						
Metasedimentary and volcaniclastic rocks									
E178	Jones Creek Sequence	15	1.01	55 (55)	2649±11	2665±5	23 (1.09)	2660	2816±13 ($n=6$)
MB	Merougil Sequence	15	1.57	66 (66)	2630±18	2664±6	18 (0.78)	2658	–
MQ	Merougil Sequence	16	1.21	66 (66)	2551±8	2671±3	30 (1.07)	2668	–
KU-1	Kurrawang Sequence	11	1.45	53 (53)	2632±17	2679±10	10 (1.17)	2669	2730±4 ($n=19$), 2828±3 ($n=2$), 3567±5 ($n=2$)
KU-2	Kurrawang Sequence	15	2.55	69 (66)	2652±17	2671±3	38 (0.96)	2668	–
KU-3	Kurrawang Sequence	13	1.60	63 (63)	2637±16	2666±5	36 (0.99)	2661	2703–2720, 2773–2867, 2961–3028
BF-1	Kalgoorlie Sequence	13	0.90	63 (63)	2648±11	2667±5	34 (0.89)	2662	2690±9 ($n=13$), *(2843±8)*, 2861±21
EMD-10	Kalgoorlie Sequence	11	1.25	26 (24)	2654±15	2669±7	16 (0.83)	2662	–
EMD-2	Kalgoorlie Sequence	11	1.10	45 (43)	2631±23	2666±6	26 (1.33)	2660	2728±9 ($n=11$), *(3368±7)*
E179	Kalgoorlie Sequence	13	0.90	38 (38)	2652±10	2676±5	9 (0.26)	2671	2904±9, *3311±5*
BF-6	Kalgoorlie Sequence	11	0.74	42 (40)	2761±11	2813±3	34 (1.37)	2810	*3480±7*
E7	Spargoville Sequence	15	1.24	30 (18)	2662±8	2686±3	18 (1.08)	2683	–
NG1	Kambalda Sequence	13	1.10	62 (62)	2858±7	2864±5	2 (1.11)	2859	2901±7 ($n=3$), 2983±3 ($n=2$), 3132±4, *(3401±4)*
Intrusive porphyries and lava lithofacies									
E75	Spargoville Sequence	12	2.60	25 (22)	–	2698±6	9 (1.36)	2692	–
KWA-1	Kalgoorlie Sequence	7	1.13	24 (23)	–	2676±4	22 (1.06)	2672	–
MGS-1	Kalgoorlie Sequence	14	0.65	19 (19)	–	2678±8	13 (0.86)	2670	*2770±8*
SHD-9	Kalgoorlie Sequence	14	0.65	30 (24)	–	2658±3	23 (0.95)	2655	–

[a] Number of SHRIMP analyses, with total number of grains analysed in parentheses.
[b] Youngest concordant (95–105%) analysis with U<350 ppm, and f206%<2.0. Quoted errors are 1σ analytical errors.
[c] Weighted mean of the youngest population (χ^2 close to or less than 1) of concordant analyses. Quoted errors are the 95% confidence interval.
[d] $^{207}Pb/^{206}Pb$ ages of older detrital zircons; populations ($\chi^2<1$), ranges ($\chi^2>1$), single grains (italics), discordant single grains (italics, bracketed).

were chosen from type field sections of those sequences and available diamond drill core. The three intrusive porphyry samples are interpreted to be subvolcanic, time-stratigraphic equivalents of strata in the Kalgoorlie Sequence (Table 2).

The aims of the geochronological programme are not only to constrain the range in ages to source rocks of those sequences, but also to constrain the maximum age of those sequences relative to the tectonomagmatic history of the Eastern Goldfields Province. The three samples of intrusive porphyry (Tables 1 and 2) test the minimum age of the Kalgoorlie Sequence, and the relationships of those porphyries to the ages of granitoids and compressive deformation.

6.2. Sample preparation and analytical techniques

Samples were crushed, and then milled in a tungsten-carbide ring mill such that all material passed through a 60 mesh nylon sieve. Powders were washed to remove very fine material, prior to heavy-mineral separation. Heavy minerals were concentrated by standard heavy-liquid and magnetic separation techniques, and zircon grains were hand-picked from the concentrate. These grains were then mounted in epoxy resin with several chips of the CZ3 Sri Lankan zircon standard ($^{206}Pb/^{238}U=0.0914$, corresponding to an age of 564 Ma) on 25 mm diameter mounts, and polished to expose zircon cross-sections. The polished

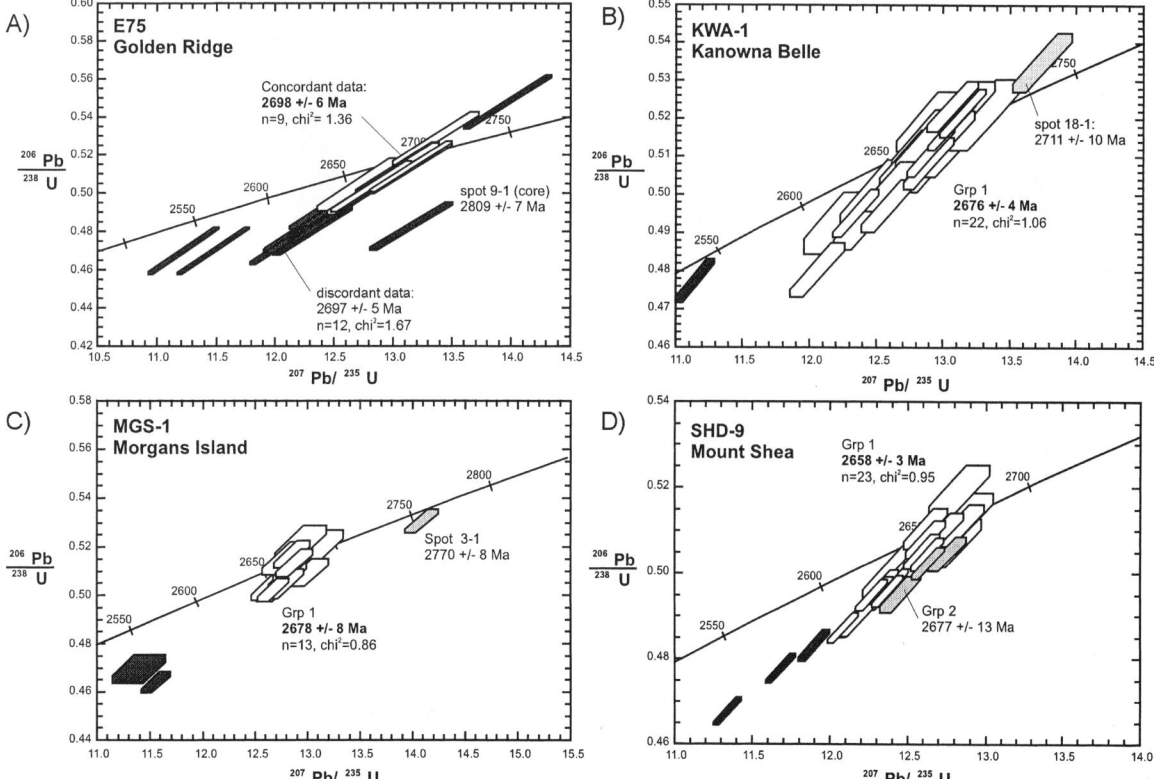

Fig. 4. Concordia diagrams showing SHRIMP zircon age data for lava-lithofacies and intrusive porphyry samples. Refer to Tables 1 and 2 for sample locations, stratigraphic affinities, and summarized statistical data. Unfilled (clear) boxes represent those analyses used to calculate mean ages, filled (black) boxes represent analyses excluded from the mean due to discordance, and filled (grey) boxes represent analyses or populations that are statistically outside the main population and considered to represent xenocrysts; error boxes are 1σ.

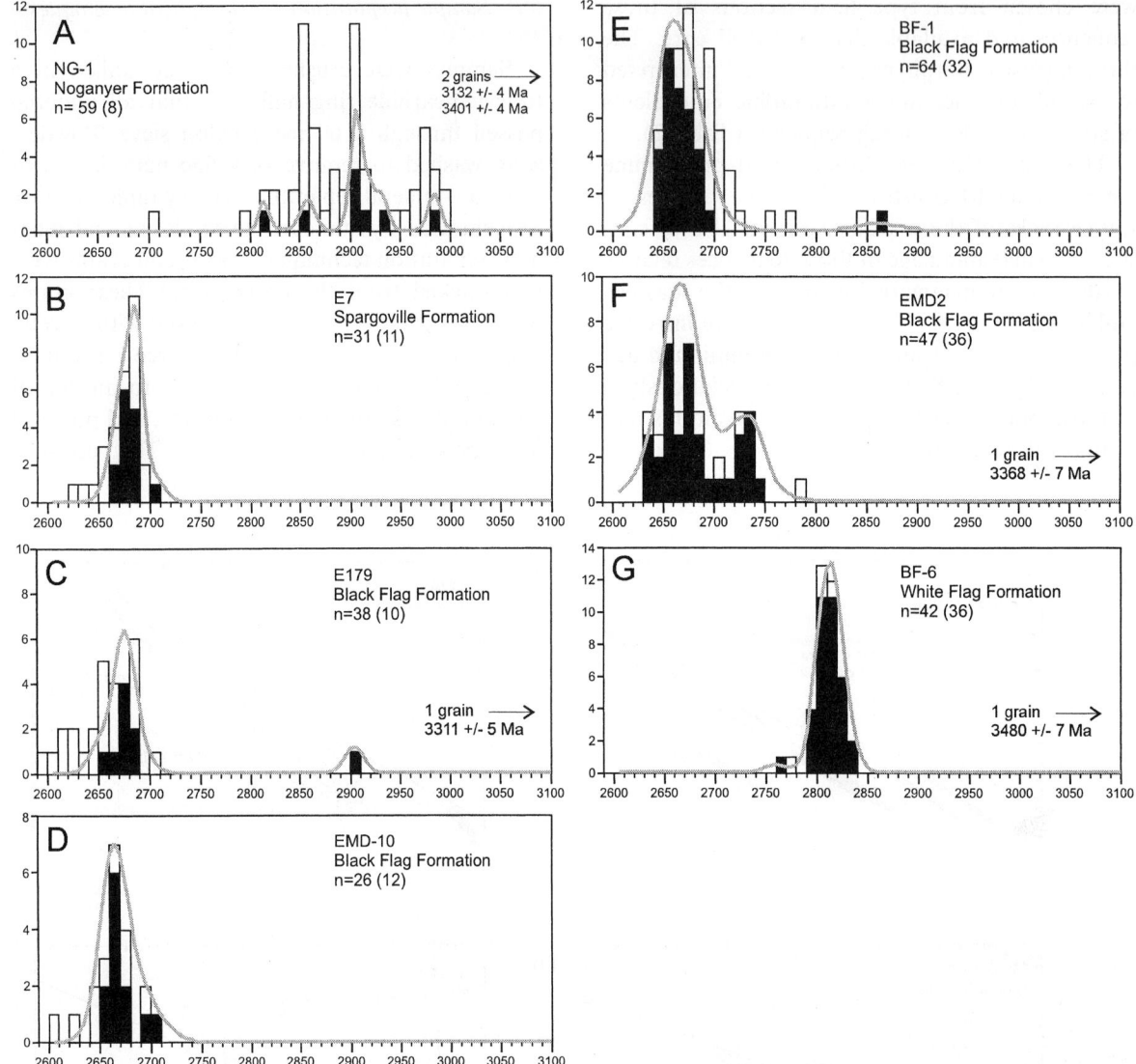

Fig. 5. Histograms showing SHRIMP zircon age distributions for detrital grains from metasedimentary samples. Refer to Tables 1 and 2 for sample locations, stratigraphic affinities, and summarized statistical data. Numbers in parentheses refer to the number of acceptable, concordant analyses. Filled (black) bars are concordant data (95–105% concordant), and unfilled (clear) bars represent data rejected due to discordance and on the basis of other criteria outlined in the text. The grey-lined curves are summed probability curves calculated from concordant analyses only.

mounts were carbon-coated prior to SEM examination. Backscattered electron (BSE) and CL images were obtained for all the mounts, and, together with transmitted-light and reflected-light photomicrographs, were used to characterise each grain in terms of colour, size, growth morphology and internal structure. The carbon coat was removed and the mount gold-coated for SHRIMP analysis.

SHRIMP analyses were performed under the general operating conditions described by Smith et al. (1998). In this study, count times (per scan) were 10 s for the ^{204}Pb, ^{206}Pb, and ^{208}Pb mass peaks and background, and 30 s for the ^{207}Pb mass

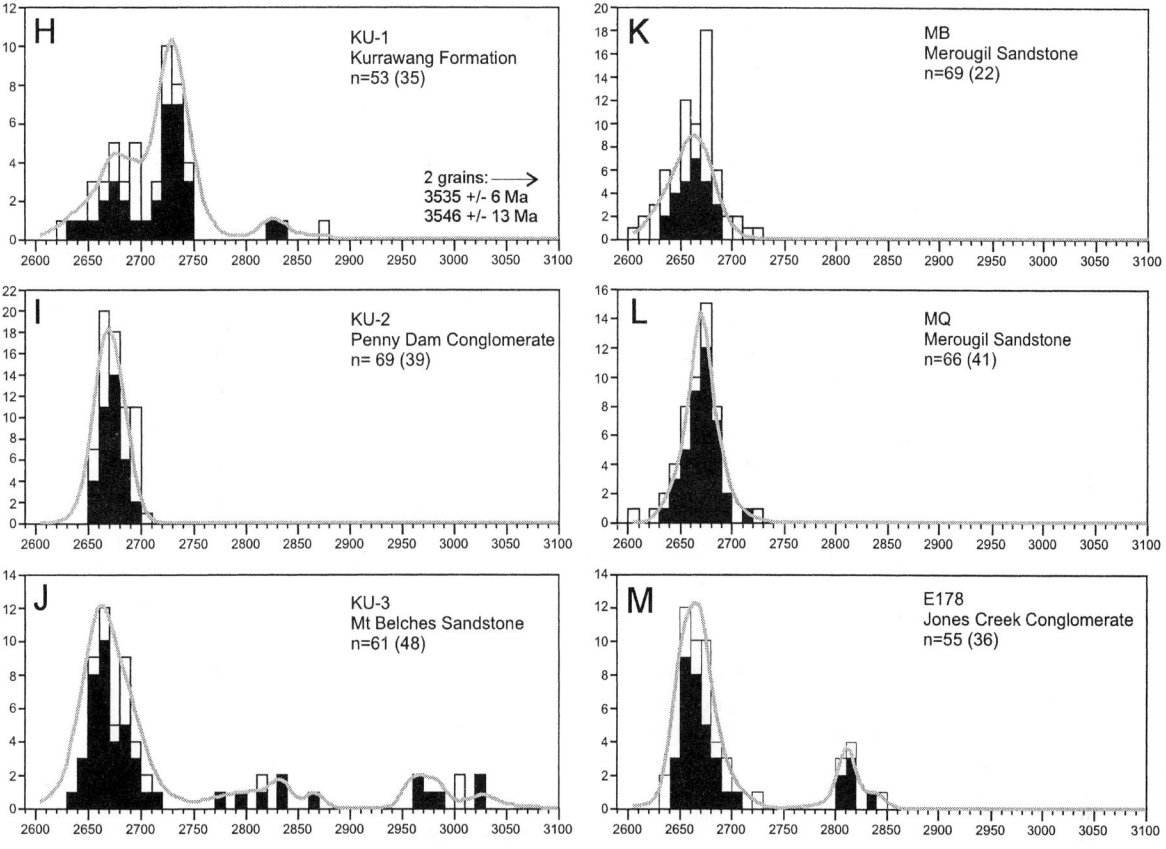

Fig. 5. (*continued*).

peak. Single analyses involved six scans for lava-lithofacies and coarse-grained volcaniclastic rocks. The number of scans was reduced to five for detrital zircon suites to reduce the total analysis time, and therefore allow a greater number of grains to be analysed in a 24 h session. Typically between 50 and 66 grains were analysed for sedimentary samples, and between 20 and 30 grains for intrusive and volcanic rocks, and monomictic volcaniclastic breccias. Analyses were referenced to multiple analyses of the CZ3 standard. Analytical errors on the standard for all samples are included in Table 2.

7. Results of SHRIMP zircon dating

A total of 778 SHRIMP analyses were obtained from the 17 samples. Because of the large number of analyses, only mean population ages and χ^2 values for age distributions are presented (Table 2). A complete list of analyses is available as supplementary data (Appendix, Table 1A). Concordia diagrams are presented for the dacite lava and the intrusive porphyry samples (Fig. 4), whereas age data for sedimentary samples are shown as histograms (Fig. 5).

7.1. Kambalda Sequence

The quartzose-sandstone sample from the base of the Noganyer Formation (Sample NG-1) yielded many zircon grains for analysis, but almost all lie below concordia, indicating some radiogenic Pb loss (refer to Sample NG-1, Table 1A). Nevertheless, near-concordant analyses span a range from 3401 to 2811 Ma (Fig. 5A), indicating either first-cycle provenance from a number of older crustal sources of felsic composition, or recycling from older sedimentary rocks. The youngest, but discordant (87%), individual age of 2702 ± 6 Ma (Sample NG-1, analysis 17-1, Table 1A) may constrain a minimum, rather than a maximum, depositional age. Although that individual age is within error of the age of a synvolcanic mafic sill (2714 ± 5 Ma) in the overlying Woolyeenyer Formation (Fig. 3), it is better equated with the hydrothermal zircon age of 2706 ± 5 Ma previously recorded from 'banded iron formation' of the Noganyer Formation (see above).

7.2. Spargoville Sequence

For the two samples from the Spargoville Sequence, the sample of dacite lava (Sample E75, Fig. 4A) provides an age for volcanic eruption and deposition, whereas the sample of dacite breccia (Sample E7, Fig. 5B) represents a syneruptive deposit and therefore provides an estimate of the age of deposition. Mean eruptive and depositional ages range from 2698 ± 6 Ma (E75; Table 2) to 2686 ± 3 Ma (Sample E7, Table 2).

7.3. Kalgoorlie Sequence

For the Kalgoorlie Sequence, interpreted syneruptive depositional ages from the Black Flag Formation are 2676 ± 5 Ma, for the volcaniclastic dacite conglomerate of Sample E179 (Table 2 and Fig. 5C), and 2669 ± 7 Ma, for the epiclastic dacite breccia of Sample EMD-10 (Table 2 and Fig. 5D). Detrital zircon grains from sandstone samples of the Black Flag Formation yield mean ages for the youngest age distribution of 2666 ± 5 Ma (Sample BF-1, Table 2 and Fig. 5E) and 2666 ± 6 Ma (Sample EMD-2, Table 2 and Fig. 5F). In detail, Samples BF-1 and EMD-2 comprise several age distributions. Concordant age data from Sample BF-1 are a unimodal age distribution with a mean of 2666 ± 5 Ma, but discordant data define a slightly older, high-U distribution (refer to Sample BF-1, Table 1A) with a minimum age of 2690 ± 9 Ma (Fig. 5E). Two older grains define a third distribution in Sample BF-1 that has a mean age of 2845 ± 8 Ma. Zircon age data from Sample EMD-2 define a bimodal population (Fig. 5F) with mean ages of 2666 ± 6 and 2728 ± 9 Ma for the two modes, and a single grain with an age of 3368 ± 7 Ma.

Detrital zircon grains from the sample of sandy matrix of the andesite-clast conglomerate in the White Flag Formation (Sample BF-6) produced an age spectrum that is significantly older than samples from the Black Flag Formation. An age distribution of 34 grains has a mean age of 2813 ± 3 Ma (Table 2 and Fig. 5G). One analysis of one grain that gave an age of 2761 ± 11 Ma defines the maximum age of deposition, whereas one analysis of another grain that gave an age of 3480 ± 7 Ma defines the oldest source (refer to Sample BF-6, analyses 35-1 and 32-1, Table 1A).

For the three intrusive rhyolite porphyries, two have similar mean ages of 2676 ± 4 Ma (Sample KWA-1, Table 2 and Fig. 4B) and 2678 ± 8 Ma (Sample MGS-1, Table 2 and Fig. 4C), whereas the third has a mean age of 2658 ± 3 Ma (Sample SHD-9, Table 2 and Fig. 4D).

7.4. Kurrawang Sequence

Two of the samples analysed from the Kurrawang Sequence are sandstone interbeds in conglomerate (Samples KU-1 and KU-2, Fig. 5H and I), and the third is from a thick sandstone succession (Sample KU-3, Fig. 5J). Samples KU-1 and KU-3 produced polymodal detrital zircon age populations that are characterized by a young age distribution (2679 ± 10 and 2666 ± 5 Ma, respectively), and by multiple older age distributions ranging back to 3567 ± 5 Ma (Table 2, Fig. 5H and J). However, in detail, the zircon age populations for these two samples are different. Sample KU-1 contains four discernible age distributions with means at 2679 ± 10, 2730 ± 4, 2828 ± 3 Ma and 3567 ± 5 Ma, for concordant or near-concordant data (Fig. 5H). Sample KU-3 comprises at least five age distributions, many of which show considerable spread outside that expected for a single source (i.e. $\chi^2>1.0$), and are best described in terms of age ranges or broad groups. Sample KU-3 contains a large age distribution at 2666 ± 5 Ma, and older distributions and age ranges of 2720–2703, 2867–2773, and 3028–2961 Ma (Fig. 5J). In contrast, Sample KU-2 has a unimodal zircon age distribution with a mean age of 2671 ± 3 Ma (Table 2, Fig. 5I) from 38 concordant analyses ($\chi^2 = 0.96$).

7.5. Merougil Sequence

Two samples of sandstone from the Merougil Sequence (Samples MB and MQ, Table 1) have slightly different zircon age distributions (Fig. 5K and L), with mean ages of 2664 ± 6 Ma (Sample MB, Table 2) and 2671 ± 3 Ma (Sample MQ, Table 2). The mean ages are similar, with overlapping errors at the 1σ level, but their younger 95 percentiles define different maximum ages of deposition at 2658 and 2668 Ma.

7.6. Jones Creek Sequence

The sample of sandstone from this sequence (Sample E178, Table 1) produced a bimodal zircon age population, with most of the analyses defining a single age distribution at 2665 ± 5 Ma (younger 95 percentile at 2660 Ma), and an older age distribution ranging between 2830 and 2780 Ma (Table 2, Fig. 5M). The youngest age distribution is interpreted to reflect the age of crystallization of a granitic source rock, and its minimum age provides a maximum depositional age. The older age distribution represents analyses of zircon cores. Those cores have younger rims, which are interpreted to be original magmatic overgrowths. The older grains are interpreted to have been xenocrysts in the granitic source rock, and therefore indicate the presence of >2800 Ma crust during intrusion of the granitoid at ~2665 Ma.

7.7. Morphology of zircon grains in all samples

Morphological aspects of analysed zircon grains are summarized in Table 3. Zircon grains typically display a wide range in morphological types and sizes (Fig. 6). Euhedral magmatic grains and broken fragments are common, and rounding, when present, varies from slightly rounded (subhedral) magmatic grains through to well-rounded, near-spherical clear grains. Zircon grains with older cores and euhedral magmatic overgrowths are rare, but have been noted in some samples (e.g. Fig. 6A). Euhedral and rounded grains of the same age in a sample indicate that the euhedral grains were most likely liberated from within lithic clasts during sample preparation. In general, most samples show no clear relationship between grain morphology and age, but larger, euhedral and slightly metamict grains from Sample KU-1 (Table 3) tend to belong to the youngest age distribution (i.e. <2700 Ma).

Backscatter electron (BSE) and CL images reveal that partially metamict zircon grains are relatively common, and typically display preferential alteration (dark areas in CL) in growth zones with particularly high U concentrations, and radial fractures. Recrystallization is also common, but may be restricted to particular growth zones, thereby enhancing original magmatic zonation (Fig. 6B and D). In other grains, recrystallization is present along fractures (Fig. 6E) that facilitated the passage of fluids into the crystal. In some instances, the entire grain appears to have been recrystallized prior to sedimentary rounding and fracturing (Fig. 6F and H). The high abundance of recrystallized grains in the samples implies that many grains were subject to high-grade metamorphism prior to erosion and incorporation into their host strata.

7.8. Significance of zircon recrystallization and ancient Pb loss

It is evident from the analytical results (Table 1A) that all samples produced some spuriously young ages. Although more research is required on this problem, it appears that spots that can produce spuriously young ages will not always be rejected by prior SEM-image analysis. The potential error associated with erroneous acceptance of spuriously young ages as detrital ages has been minimized by rejecting all discordant and high-U data from mean age calculations.

It is now obvious that the youngest zircon age may not necessarily provide an estimate of a maximum depositional age for two main reasons: (1) the youngest concordant mean age may be a composite of more than one age distribution (mode), distinctions of which cannot be resolved

Table 3
Summary of zircon morphology and composition for metasedimentary and volcaniclastic samples

Sample	Unit	Morphology	Th and U contents
E178	Jones Creek Conglomerate	Zircons range from small (50–120 μm), relatively clear euhedral grains, to larger (100–300 μm) euhedral to broken subhedral grains which are often variably metamict. BSE imaging of the grains shows that oscillatory zoned outer rims are often variably metamict, while interior zones commonly show evidence for recrystallization. Curvilinear and radial fractures are common in partially metamict grains. Some grains (e.g. 13, 42) clearly show rounded cores with euhedral magmatic rims, which yield significantly older ages	Th: 7–380 ppm (1 analysis 620 ppm) U: 10–245 ppm (1 analysis 610 ppm) Th/U: 0.4–1.7 (older population <0.8)
MQ	Merougil Sandstone	Grains range from 50 to 400 μm, and are typically euhedral to broken. Most grains are clear to slightly metamict	U: most 20–400 ppm (5 analyses >1000 ppm) Th: most 20–300 ppm (4 analyses >800 ppm) Th/U: most 0.2–0.8, up to 1.4. High-U grains (>250 ppm tend to yield young, discordant ages
MB	Merougil Sandstone	Mostly small, equant (typically 120 × 150 μm, up to 350 μm) colourless to pale olive, euhedral to slightly rounded grains. Some grains are partly to extensively metamict. Several highly rounded grains	U: most 50–230 ppm (2 analyses >400 ppm) Th: most 40–250 ppm (1 analysis >400 ppm) Th/U: most 0.3–1.5, up to 2.1
KU-3	Mt Belches Sandstone	Zircons from KU-3 range from 60 to 350 μm in size. They range in morphology from euhedral prismatic to anhedral, broken and resorbed grains. Many grains are slightly metamict. Although some of the oldest age analyses come from zircon cores or rounded/resorbed grains, many also come from euhedral grains, and there is no clear relationship between age and morphology	U: most 15–250 ppm Th: 5–280 ppm Th/U: most 0.1–1.20 (up to 2.3)
KU-2	Penny Dam Conglomerate	One main morphological type; grains are typically large (100–380 μm), and variably metamict. Morphologically, grains range from large equant, euhedral or subhedral zoned magmatic grains, to more elongate or broken zoned grains. Some grains are slightly rounded	U: most 40–200 ppm (1 analysis >200 ppm) Th: 10–120 ppm Th/U: most 0.2–0.85 (narrow range)
KU-1	Kurrawang Formation	Range in zircon size (50–350 μm) morphology, three main types: (A) colourless euhrdral magmatic grains, either elongate or more equant (120 × 150 μm), (B) colourless fractured and/or slightly rounded grains, (C) dark, variably metamict, euhedral grains with fine magmatic zonation. Many type C grains are <2700 Ma, but no clear relationship exists between morphology and age. Many older zircons (>2700 Ma) show evidence of recrystallization in BSE images	U: most 20–160 ppm (3 analyses >300 ppm) Th: most 10–200 ppm (4 analyses >300 ppm) Th/U: most 0.4–1.8, up to 3.8; there is a tendancy for the youngest grains (<2700 Ma) to have lower Th/U (<1.1), than the 2730 Ma population (Th/U for most grains: 0.9–1.8)
BF-6	White Flag Formation	Single morphological population; small (50–120 μm), euhedral or broken pale olive–brown grains	U: most 40–240 ppm (1 analysis 780 ppm) Th: most 30–180 ppm (1 analysis 790 ppm) Th/U: 0.5–1.0
EMD-2	Black Flag Formation	Wide range in zircon grain morphologies. Many are large (100–150 μm), equant to slightly elongate euhedral magmatic grains with oscillatory zoning and common metamictization of high U zones. Some of these grains contain recrystallized patches, and have common <5 μm mineral inclusions. Other grains can be anhedral or subhedral, with broken or resorbed margins and extensively recrystallized interiors and may have had a meta-volcanic origin. Euhedral magmatic grains tend to be associated with the younger ages (<2670 Ma), whilst the subhedral, recrystallized grains tend to yield ages within the older population (>2710 Ma)	U: most 20–200 ppm (2 analyses >300 ppm) Th: most 10–180 ppm (1 analysis >500ppm) Th/U: most 0.2–1.2 (up to 1.5), All of the older grains (>2700 Ma) tend to have relatively low U contents (<120 ppm)

Table 3 (*continued*).

Sample	Unit	Morphology	Th and U contents
BF-1	Killaloe Formation	Zircons range from >50 µm to 300 µm. Most are euhedral magmatic grains, but subhedral and broken morphologies are also common. Grains range from clear to variably or completely metamict and dark in transmitted light	U: most 20–200 ppm (13 analyses 200–750 ppm) Th: most 10–500 ppm (13 analyses 800–2500 ppm) Th/U: most 0.2–2.0 (up to 6.5), high U, and high Th/U grains are typically discordant
EMD-10	Black Flag Formation	Zircons were mostly small (<100 µm) euhedral grains or anhedral broken crystal fragments. Many were fractured and/or metamict	U: most 40–200 ppm (3 analyses >300 ppm) Th: most 10–200 ppm Th/U: most 0.2–1.5
E179	Black Flag Formation	Zircon grains range from 60 to 250 µm in size, and range from euhedral to broken anhedral grains. Most are metamict, and this is reflected in the high number of discordant analyses	U: most 50–250 ppm (7 analyses 300–1100 ppm) Th: most 20–300 ppm (6 analyses 400–1600 ppm) Th/U: most 0.4–1.5 (up to 2.5). High U and Th/U grains typically have discordant younger ages
E7	Spargoville Formation	Sample E7 yielded a limited number of small (50–150 µm) subhedral, and often fractured zircons. Backscattered electron imaging and cathodoluminescence reveal resorbed cores and oscillatory zoned rims in several analysed grains.	U: 140–730 ppm (wide range) Th: 40–520 ppm Th/U: 0.3–0.9. High U analyses (>300) tend towards younger ages
NG1	Noganyer Formation	Mostly small, dark (metamict) grains. Nearly all grains have angular broken morphologies.	U: 110–900 ppm (wide range) Th: most 80–1000 ppm (up to 6770 ppm) Th/U: most 0.3–2.3 (up to 12.1). Analyses with >350 ppm U produce younger and discordant ages

using standard SHRIMP U–Pb dating; and (2) the youngest age distribution may be skewed towards spuriously young ages. Age distribution skewing is most likely linked with recrystallization and radiogenic Pb loss during post-depositional thermal events. Examination of all the age data presented here suggests that such an event may have occurred between 2650 and 2630 Ma, whereas other thermal events occurred at <2600 Ma (refer to Table 1A and Fig. 5), although there is also concern as to the significance of analyses with ages of ∼2660 Ma (Sample E7, Table 2, for instance).

Zircon grains from the older sources also appear to have been recrystallized, and to have recrystallized during an event prior to their incorporation as detrital grains. For example, Sample EMD-2 has many recrystallized ages at ∼2730 Ma (refer to Tables 2 and 4 and Fig. 5F). This aspect of provenance dating has, therefore, established that many zircon grains record tectonic recycling through previously unrecognized thermotectonic events.

Further inspection of the age-data (Table 1A and Figs. 4 and 5) shows that for most analyses that are not considered to be representative of zircon ages in sources, the non-representative analyses are those that are >5% discordant. In many, but not all, cases, these are analyses that have high U or high common-Pb contents. For the sample of the Noganyer Formation (NG-1, Table 1A), discordance may be due to modern processes (i.e. surface weathering). This is corroborated by SHRIMP U–Pb zircon dating of a drill-core sample from the Upper Penneshaw Formation, at close to the same sampling site and in rocks of the same metamorphic grade (see Nelson, 1995, table 37). In that sample, almost all analyses are concordant, and there are no spuriously young ages. Samples that have the largest number of discordant analyses within error of synvolcanic and depositional ages are those of intrusive porphyry (Samples KWA-1, MGS-1 and SHD-9; Table 1A and Fig. 4), and those from the Spargoville and Kalgoorlie Sequences (Samples E7, E179, EMD-10, BF-1 and EMD-2, Table 1A and Fig. 5B–F), but not Sample BF-6 (Fig. 5G).

One conclusion from this study is that there may be a link between potential for zircon-age

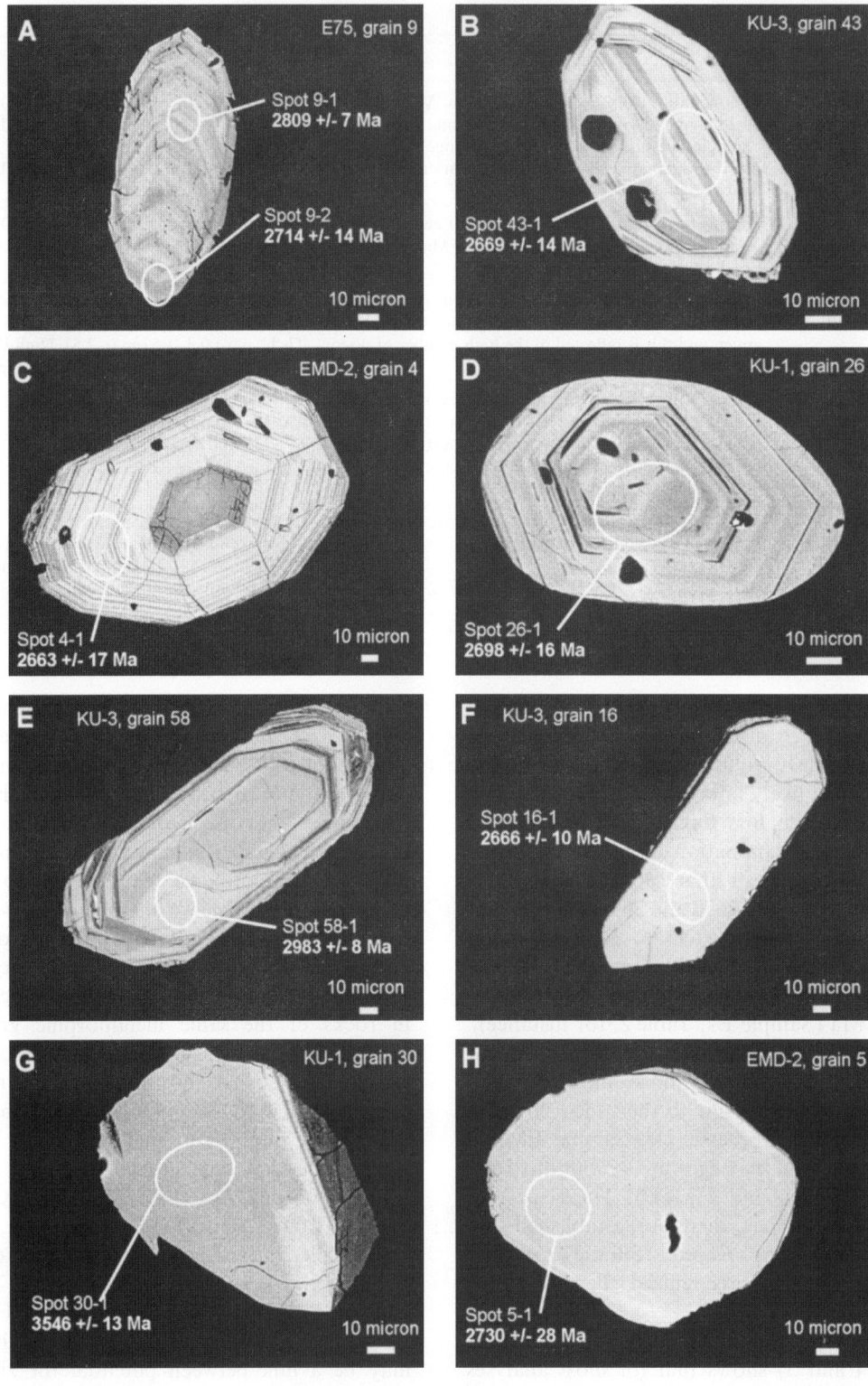

resetting, source-rock ages, depositional style and basin type. As stated above, the problem of spuriously young zircon ages within the error of interpreted maximum depositional ages is not present in sequences that are much younger than their zircon sources (e.g. Samples NG-1 and BF-6, Fig. 5A and G). Neither is it a problem with the Kurrawang, Merougil or Jones Creek Sequences, despite the presence of relatively young zircon grains, for which probability curves reveal detrital populations (Fig. 5H–M). We believe the problem is associated with the preservation, in detrital suites, of zircon grains that were susceptible to age resetting, particularly those with high U contents. Those types of grains were mostly removed in siliciclastic depositional systems, but the potential for survival in volcaniclastic systems, and intrusive porphyries, was much greater. The problem is further exasperated by the interpretation that the sequences that received zircon grains susceptible to age-resetting (i.e. the Spargoville and Kalgoorlie Sequences) were deposited in basins associated with inherently high heat-flow and coeval intrusion of granitoids (see below). It may have been the association of volcaniclastic systems and syndepositional high heat flow that produced spuriously young zircon ages within the error of depositional ages.

8. Synthesis and discussion

8.1. Stratigraphic implications

8.1.1. Kambalda Sequence

The range in detrital zircon ages from the basal quartzose–sandstone of the Noganyer Formation (Fig. 5A) is comparable to that obtained from the Upper Penneshaw Formation (cf. Hill et al., 1992; Nelson, 1995), which implies that both formations had a similar provenance. Synchrony of basalt–komatiite volcanism in a deep-marine basin (i.e. Kambalda Sequence; Fig. 3) and calc-alkalic magmatism in a volcanoplutonic arc is implied by the interpreted backarc setting of the Kambalda Sequence (Krapez, 1997, fig. 10). However, the lack of detrital zircon grains with ages similar to the interpreted depositional age of the Kambalda Sequence (i.e. ≤2715 Ma; see above) implies that sedimentary rocks in the Upper Penneshaw Formation and the Noganyer Formation may not have been derived from coeval, arc-related, felsic rocks. Rather, the detrital zircon dating implies that felsic detritus in those sedimentary rocks was derived from older sources. Detrital zircon grains in thinly bedded turbidites ('banded iron formations' in Hill et al., 1992) yield ages that imply recycling of Early Archaean rocks. Feldpathochloritic sandstones were most likely derived from an andesitic source, but samples of those sandstones are presently undated.

This apparent lack of provenance from an adjacent arc may have a sedimentological explanation. By analogy with the provenance of (and stratigraphic similarity to) turbidites intercalated with spreading-related basalts in the modern South China and Andaman Seas (see Curray et al., 1978), sediment dispersal systems may have been longitudinal, rather than arc-derived transverse. If so, sediment sources were more-likely distal basement rocks. In addition, transverse (andesitic) sources may have been either zircon poor or cut off by post-rift drowning of the adjacent arc. 'Old' detri-

---

Fig. 6. Representative back-scattered electron (BSE) images for zircons showing the range in morphological types, internal structure, and variable (but common) effects of recrystallization and metamictization. The zircon grain in (A) is from a sample of dacite lava-lithofacies (Sample E75), whereas zircon grains in (B)–(H) are detrital zircon grains from samples of metasedimentary rocks. Refer to Tables 1 and 2 for sample locations and stratigraphic affinities. (A) Grain with a rounded (and older) xenocrystic core and a euhedral magmatic overgrowth. (B) Euhedral grain with a euhedral core, showing areas of zone- and sector-specific recrystallization, and patchy recrystallization around inclusions. (C) Large, zoned euhedral grain with metamict central core, radiating fractures and abundant inclusions; this grain has also undergone zone-specific and patchy recrystallization (top-right of grain). (D) Recrystallized and well-rounded detrital grain; erosional truncation of recrystallized zones indicates that recrystallization occurred prior to erosion and deposition. (E) Euhedral grain showing both local metamictization and at least two phases of recrystallization; the light area in the lower left of the grain represents recrystallization associated with a series of fractures that cut across original zonation. (F) Radial fractures in a euhedral grain that have been sealed in the core region. (G) Broken fragment of recrystallized grain with a spot age of 3546 ± 13 Ma. (H) Well-rounded and thoroughly recrystallized grain with a spot age of 2730 ± 28 Ma.

tal zircon grains recorded from 'interflow' sedimentary rocks (e.g. Claoué-Long et al., 1988; Nelson, 1996) may also indicate deposition on longitudinal depositional systems and derivation from distant basement rocks. Sediment sources for turbidites of the Kambalda Sequence appear to have changed upwards from basement only in the Noganyer Formation, to mixed basement and contemporaneous arc in the Kapai Slate (cf. Claoué-Long et al., 1988), possibly reflecting increasing contributions from transverse systems.

8.1.2. Spargoville and Kalgoorlie Sequences

The range in zircon ages obtained from samples of the Spargoville and Kalgoorlie Sequences implies a long period (~30 million years) of volcanic eruption and syneruptive deposition (Fig. 7). However, two depositional basins are interpreted, with the boundary between deposition of the two sequences placed statistically between 2683 and 2681 Ma (the minimum synvolcanic age from Sample E7 and the maximum synvolcanic age from Sample E179, respectively; refer to Table 2 and Fig. 5B and C).

The oldest synvolcanic age for the Spargoville Sequence (2698 Ma from Sample E75, Table 2 and Fig. 4A) is older than the mean age of the Kapai Slate (2692 Ma, Claoué-Long et al., 1988), which is a stratigraphic unit in the Kambalda Sequence (Fig. 3). From statistical analysis alone, a syneruptive age of 2692 Ma for the Kapai Slate would make it a genetic–stratigraphic unit of the Spargoville Sequence. However, there is no genetic–stratigraphic link between strata of the Kambalda and Spargoville Sequences. This lack of stratigraphic relationship, and new age constraints on the Spargoville Sequence, suggests that the high proportion of discordant ages, and the spread of young ages (younger than 2700 Ma), used to define the mean age of 2692 ± 4 Ma for the Kapai Slate (Claoué-Long et al., 1988), could be post-depositional in origin.

Sample E7 contains similarly spuriously young zircon ages (refer to Table 1A and Fig. 5B) that may indicate either the amphibolite–facies prograde metamorphism or lower-greenschist facies retrograde metamorphism established from field exposure (Barley et al., 1998b). The younger analyses tend to be more discordant and are associated with grains with higher U content than the syneruptive distribution (refer to Sample E7, Table 1A). Nevertheless, they define an age distribution with a mean of 2661 ± 4 Ma. On-going analyses, at the same field location, are focusing on additional dating of samples from the Spargoville and Kalgoorlie Sequences and a porphyritic-granitoid that intrudes the Spargoville Sequence, in order to test whether the 2661 Ma zircon age distribution can be linked to the intrusion, and thereby define a metamorphic age.

Detrital zircon ages from the two sandstone samples of the Kalgoorlie Sequence (Samples BF-1 and EMD-1, Table 2 and Fig. 5E and F) indicate two felsic sources: one intrabasinal and volcanic (as defined petrographically by Hand, 1998), and the other extrabasinal. Based on the statistical convention advocated here, the maximum ages of deposition defined by Samples BF-1 and EMD-2 are 2661 and 2660 Ma respectively (Table 2). Those ages overlap with the ages of granitoids that are considered to have intruded synchronously with D_2 compression (2675–2657 Ma), and with post-D_2 extension that is linked to the uplift of granitoid–gneiss complexes at ~2660 Ma (see above).

Because strata of the Kalgoorlie Sequence are deformed by D_2, there is clearly a problem linking 2675–2657 Ma granitoids, and their post-intrusion uplift, to D_2 compression. Rather, those granitoids were coeval with the Kalgoorlie Sequence, and therefore emplaced during regional extension associated with basin formation. However, it is unlikely that presently exposed granitoids of that age were sources to the Kalgoorlie Sequence, whereas few of them have a TTG geochemistry. It is more likely that volcanic sources of the late stage of the Kalgoorlie Sequence (Hand, 1998) were high-level equivalents of extrabasinal TTG granitoids. Although the study of crustal elements in the eastern domains of the Eastern Goldfields Province is incomplete, an extensive TTG volcanoplutonic source has not been recognized.

Porphyry intrusions present the same timing problem with D_2 compression. For instance, the Mount Shea porphyry (Sample SHD-1, Table 2 and Fig. 4D), which intruded at 2658 ± 3 Ma, was

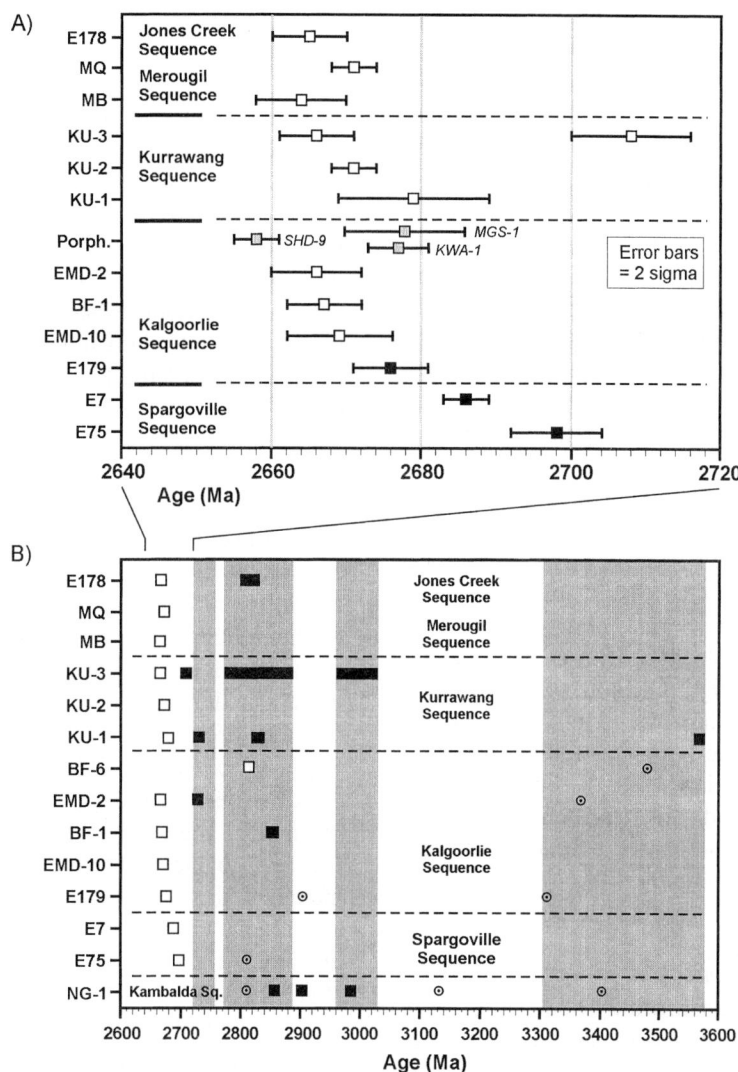

Fig. 7. Summary of SHRIMP age data for samples reported here. (A) Summary of the youngest detrital age distributions (modes) and hence maximum depositional ages (unfilled boxes) for samples of metasedimenatry rocks. Depositional ages of the Spargoville and Kalgoorlie Sequences are further constrained by the ages of rhyolite porphyry intrusions (grey filled boxes) and by lava-lithofacies and coarse, monomict volcaniclastic breccias (black filled boxes). (B) Summary of older detrital age distributions (modes); unfilled (clear) boxes are youngest detrital age distributions, filled (black) boxes are older detrital age distributions and detrital age ranges. Single grain analyses are shown as circles. The solid-grey vertical blocks show age ranges of common detrital zircon grains, and reflect the existence of older felsic source-rocks within those age ranges. Refer to Tables 1 and 2 for stratigraphic nomenclature and localities of samples.

deformed by D_2. Based on the 95 percentile, it is therefore unlikely that D_2 occurred until at least 2655 Ma, which is also the most likely maximum depositional age of the Kurrawang Sequence. The ages of the porphyry intrusions also define broadly early and late tectonic stages of the Kalgoorlie Sequence, with a boundary age between 2670 and 2661 Ma (respectively, the minimum age of Sample

MGS-1 and the maximum age of Sample SHD-1; Table 2).

Further complications with provenance of the Kalgoorlie Sequence arise from the age spectrum obtained from Sample BF-6 of the White Flag Formation (Table 2 and Fig. 5G). The composition of the White Flag Formation has been considered to record contemporaneous andesitic volcanism (e.g. Morris and Witt, 1997). However, the dominant detrital zircon age distribution (mean age 2813 ± 3 Ma) of Sample BF-6 does not indicate a contemporaneous andesitic source, but rather a derivation from an unknown andesitic terrane, unless all the zircon grains were xenocrysts in the source andesite.

Vertically and laterally changing, and vanished sources, as well as longitudinal basin filling, characterize the Kalgoorlie Sequence (Hand, 1998). These characteristics, when viewed collectively, are typical of strike-slip basins (e.g. Nilsen and McLaughlin, 1985). Although there are insufficiently preserved stratigraphic data to establish an unequivocal relationship between sedimentation patterns and strike-slip boundary faults, it is possible that at least the late stage of the Kalgoorlie Sequence developed during transtension. If so, coeval granitoids were also emplaced during transtension.

8.1.3. Kurrawang Sequence

The erosional unconformity beneath basal strata of the Kurrawang Sequence implies considerable subaerial uplift and erosion of the Kalgoorlie Sequence, whereas turbidite lithofacies in the Kurrawang Sequence imply subsequent rapid drowning. Strata of the Kurrawang Sequence were clearly not genetically associated with the Kalgoorlie Sequence. As such, the maximum age of the Kurrawang Sequence must be ≤ 2655 Ma (see above). However, strata of the Kurrawang Sequence were deformed during D_2.

This estimated maximum age of the Kurrawang Sequence is in conflict with previous interpretations. For instance, the northwestern part of the Kurrawang Syncline (the type area of the Kurrawang Formation; Fig. 2) is considered to have been laterally displaced during intrusion of a post-D_2 granitoid, which has a SHRIMP zircon age of 2660 ± 3 Ma (Nelson, 1997). The implications are that folding predated granitoid intrusion, and the granitoid constrains a minimum age for the Kurrawang Formation. The interpretations are possible if based solely on the youngest detrital zircon age distribution (mean 2669 Ma) of Sample KU-1 (Table 2, Fig. 5H), but conflicts with stratigraphic relationships and provenance age data from the Kalgoorlie Sequence, which imply that the same granitoid must have been intruded during the late stage of the Kalgoorlie Sequence. Lateral displacement of the Kurrawang Syncline against the granitoid may have been a post-D_2 structural event, but intrusion of that granitoid must have predated that event.

There is also no evidence for volcanism during deposition of the Kurrawang Sequence, and no coeval granitoid plutons have been recognized. Rather, deposition on a deep-marine, longitudinal submarine-fan in a basin lacking syndepositional intrabasinal magmatism, with sediments derived from synorogenic sources, is envisaged (Krapez, 1997; Krapez et al., 1997). Those interpretations are consistent with deposition in a remnant-ocean basin (e.g. Ingersoll et al., 1995). In contrast to the Spargoville and Kalgoorlie Sequences, sediments of the Kurrawang Sequence were derived exclusively from extrabasinal supracrustal and granitoid rocks (Krapez, 1997; Barley et al., 1998b).

Recycling of extrabasinal sources is proven by the age range of detrital zircon grains in Samples KU-1 and KU-3 (Fig. 5H and J), with grains as old as the Early Archaean, but supracrustal and crustal rocks as young as the Kalgoorlie Sequence supplied significant detritus to the basin (Fig. 7). The flux of detritus from older extrabasinal sources implies uplift of distant basement rocks, which is consistent with a remnant-ocean basin setting. An estimated maximum depositional age of 2661 Ma for the Kurrawang Sequence is provided by Sample KU-3 (Table 2).

In contrast to Samples KU-1 and KU-3, the unimodal detrital zircon age population for Sample KU-2 (Table 2 and Fig. 5I) implies derivation from a felsic source with no basement rocks. In fact, the mean detrital zircon age of 2671 ± 3 Ma (Table 2) implies recycling of source rocks the same age as the Kalgoorlie Sequence. Because palaeocurrent data from the sample site of Sample KU-2 (i.e. Penny Dam Conglomerate, Fig. 3) have the same trend as the entire Kurrawang Sequence (Barley et al., 1998b), the unimodal detrital zircon age population is interpreted to represent longitudinal reworking of a transverse system. The contrast to Sample KU-1 (Kurrawang Formation) and Sample KU-3 (Mt Belches Sandstone) is possibly because those samples come from more-central depositional sites in the original basin, and therefore regionally amalgamated sources, although there are no apparent differences in clast lithotypes or sandstone composition (Barley et al., 1998b). Longitudinal reworking of a transverse system implies that there was an eastern or northeastern margin to the original basin, in the direction of the interpreted Kalgoorlie-aged volcanoplutonic arc.

8.1.4. Merougil and Jones Creek Sequences

The unconformable relationship between the Merougil and Kalgoorlie Sequences implies that the maximum age of the Merougil Sequence is ≤ 2655 Ma, whereas the interpreted maximum depositional age from detrital zircon age data is 2658 Ma (Sample MB, Table 2 and Fig. 5K). The maximum depositional age of the Jones Creek Sequence is approximately the same (Sample E178, Table 2 and Fig. 5M). The Merougil and Jones Creek Sequences were deformed by D_2. Taken in isolation, detrital zircon ages establish that the Merougil and Jones Creek Sequences were younger than 2675–2657 Ma granitoid intrusions interpreted to have been synchronous with D_2 (e.g. Nelson, 1997).

Source rocks of sediments deposited in the Merougil–Jones Creek Sequences were similar to those of the Kurrawang Sequence in that there is no evidence for a coeval volcanoplutonic source. Similarly, the sequences contain no evidence for intrabasinal volcanism. Rather, source terrains comprised extrabasinal supracrustal and crustal rocks, and the sequences are exclusively epiclastic sedimentary rocks. However, in contrast to the Kurrawang Sequences, polymodal zircon age populations are not recorded (Figs. 5K–M and 7), thereby implying that detritus was not recycled from the Kurrawang Sequence. Rather, all three samples from the Merougil and Jones Creek Sequences imply that detrital zircon grains were derived from single provenances, but not necessarily the same provenances, which is consistent with the interpreted fault-bound braid–plain depositional systems (e.g. Krapez, 1997) and derivation from source rocks the same age as the Kalgoorlie Sequence. Because there is no evidence for eroded remnants of the Kurrawang Sequence below the Merougil or Jones Creek Sequences, it is likely that Kurrawang strata were subaerially eroded and recycled into a distant sedimentary basin prior to Merougil–Jones Creek sedimentation.

8.2. Recycled provenances

Arguably the most significant concepts that modern basin analysis brings to studies of Archaean tectonics are recognition of the geographic scale of the links between source areas and depositional basins, and between dispersal and depositional systems. For instance, it is now well-established that deep-marine submarine fans are supplied during sea-level lowstands by subaerial erosion and resedimentation of detritus stored in fluvial to shallow-marine depositional systems during preceding sea-level highstands (e.g. Krapez, 1996, 1997). Therefore, palaeogeographic reconstructions must allow for not only the scale of deep-marine submarine–fan systems (some of which are more than 1000 km long) but also the scale of the shallow-marine to fluvial systems that supplied the resedimented detritus. Similarly, allowance must be made for the scale of the drainage systems that supplied detritus to the subaerial and shallow-marine systems. It is only in fault-flank, fault-bound and volcano-bound depositional systems that a local physical link between source and basin can be interpreted, and even that link relies on the preservation of proximal boulder deposits. In deep-marine submarine–fan systems, and longitudinal systems in particular, source areas and depositional basins may have large-scale geographic separations.

Detrital zircon dating of the Jones Creek Sequence supports a close physical link between the source area and the depositional basin. The implication is that the source area of the Jones Creek Sequence was probably within the Eastern Goldfields Province. A similar link may be made for the Merougil and Spargoville Sequences. However, no such link can be postulated for deep-marine deposited sedimentary rocks of the Kambalda, Kalgoorlie or Kurrawang Sequences, regardless of the coarse-grain size of sedimentary strata. The SHRIMP zircon provenance age data from the Kambalda, Kalgoorlie and Kurrawang Sequences therefore provide information on the regional-scale distribution of source rocks, and imply sources outside the Eastern Goldfields Province. Implied sources include contemporaneous volcanoplutonic-arc and rifted-arc rocks, same tectonic–megacycle arc and basin rocks, and older crustal and supracrustal rocks.

Presently known SHRIMP zircon dates of Eastern Goldfields rocks, which have similar ages to the postulated sources for the unconformity bounded sequences reported here, range from 2719 ± 5 to 2657 ± 5 Ma for granitoids, and from 2713 ± 4 to 2658 ± 6 Ma for volcanic and subvolcanic rocks (e.g. Nelson, 1997), although Kent and Hagemann (1996) have dated a subvolcanic felsic sill from the Wiluna Domain (Fig. 3) at 2749 ± 7 Ma. Supracrustal rocks older than 2720 Ma have yet to be recognized, although likely candidates include Lower Penneshaw tholeiite from Norseman, Association I from Laverton (see Hallberg, 1985), and much of the succession from Wiluna (see Fig. 3). Although those successions probably did not provide detritus to the sequences reported here, they imply that pre-2720 Ma sources existed in adjacent tectonic settings.

Detrital zircon ages that fall into the age ranges 2730, 2830–2800, 2950–2900 and >3000 Ma (Fig. 7) establish not only the existence of those pre-2720 Ma sources, but also that they were subaerially exposed in order to have been contributors of sediment to the unconformity bounded sequences. The implication is that pre-Merougil–Jones Creek sequences received detritus from sources external to contemporaneous arc and orogen tectonic settings, which is consistent with the longitudinal, deep-marine settings of the interpreted depositional basins. Crustal and supracrustal rocks of pre-2720 Ma age are preserved in other parts of the Yilgarn Craton. In particular, felsic volcanic rocks with formation ages of 2760–2720 and 3000–2950 Ma are present in the Murchison Province (Pidgeon, 1986; Pidgeon and Wilde, 1990; Pidgeon and Hallberg, 2000) and the Southern Cross Province (Savage et al., 1996). Early Archaean source rocks are preserved in the Narryer Terrane (e.g. Kinny et al., 1990).

Because of the orogen scale of the interpreted depositional basins, it is suggested that the former extent of Yilgarn-type crustal and supracrustal source-rocks, with ages back to the Early Archaean, was much greater than the preserved Yilgarn Craton. If so, the Yilgarn Craton is merely a preserved portion of a once much larger slice of Archaean crust, most of which has been removed through a combination of tectonic and sedimentary recycling, and tectonic severance. In fact, interpreted depositional basins and tectonic settings similar in age to those documented in the Eastern Goldfields Province are present in the Superior and Slave Provinces of Canada (Barley et al., 1998a, fig. 2), identifying a tentative correlation, at least by geotectonic settings. Corollaries of those interpretations are that detrital zircon suites in Archaean sedimentary sequences not only identify sediment derivation from vanished source-rocks, but also indicate the degree of sedimentary recycling of Archaean crustal and supracrustal terrains. It is therefore concluded that a greater use of detrital zircon suites can document a balance between sedimentary recycling of coeval volcanoplutonic and orogenic sources, which themselves are indicators of crustal growth, and vanished older sources, which are indicators of crustal recycling.

A significant conclusion from SHRIMP detrital zircon ages of the Kalgoorlie Sequence, which impacts on previous provenance models defined by sedimentary geochemistry (e.g. Nance and Taylor, 1977; Taylor and McLennan, 1985), is that, despite its interpreted intra-arc setting and the dominance of intrabasinal sources, it contains detritus recycled from much older sources. Basin studies that have proposed that depositional systems of the Kalgoorlie Sequence were longitudinal in basin orientation (Barley et al., 1998b; Hand, 1998), suggest that old detritus was provided from sources external to an adjacent volcanoplutonic arc. Such provenance interpretations imply that provenance models based on sedimentary geochemistry need to link interpreted proportions of intrabasinal and extrabasinal sources, defined by petrographic data and detrital zircon age data, to geochemical analyses for each sample, and for each unconformity bounded sequence.

9. Conclusions

SHRIMP U–Pb zircon dating of sedimentary, volcanic and subvolcanic-intrusive rocks from the Eastern Goldfields Province, defines maximum depositional ages from a set of unconformity bounded sequences, and provides age constraints on crustal and supracrustal sources. The Kambalda Sequence, which records a backarc basin, was deposited prior to 2700 Ma and from at least 2715 Ma. The Spargoville Sequence, which records an arc-adjacent, volcano-bound basin, was deposited between 2700 and 2683 Ma. The Kalgoorlie Sequence, which records an intra-arc rift basin, comprises two tectonic stages with depositional ages of 2681–2670 and 2661–2655 Ma. Younger, submarine-fan (Kurrawang Sequence) and braided-fluvial (Merougil and Jones Creek Sequences) deposits of a remnant-ocean basin were deposited at <2655 Ma, prior to regional compression and metamorphism, which apparently occurred between 2650 and 2630 Ma.

Many detrital zircon grains, particularly those high in U and Th, have undergone radiation damage, and subsequent Pb loss and recrystallization during post-depositional metamorphic events. While recent Pb loss can be identified by significant discordance (on a concordia plot), if the Pb-loss event is close in age to the magmatic age of the zircon, ages may still appear concordant despite partial or complete isotopic resetting. This has important implications for determining the maximum depositional age of a unit (from its youngest detrital population), because apparently concordant populations (and the mean) may be skewed towards a younger reset age. To minimize this problem, we suggest the use of various criteria to exclude data from zircons that may be susceptible to Pb loss (high-U, high common Pb, and <95% concordant analyses, and metamict grains), and to identify and interpret accordingly those grains that have undergone partial or complete recrystallization.

Deposition of the Spargoville and Kalgoorlie Sequences (2700–2655 Ma) overlapped with the intrusion of voluminous granitoids, and subvolcanic porphyries, thereby indicating that those granitoids were emplaced during crustal extension, and not, as commonly advocated, during regional compression and metamorphism.

Most felsic detritus was recycled from coeval volcanoplutonic arcs and the orogen to the remnant-ocean basin sequences, with zircon ages ranging from ~2730 to ~2660 Ma. Older felsic sources, with age ranges of 3570–3130, 3030–2900, 2870–2770 and ~2760 Ma, could have been structural enclaves within the volcanoplutonic arcs or orogen, but were more likely within distant cratonic basement. When coupled to palaeocurrent data, the detrital ages fingerprint southeast-oriented amalgamation of felsic detritus from multiple sources within longitudinal depositional systems, and longitudinal reworking of felsic detritus from single-sourced transverse systems. Mafic and intermediate sources are not represented in the detrital zircon populations due to the inherent zircon-poor nature of these lithologies, but are evident from petrographic studies.

Detrital-zircon age distributions preserve evidence for the sedimentary recycling of coeval volcanoplutonic arc or orogenic sources and cratonic basement. Most of these source terrains are no longer preserved within, or adjacent to, the Eastern Goldfields Province, presumably destroyed or translated through a combination of sedimentary recycling and tectonic severance. It is therefore important in models for crustal evolution during the Archaean to consider increments of crustal growth, as recorded by sedimentary provenance, that are no longer preserved. Increased and more widespread use of detrital zircon suites from Archaean sedimentary sequences is advocated in order to help quantify the balance between

Appendix A

Table 1A
SHRIMP zircon data (Pb204-corrected) — Eastern Goldfields Province, Western Australia

Mount	Analysis	U/ppm	Th/ppm	Th/U	f206%	207/206	±	208/206	±	206/238	±	207/235	±	%conc	Age (206/238)	±	Age (207/206)	±
E75, Dacite, Spargoville Formation, Golden Ridge
U637B	29-1	860	890	1.03	0.06	0.10008	0.00026	0.11070	0.00048	0.41051	0.01034	5.66471	0.14526	136	2217	47	1626	5
U637B	34-1	174	62	0.36	0.35	0.14043	0.00085	0.06216	0.00154	0.56672	0.01446	10.97271	0.29554	130	2894	59	2232	10
U637B	12-1	417	170	0.41	0.40	0.17334	0.00055	0.11519	0.00098	0.47344	0.01194	11.31517	0.29216	96	2499	52	2590	5
U637B	31-1	419	261	0.62	0.14	0.17721	0.00048	0.16175	0.00081	0.47348	0.01195	11.56849	0.29757	95	2499	52	2627	5
U637B	1-1	675	497	0.74	0.18	0.17863	0.00045	0.20062	0.00082	0.43875	0.01106	10.80596	0.27716	89	2345	50	2640	4
U637B	19-1	708	514	0.73	0.03	0.18232	0.00061	0.19123	0.00105	0.39872	0.01004	10.02334	0.25870	81	2163	46	2674	6
U637B	5-1	284	74	0.26	0.16	0.18302	0.00062	0.06746	0.00087	0.49783	0.01262	12.56229	0.32652	97	2604	54	2680	6
U637B	18-1	94	20	0.21	0.26	0.18309	0.00128	0.05443	0.00216	0.50816	0.01312	12.82812	0.35392	99	2649	56	2681	6
U637B	27-1	160	47	0.29	0.13	0.18353	0.00082	0.07906	0.00116	0.53266	0.01361	13.47877	0.35680	103	2753	57	2685	12
U637B	20-1	372	156	0.42	0.14	0.18418	0.00055	0.11094	0.00082	0.49914	0.01262	12.67586	0.32724	97	2610	54	2691	5
U637B	13-1	203	62	0.31	0.13	0.18436	0.00070	0.08243	0.00100	0.51726	0.01314	13.14821	0.34382	100	2688	56	2692	6
U637B	35-1	174	49	0.28	3.56	0.18447	0.00191	0.07749	0.00408	0.53308	0.01360	13.55869	0.38914	102	2754	57	2693	17
U637B	2-1	274	63	0.23	0.12	0.18460	0.00063	0.06132	0.00088	0.51672	0.01310	13.15191	0.34175	100	2685	56	2695	6
U637B	26-1	133	34	0.26	0.28	0.18461	0.00107	0.06727	0.00171	0.49094	0.01257	12.49619	0.33674	96	2575	54	2695	10
U637B	26-3	199	62	0.31	0.15	0.18463	0.00094	0.08503	0.00143	0.48541	0.01240	12.35720	0.32934	95	2551	54	2695	8
U637B	10-1	289	103	0.36	0.09	0.18502	0.00061	0.09665	0.00085	0.47873	0.01213	12.21281	0.31696	93	2522	53	2698	5
U637B	14-1	181	52	0.29	0.12	0.18494	0.00078	0.07799	0.00111	0.50580	0.01287	12.89777	0.33942	98	2639	55	2698	7
U637B	6-1	371	196	0.53	0.13	0.18523	0.00051	0.14316	0.00083	0.55192	0.01396	14.09556	0.36352	105	2833	58	2700	5
U637B	8-1	314	146	0.46	0.11	0.18553	0.00056	0.12830	0.00086	0.49059	0.01241	12.54956	0.32448	95	2573	54	2703	5
U637B	26-2	383	230	0.60	0.13	0.18577	0.00052	0.16473	0.00087	0.51789	0.01308	13.26550	0.34146	99	2690	56	2705	5
U637B	3-1	475	253	0.53	0.09	0.18589	0.00047	0.14180	0.00070	0.49089	0.01239	12.58181	0.32305	95	2575	54	2706	4
U637B	33-1	175	47	0.27	0.08	0.18621	0.00084	0.07243	0.00122	0.48386	0.01235	12.42280	0.32889	94	2544	54	2709	7
U637B	9-2	214	100	0.47	1.80	0.18680	0.00161	0.11796	0.00332	0.44771	0.01147	11.53115	0.32315	88	2385	51	2714	14
U637B	9-1	259	119	0.46	0.13	0.19792	0.00079	0.12651	0.00118	0.48625	0.01230	13.26951	0.34614	91	2554	53	2809	7
U637B	36-1	61	36	0.59	1.46	0.19901	0.00235	0.15685	0.00484	0.54178	0.01424	14.86567	0.44800	99	2791	60	2818	19
KWA-1, Intrusive porphyritic rhyolite, Black Flag Formation, Kanowna Belle
u702	6-1	567	215	0.38	0.99	0.16938	0.00065	0.10237	0.00127	0.47733	0.00572	11.14730	0.14527	99	2516	25	2551	6
u702	2-1	110	44	0.40	0.65	0.17999	0.00146	0.10266	0.00267	0.51911	0.00779	12.88265	0.23085	102	2695	33	2653	13
u702	19-1	39	14	0.36	1.77	0.18056	0.00287	0.09687	0.00592	0.49237	0.00821	12.25783	0.29931	97	2581	35	2658	26
u702	16-1	256	121	0.47	0.15	0.18100	0.00070	0.12688	0.00110	0.49491	0.00628	12.35109	0.16950	97	2592	27	2662	6
u702	17-1	43	22	0.53	0.33	0.18124	0.00189	0.14225	0.00346	0.50030	0.00870	12.50204	0.26711	98	2615	37	2664	17
u702	25-1	61	23	0.38	0.29	0.18121	0.00142	0.09814	0.00225	0.51443	0.00794	12.85308	0.23324	100	2676	34	2664	13
u702	29-1	193	66	0.34	0.47	0.18148	0.00128	0.09192	0.00147	0.52318	0.00671	13.07641	0.18568	102	2713	28	2665	8
u702	4-1	104	57	0.54	0.27	0.18169	0.00128	0.14543	0.00229	0.50952	0.00749	12.74943	0.21775	100	2655	32	2666	12
u702	25a-1	82	50	0.61	0.14	0.18196	0.00120	0.16288	0.00215	0.51355	0.00745	12.86501	0.21444	100	2672	32	2668	11
u702	20-1	149	41	0.27	0.91	0.18196	0.00116	0.07334	0.00217	0.50210	0.00653	12.59720	0.19130	98	2623	28	2671	11
u702	5-1	121	44	0.37	0.04	0.18220	0.00116	0.09948	0.00174	0.48015	0.00719	12.06211	0.20484	95	2528	31	2673	11
u702	7-1	85	38	0.45	0.37	0.18255	0.00146	0.11748	0.00253	0.49319	0.00767	12.41391	0.22757	97	2584	33	2676	13
u702	14-1	205	74	0.36	3.35	0.18266	0.00152	0.10301	0.00323	0.52317	0.00661	13.17638	0.21063	101	2713	28	2677	14
u702	6-2	97	29	0.30	0.99	0.18275	0.00154	0.08191	0.00295	0.49775	0.00706	12.54226	0.21811	97	2604	30	2678	14
u702	15-1	225	67	0.30	0.07	0.18296	0.00061	0.07918	0.00075	0.51800	0.00649	13.06711	0.17449	100	2691	28	2680	5
u702	26-1	358	206	0.58	0.26	0.18293	0.00055	0.15744	0.00096	0.52138	0.00633	13.15060	0.16938	101	2705	27	2680	5
u702	31-1	256	290	1.14	0.03	0.18320	0.00068	0.30603	0.00140	0.51122	0.00662	12.91315	0.17955	99	2662	28	2682	6
u702	1-1	45	17	0.39	0.24	0.18349	0.00217	0.10116	0.00380	0.51025	0.00976	12.90899	0.30620	99	2658	42	2685	20
u702	10-1	93	17	0.18	0.43	0.18352	0.00122	0.05200	0.00200	0.51474	0.00727	13.02506	0.21263	100	2677	31	2685	11
u702	11-1	125	76	0.61	0.28	0.18375	0.00107	0.16901	0.00195	0.49682	0.00686	12.58736	0.19688	97	2600	30	2687	10
u702	33-1	49	17	0.34	0.27	0.18411	0.00205	0.08791	0.00350	0.51429	0.00967	13.05568	0.30047	99	2675	41	2690	18
u702	13-1	136	70	0.51	0.09	0.18409	0.00088	0.13771	0.00140	0.50816	0.00681	12.89834	0.19047	98	2649	29	2690	8

u702	12-1	56	27	0.48	0.21	0.18505	0.00188	0.13206	0.00335	0.52096	0.00935	13.29219	0.28816	2703	40	2699	17
u702	18-1	79	68	0.85	0.16	0.18648	0.00112	0.22753	0.00208	0.53476	0.00781	13.74921	0.22648	2761	33	2711	10

MGS1, Intrusive porphyritic rhyolite, Black Flag Formation, Morgans Island

u702A	22-1	59	58	0.97	2.00	0.17623	0.00290	0.13985	0.00610	0.46923	0.00589	11.40191	0.24985	2480	26	2618	27
u702A	28-1	41	16	0.39	0.15	0.18043	0.00212	0.10665	0.00367	0.52061	0.00721	12.95132	0.24911	2702	31	2657	19
u702A	30-1	47	1	0.02	0.05	0.18046	0.00166	0.00453	0.00185	0.51457	0.00690	12.80341	0.22003	2676	29	2657	15
u702A	1-1	142	93	0.65	0.36	0.18079	0.00116	0.17023	0.00216	0.46366	0.00424	11.55758	0.13658	2456	19	2660	11
u702A	5-1	80	35	0.43	0.10	0.18085	0.00125	0.11609	0.00191	0.51607	0.00550	12.86851	0.17254	2682	23	2661	11
u702A	14-1	6	4	0.75	1.03	0.18132	0.01474	0.08510	0.03189	0.46210	0.01659	11.55279	1.07294	2449	73	2665	135
u702A	25-1	79	29	0.38	0.00	0.18243	0.00114	0.10029	0.00124	0.51239	0.00567	12.88854	0.17246	2667	24	2675	10
u702A	26-1	102	40	0.39	0.52	0.18252	0.00143	0.10583	0.00257	0.50232	0.00509	12.64166	0.17152	2624	22	2676	13
u702A	2-1	84	33	0.40	0.01	0.18262	0.00128	0.10419	0.00143	0.51059	0.00509	12.85662	0.17336	2659	23	2677	12
u702A	8-1	110	81	0.73	0.81	0.18295	0.00173	0.14685	0.00192	0.50527	0.00545	12.74533	0.19214	2636	23	2680	16
u702A	29-1	30	17	0.55	0.14	0.18312	0.00240	0.15249	0.00335	0.51797	0.00537	13.07787	0.28152	2691	23	2681	22
u702A	23-1	137	59	0.43	0.45	0.18344	0.00117	0.11171	0.00431	0.50154	0.00803	12.68548	0.15312	2620	34	2684	11
u702A	11-1	85	36	0.42	0.43	0.18352	0.00156	0.10614	0.00203	0.50208	0.00476	12.70452	0.18221	2623	20	2685	14
u702A	7-1	84	36	0.43	0.25	0.18349	0.00138	0.11389	0.00281	0.50685	0.00528	12.82341	0.17847	2643	23	2685	12
u702A	13-1	62	27	0.44	0.00	0.18410	0.00127	0.11991	0.00230	0.50402	0.00547	12.79366	0.18328	2631	25	2690	11
u702A	20-1	56	36	0.63	3.97	0.18522	0.00433	0.21232	0.00150	0.41640	0.00592	10.63397	0.29806	2244	24	2700	39
u702A	16-1	79	45	0.57	0.36	0.18573	0.00159	0.12865	0.00952	0.50855	0.00534	13.02313	0.19278	2650	24	2705	14
u702A	17-1	12	0	0.00	1.18	0.19034	0.00606	-0.02021	0.00288	0.52490	0.00561	13.77524	0.58724	2720	55	2745	52
u702A	3-1	138	82	0.59	0.09	0.19322	0.00098	0.15854	0.01178	0.52951	0.01292	14.10689	0.15605	2739	21	2770	8

SHD-9, Intrusive porphyritic rhyolite, Black Flag Formation, Mount Shae

u702B	1-1	850	1361	1.60	0.03	0.13539	0.00038	0.44168	0.00156	0.35227	0.00254	6.57613	0.05285	1945	12	2169	5
u702B	22-1	621	127	0.20	0.00	0.17598	0.00042	0.05670	0.00115	0.46751	0.00338	11.34364	0.08950	2473	15	2615	5
u702B	13-1	645	1074	1.67	0.01	0.17739	0.00041	0.44550	0.00042	0.47761	0.00347	11.68146	0.09224	2517	15	2629	4
u702B	21-1	385	532	1.38	0.01	0.17865	0.00054	0.37183	0.00110	0.48299	0.00372	11.89686	0.10257	2540	16	2640	4
u702B	15-1	68	33	0.48	0.09	0.17929	0.00142	0.12812	0.00128	0.51909	0.00594	12.83239	0.18866	2695	25	2646	5
u702B	32-1	134	97	0.72	0.01	0.17929	0.00091	0.19267	0.00226	0.51038	0.00474	12.61670	0.14028	2658	20	2646	13
u702B	23-1	507	564	1.11	0.00	0.17994	0.00045	0.30452	0.00153	0.48571	0.00380	12.08529	0.09864	2558	16	2652	8
u702B	29-1	413	390	0.94	0.01	0.17987	0.00049	0.25629	0.00094	0.49472	0.00363	12.08529	0.10197	2591	16	2652	4
u702B	16-1	346	389	1.12	0.04	0.18003	0.00057	0.30111	0.00093	0.49686	0.00372	12.26957	0.10779	2600	17	2653	5
u702B	30-1	352	340	0.97	0.00	0.18007	0.00053	0.26469	0.00118	0.49540	0.00386	12.33347	0.10557	2594	16	2654	5
u702B	14-1	429	292	0.68	0.02	0.18025	0.00052	0.18415	0.00102	0.49390	0.00381	12.30008	0.10297	2588	16	2655	5
u702B	11-1	406	389	0.96	0.10	0.18024	0.00055	0.25959	0.00084	0.49265	0.00372	12.27523	0.10469	2582	16	2655	5
u702B	19-1	182	183	1.00	0.00	0.18021	0.00075	0.27474	0.00109	0.50787	0.00375	12.24283	0.12823	2648	19	2655	7
u702B	20-1	324	247	0.76	0.09	0.18038	0.00061	0.20635	0.00146	0.48874	0.00445	12.61939	0.10765	2565	16	2656	7
u702B	2-1	43	22	0.51	0.02	0.18039	0.00184	0.14162	0.00061	0.49989	0.00380	12.15538	0.22028	2613	29	2656	17
u702B	6-1	407	378	0.93	0.00	0.18059	0.00051	0.25410	0.00304	0.48889	0.00664	12.43327	0.10279	2566	16	2658	5
u702B	19-2	204	180	0.88	0.00	0.18061	0.00076	0.24417	0.00095	0.49557	0.00372	12.17282	0.12439	2595	17	2658	7
u702B	24-1	162	131	0.81	0.03	0.18076	0.00080	0.21949	0.00142	0.50182	0.00428	12.34089	0.13008	2622	19	2660	7
u702B	31-1	230	121	0.52	0.02	0.18088	0.00066	0.14377	0.00139	0.50494	0.00446	12.50726	0.11876	2635	18	2661	6
u702B	13-2	607	881	1.45	0.00	0.18101	0.00046	0.38947	0.00092	0.49570	0.00416	12.59314	0.10085	2595	16	2662	4
u702B	12-1	481	52	0.11	0.01	0.18095	0.00047	0.02973	0.00110	0.49726	0.00369	12.37142	0.10135	2602	16	2662	4
u702B	28-1	156	128	0.82	0.01	0.18112	0.00053	0.22380	0.00036	0.50948	0.00372	12.40595	0.10619	2654	19	2663	8
u702B	20-2	255	136	0.53	0.24	0.18130	0.00076	0.14648	0.00147	0.50334	0.00428	12.72332	0.13321	2628	18	2665	7
u702B	23-2	414	391	0.95	0.00	0.18163	0.00053	0.25646	0.00102	0.50432	0.00453	12.58193	0.12128	2632	17	2668	5
u702B	27-1	73	21	0.29	0.34	0.18179	0.00151	0.07710	0.00131	0.51311	0.00411	12.62991	0.10787	2670	24	2669	14
u702B	9-1	126	105	0.83	0.08	0.18224	0.00098	0.22980	0.00093	0.51108	0.00387	12.86099	0.18728	2661	21	2673	9
u702B	5-1	424	494	1.17	0.02	0.18242	0.00052	0.31754	0.00249	0.50229	0.00563	12.84187	0.14896	2624	16	2675	5
u702B	8-2	154	90	0.58	0.00	0.18239	0.00082	0.16171	0.00180	0.49531	0.00492	12.63338	0.10619	2594	20	2675	7
u702B	9-2	78	51	0.66	0.47	0.18268	0.00175	0.18889	0.00116	0.50709	0.00380	12.45630	0.13386	2644	24	2677	16
u702B	8-1	272	180	0.66	0.17	0.18341	0.00070	0.18182	0.00346	0.50444	0.00568	12.77213	0.19953	2633	17	2684	6

NG-1, quartzose sandstone, Noganyer Formation, South Scotia

u725A	17-1	407	223	0.55	0.36	0.185475	0.000638	0.149328	0.001117	0.44197	0.004792	11.302635	0.133099	2359	21	2702	6
u725A	39-1	414	285	0.69	0.67	0.196126	0.000962	0.079259	0.001702	0.279697	0.003067	7.563504	0.094994	1590	15	2794	8
u725A	31-1	884	614	0.69	0.03	0.198142	0.000342	0.187324	0.000523	0.511682	0.005356	13.979059	0.151264	2664	23	2811	3

Table 1A (continued).

Mount	Analysis	U/ppm	Th/ppm	Th/U	f206%	207/206	±	208/206	±	206/238	±	207/235	±	%conc	Age (206/238)	±	Age (207/206)	±
u725A	21-1	899	1465	1.63	1.19	0.199089	0.000969	0.149939	0.001921	0.168915	0.001795	4.636779	0.056712	36	1006	10	2819	8
u725A	25-1	544	1945	3.57	1.31	0.20002	0.001136	0.14068	0.002248	0.199655	0.002159	5.506233	0.070625	42	1173	12	2826	9
u725A	61-1	714	254	0.36	0.03	0.200415	0.000385	0.089474	0.000422	0.513602	0.005429	14.192508	0.15578	94	2672	23	2830	3
u725A	41-1	667	279	0.42	0.01	0.202692	0.00039	0.109435	0.000452	0.505687	0.005343	14.132488	0.155081	93	2638	23	2848	3
u725A	57-1	262	261	1.00	0.07	0.202647	0.000685	0.083389	0.000855	0.49661	0.005567	13.875786	0.167973	91	2599	24	2848	6
u725A	40-1	274	545	1.99	0.26	0.203231	0.000867	0.162174	0.001467	0.366168	0.004131	10.26056	0.12876	71	2011	19	2852	7
u725A	48-1	653	1320	2.02	1.39	0.203227	0.001046	0.105286	0.002039	0.2177	0.002331	6.100177	0.07593	45	1270	12	2852	8
u725A	45-1	785	4535	5.78	0.48	0.203348	0.000769	0.117801	0.001315	0.185988	0.001982	5.214662	0.061237	39	1100	11	2853	6
u725A	49-1	667	4418	6.62	0.45	0.203318	0.000913	0.128419	0.001593	0.156132	0.00168	4.376909	0.053267	33	935	9	2853	7
u725A	29-1	639	923	1.45	1.99	0.203553	0.001316	0.142865	0.002693	0.173414	0.001864	4.867012	0.064326	36	1031	10	2855	11
u725A	24-1	689	302	0.44	0.04	0.20371	0.000376	0.125855	0.000481	0.529015	0.005573	14.858831	0.162264	96	2737	23	2856	3
u725A	49-1	124	467	3.76	0.40	0.203848	0.001286	0.17247	0.002285	0.422777	0.005207	11.882823	0.172581	80	2273	24	2857	10
u725A	46-1	224	138	0.62	0.05	0.203885	0.000861	0.159876	0.001247	0.548376	0.006285	15.415764	0.195796	99	2818	26	2858	7
u725A	30-1	177	166	0.94	0.05	0.204122	0.000861	0.235403	0.001522	0.467528	0.005448	13.158261	0.169456	86	2473	24	2859	7
u725A	15-1	436	293	0.67	0.08	0.20424	0.000505	0.180819	0.00079	0.522498	0.005625	14.713846	0.166902	95	2710	24	2860	4
u725A	18-1	187	241	1.28	0.04	0.204339	0.000854	0.213579	0.001477	0.446027	0.005137	12.566482	0.160027	83	2378	23	2861	7
u725A	9-1	167	84	0.50	0.00	0.204568	0.000791	0.070604	0.000656	0.449534	0.005239	12.679462	0.161413	84	2393	23	2863	6
u725A	43-1	738	441	0.60	0.05	0.204545	0.000381	0.156943	0.000525	0.517544	0.00545	14.596104	0.1594	94	2689	23	2863	3
u725A	36-1	708	5698	8.04	2.85	0.204652	0.001487	0.145494	0.003119	0.167633	0.001803	4.730165	0.064855	35	999	10	2864	12
u725A	2-1	365	298	0.82	0.00	0.205089	0.000481	0.218743	0.000747	0.543436	0.005895	15.367133	0.174884	98	2798	25	2867	4
u725A	22-1	560	6767	12.08	0.90	0.205594	0.000909	0.184515	0.001767	0.23409	0.002513	6.635802	0.080373	47	1356	13	2871	7
u725A	51-1	405	2471	6.09	0.19	0.206983	0.000711	0.107939	0.001069	0.354466	0.003841	10.116054	0.119065	68	1956	18	2882	6
u725A	8-1	434	675	1.56	0.68	0.20746	0.000861	0.132112	0.001571	0.318519	0.003458	9.111099	0.110224	62	1782	17	2886	6
u725A	53-1	400	300	0.75	0.08	0.207575	0.000543	0.198462	0.000871	0.540284	0.005845	15.46318	0.176958	96	2785	24	2887	4
u725A	4-1	430	751	1.75	0.44	0.208228	0.000864	0.131886	0.001492	0.263443	0.002848	7.563564	0.091174	52	1507	15	2892	7
u725A	35-1	500	299	0.60	0.01	0.20897	0.000441	0.1611	0.000601	0.560258	0.005981	16.142609	0.179811	99	2868	25	2898	3
u725A	3-1	490	561	1.15	0.05	0.209446	0.000476	0.140223	0.000642	0.502977	0.005368	14.525173	0.162471	91	2627	23	2901	4
u725A	33-1	194	50	0.26	0.03	0.209685	0.000769	0.060077	0.000849	0.562993	0.006468	16.276942	0.203286	99	2879	27	2903	6
u725A	47-1	530	1184	2.23	0.15	0.209611	0.000621	0.095818	0.000865	0.338049	0.003633	9.769985	0.112332	65	1877	18	2903	5
u725A	6-1	153	176	1.15	0.07	0.209784	0.00091	0.144864	0.001394	0.520176	0.006123	15.046072	0.196278	93	2700	26	2904	7
u725A	10-1	288	360	1.25	0.03	0.209833	0.000656	0.144322	0.000878	0.502652	0.0056	14.542604	0.173672	90	2625	24	2904	6
u725A	58-1	220	218	0.99	0.04	0.209864	0.000775	0.148551	0.001101	0.467617	0.005332	13.530983	0.167999	85	2473	23	2904	6
u725A	42-1	311	294	0.94	0.06	0.209908	0.000683	0.069461	0.000757	0.432803	0.004771	12.526234	0.148772	80	2318	21	2905	5
u725A	13-1	249	80	0.32	0.03	0.210018	0.000657	0.143092	0.000729	0.558889	0.006218	16.183912	0.194886	98	2862	26	2906	5
u725A	28-1	300	328	1.09	0.26	0.210015	0.000688	0.078509	0.000943	0.440785	0.005393	12.76372	0.15343	81	2354	22	2906	5
u725A	54-1	267	142	0.53	0.00	0.211678	0.000599	0.140143	0.000705	0.555634	0.006205	16.216817	0.192304	98	2849	26	2918	5
u725A	5-1	164	121	0.73	0.06	0.213389	0.000844	0.177612	0.001338	0.558431	0.006504	16.430217	0.209673	98	2860	27	2931	6
u725A	7-1	490	244	0.50	0.33	0.215047	0.000683	0.083917	0.001022	0.347666	0.003741	10.308538	0.119499	65	1923	18	2944	5
u725A	38-1	330	1387	4.21	0.97	0.217234	0.001116	0.204653	0.002182	0.357329	0.003995	10.702774	0.137785	67	1970	19	2960	8
u725A	11-1	337	716	2.12	0.51	0.217816	0.000774	0.110489	0.001282	0.477633	0.005228	14.344481	0.171013	85	2517	23	2965	5
u725A	23-1	443	939	2.12	0.15	0.218046	0.000777	0.135667	0.00111	0.277561	0.00304	8.34465	0.099602	53	1579	15	2966	6
u725A	19-1	419	788	1.88	0.30	0.218676	0.000657	0.107945	0.000988	0.431752	0.004678	13.017798	0.151015	78	2314	21	2971	5
u725A	32-1	252	239	0.76	0.02	0.218642	0.000623	0.079262	0.000696	0.483366	0.005298	14.583366	0.169977	86	2544	23	2971	5
u725A	50-1	420	792	1.88	0.12	0.219007	0.000623	0.106635	0.000804	0.452915	0.00492	13.676519	0.158227	81	2408	22	2973	5
u725A	55-1	475	961	2.02	0.28	0.21931	0.000711	0.153612	0.001161	0.345834	0.003747	10.4575	0.122267	64	1915	18	2976	5
u725A	26-1	293	668	2.28	0.01	0.219475	0.000751	0.189277	0.00111	0.384288	0.004266	11.629027	0.139741	70	2096	20	2977	5
u725A	1-1	305	612	2.01	0.21	0.220324	0.000781	0.170012	0.001259	0.375834	0.004133	11.417166	0.136644	69	2057	19	2983	6
u725A	52-1	488	183	0.37	0.00	0.220375	0.000474	0.099734	0.000511	0.573154	0.006149	17.415419	0.195078	98	2921	25	2983	3
u725A	14-1	195	155	0.79	0.06	0.220416	0.000786	0.204322	0.001244	0.577027	0.006651	17.536424	0.21891	98	2937	27	2984	6

u725A	60-1	249	211	0.85	0.00	0.22138	0.000799	0.169069	0.001014	0.374307	0.00424	11.425331	0.140662	69	2050		20	2991	6
u725A	34-1	239	167	0.70	0.05	0.221747	0.000749	0.116863	0.000988	0.531591	0.005994	16.253115	0.197752	92	2748		25	2993	5
u725A	16-1	254	127	0.50	0.03	0.241818	0.000653	0.130925	0.000747	0.627322	0.006998	20.916047	0.24682	100	3139		28	3132	4
u725A	59-1	213	187	0.88	0.01	0.287012	0.000813	0.170093	0.000916	0.622321	0.00707	24.627212	0.296696	92	3119		28	3401	4
E7, Dacite breccia, Spargoville Formation, Widgiemooltha																			
U968A	2.2	685	319	0.46	0.46	0.17701	0.00051	0.12943	0.00093	0.45842	0.01026	11.18819	0.25651	93				2625	5
U968A	13.1	589	531	0.90	0.12	0.17784	0.00048	0.24048	0.00091	0.48614	0.00610	11.92061	0.15676	97				2633	4
U968A	13.2	717	386	0.54	0.06	0.17878	0.00039	0.14509	0.00058	0.48981	0.01097	12.07368	0.27484	97				2642	4
U968A	4.3	356	180	0.51	0.26	0.18035	0.00065	0.12137	0.00105	0.48600	0.01093	12.08515	0.28051	96				2656	6
U968A	8.1	615	427	0.69	0.32	0.18057	0.00050	0.18979	0.00094	0.48792	0.00611	12.14763	0.15990	96				2658	5
U968A	2.1	588	316	0.54	0.09	0.18075	0.00040	0.14447	0.00060	0.53210	0.00668	13.26085	0.17261	103				2660	4
U968A	14.1	610	425	0.70	0.33	0.18094	0.00048	0.18638	0.00088	0.50828	0.00638	12.68027	0.16668	100				2661	4
U968A	3.2	300	159	0.53	0.72	0.18104	0.00084	0.09138	0.00153	0.48857	0.01100	12.19522	0.28709	96				2662	8
U968A	12.1	475	268	0.56	0.46	0.18149	0.00061	0.15297	0.00111	0.50388	0.00634	12.60871	0.16935	99				2666	6
U968A	16.1	238	115	0.48	0.66	0.18145	0.00094	0.10092	0.00171	0.43552	0.00981	10.89599	0.25864	87				2666	9
U968A	3.1	268	164	0.61	0.57	0.18202	0.00087	0.09963	0.00155	0.49879	0.00636	12.51801	0.17722	98				2671	8
U968A	5.1	151	60	0.40	3.56	0.18238	0.00214	0.12540	0.00462	0.50590	0.00664	12.72128	0.23732	99				2675	19
U968A	18.1	254	94	0.37	0.42	0.18247	0.00078	0.10335	0.00133	0.50307	0.00533	12.65698	0.29677	98				2675	7
U968A	3.3	145	45	0.31	0.14	0.18252	0.00097	0.08403	0.00156	0.50849	0.01133	12.79643	0.30678	99				2676	9
U968A	1.2	408	164	0.40	2.96	0.18276	0.00117	0.12739	0.00249	0.50625	0.01138	12.75679	0.30775	99				2678	11
U968A	11.2	165	66	0.40	0.90	0.18271	0.00113	0.10207	0.00214	0.50448	0.01140	12.70887	0.30707	98				2678	10
U968A	6.1	440	240	0.55	0.01	0.18297	0.00050	0.15122	0.00072	0.50509	0.00637	12.74260	0.16856	98				2680	5
U968A	4.1	257	107	0.42	0.11	0.18304	0.00066	0.10603	0.00094	0.52598	0.00671	13.27419	0.18170	102				2681	6
U968A	5.2	386	164	0.42	1.20	0.18314	0.00081	0.12438	0.00160	0.51533	0.01157	13.01263	0.30457	100				2681	7
U968A	1.1	399	170	0.43	0.35	0.18347	0.00063	0.12065	0.00108	0.51496	0.00651	13.02684	0.17586	100				2685	6
U968A	8.2	238	107	0.45	0.61	0.18351	0.00084	0.12454	0.00154	0.50771	0.01144	12.84588	0.30250	99				2685	8
U968A	10.1	510	298	0.58	0.06	0.18350	0.00046	0.15620	0.00072	0.52746	0.00664	13.34541	0.17527	102				2685	4
U968A	15.1	625	295	0.47	0.01	0.18361	0.00041	0.12796	0.00053	0.51690	0.00648	13.08614	0.17012	100				2686	4
U968A	12.2	571	254	0.44	0.09	0.18379	0.00042	0.12374	0.00059	0.50937	0.01141	12.90768	0.29423	99				2687	5
U968A	7.1	311	173	0.56	0.04	0.18384	0.00060	0.14999	0.00090	0.52322	0.00665	13.26236	0.17932	101				2688	4
U968A	17.1	504	226	0.45	0.03	0.18400	0.00041	0.12570	0.00053	0.51401	0.01151	13.04014	0.29697	99				2689	4
U968A	4.2	303	168	0.56	0.36	0.18406	0.00063	0.15843	0.00112	0.51094	0.01148	12.96692	0.30019	99				2690	6
U968A	6.2	303	136	0.45	0.10	0.18411	0.00062	0.11222	0.00085	0.50930	0.01146	12.92852	0.29961	99				2690	6
U968A	9.1	484	280	0.58	0.01	0.18458	0.00046	0.15644	0.00067	0.51387	0.00647	13.07815	0.17165	99				2694	4
U968A	17.2	190	80	0.42	0.04	0.18477	0.00069	0.13414	0.00093	0.62151	0.01408	15.83326	0.37061	116				2696	6
U968A	11.1	149	51	0.34	0.25	0.18576	0.00100	0.09557	0.00157	0.51645	0.00676	13.22778	0.19503	99				2705	9
E179, Dacite conglomerate, Black Flag Formation, Gibson Honman Rock																			
u717A	4-1	1093	1563	1.43	1.53	0.14975	0.00129	0.34465	0.00304	0.10178	0.00098	2.10138	0.02882	27	625	6	2343	15
u717A	13-1	522	522	1.00	0.39	0.15680	0.00052	0.26585	0.00114	0.41035	0.00394	8.87180	0.09359	92	2217	18	2421	6
u717A	14-1	137	158	1.15	16.35	0.16717	0.00500	0.28138	0.01140	0.43926	0.00535	10.12495	0.34104	93	2347	24	2530	50
u717A	1-1	344	838	2.44	10.43	0.16782	0.00286	0.52300	0.00678	0.30730	0.00321	7.11043	0.14982	68	1727	16	2536	29
u717A	20-1	297	521	1.76	22.18	0.16840	0.00621	0.31289	0.01421	0.21161	0.00257	4.91333	0.19776	49	1237	14	2542	62
u717A	9-1	407	574	1.41	16.83	0.17066	0.00391	0.53700	0.00922	0.25059	0.00268	5.89646	0.15598	56	1442	14	2564	38
u717A	31-1	217	178	0.82	23.66	0.17153	0.00537	0.33928	0.01231	0.45592	0.00545	10.78292	0.37597	94	2422	24	2573	52
u717A	26-1	321	403	1.25	28.09	0.17392	0.00568	0.42177	0.01314	0.31695	0.00368	7.60059	0.27367	68	1775	18	2596	55
u717A	17.1	504	226	1.07	20.80	0.17500	0.00564	0.17103	0.01270	0.28671	0.00341	6.91794	0.24708	62	1625	17	2606	54
u717A	12-1	228	244	1.07	20.80	0.17500	0.00564	0.17103	0.01270	0.28671	0.00341	6.91794	0.24708	62	1625	17	2606	54
u717A	25-1	177	98	0.55	6.15	0.17588	0.00277	0.14290	0.00610	0.37684	0.00412	9.13815	0.18514	79	2061	19	2614	26
u717A	32-1	246	625	2.54	3.15	0.17636	0.00203	0.28281	0.00455	0.33340	0.00362	8.10718	0.13609	71	1855	18	2619	19
u717A	38-1	167	193	1.16	4.85	0.17670	0.00272	0.25621	0.00609	0.40829	0.00463	9.94724	0.20123	84	2207	21	2622	26
u717A	3-1	261	287	1.10	22.37	0.17687	0.00490	0.34476	0.01123	0.35362	0.00400	8.62365	0.26928	74	1952	19	2624	46
u717A	30-1	279	230	0.82	1.16	0.17786	0.00119	0.18152	0.00246	0.39048	0.00405	9.57601	0.12493	81	2125	19	2633	11
u717A	19-1	173	200	1.16	6.05	0.17875	0.00292	0.22638	0.00652	0.37562	0.00426	9.25740	0.19476	78	2056	20	2641	27
u717A	34-1	150	98	0.65	3.57	0.17887	0.00247	0.13111	0.00533	0.43734	0.00510	10.78569	0.20668	89	2339	23	2642	23
u717A	7-1	266	285	1.07	3.96	0.17988	0.00172	0.21635	0.00380	0.43490	0.00446	10.78626	0.16038	88	2328	20	2652	16
u717A	29-1	193	132	0.69	0.66	0.17992	0.00111	0.17074	0.00219	0.49138	0.00537	12.18947	0.16115	97	2577	23	2652	10
u717A	2-1	68	88	1.30	6.48	0.18023	0.00418	0.31409	0.00949	0.51422	0.00703	12.77815	0.36222	101	2675	30	2655	38
u717A	18-1	184	132	0.72	14.62	0.18022	0.00394	0.24833	0.00891	0.49774	0.00564	12.36800	0.31990	98	2604	24	2655	36

Table 1A (continued).

Mount	Analysis	U/ppm	Th/ppm	Th/U	f206%	207/206	±	208/206	±	206/238	±	207/235	±	%conc	Age (206/238)	±	Age (207/206)	±
u717A	22-1	387	212	0.55	0.34	0.18047	0.00065	0.14491	0.00113	0.50156	0.00499	12.48050	0.13710	99	2620	21	2657	6
u717A	35-1	370	182	0.49	0.63	0.18110	0.00078	0.13841	0.00147	0.47209	0.00470	11.78834	0.13364	94	2493	21	2663	7
u717A	36-1	238	230	0.97	4.99	0.18148	0.00247	0.26390	0.00552	0.36662	0.00398	9.17389	0.16903	76	2013	19	2666	23
u717A	10-1	101	56	0.56	1.28	0.18151	0.00177	0.13459	0.00358	0.50880	0.00612	12.73363	0.20881	99	2652	26	2667	16
u717A	28-1	208	131	0.63	0.97	0.18150	0.00118	0.15232	0.00233	0.46328	0.00498	11.59382	0.16616	92	2454	22	2667	11
u717A	21-1	146	71	0.49	0.09	0.18213	0.00090	0.13437	0.00137	0.51890	0.00580	13.03076	0.16616	101	2695	25	2672	8
u717A	6-1	150	76	0.51	0.65	0.18239	0.00117	0.14319	0.00221	0.50259	0.00554	12.63932	0.16950	98	2625	24	2675	11
u717A	11-1	188	163	0.87	0.21	0.18254	0.00087	0.23522	0.00168	0.49250	0.00528	12.39554	0.15180	96	2581	23	2676	8
u717A	17-1	159	108	0.68	0.57	0.18267	0.00110	0.18064	0.00213	0.49404	0.00540	12.44284	0.16338	97	2588	23	2677	10
u717A	37-1	161	96	0.59	0.82	0.18323	0.00135	0.15973	0.00266	0.49445	0.00556	12.49183	0.17722	97	2590	24	2682	12
u717A	24-1	203	93	0.46	3.32	0.18329	0.00188	0.09144	0.00398	0.43725	0.00465	11.04997	0.17317	87	2338	21	2683	17
u717A	5-1	229	128	0.56	0.14	0.18359	0.00085	0.15196	0.00140	0.34802	0.00362	8.80985	0.10479	72	1925	17	2686	8
u717A	15-1	153	77	0.51	1.23	0.18377	0.00143	0.14781	0.00287	0.50440	0.00558	12.78086	0.18264	98	2633	24	2687	13
u717A	16-1	127	91	0.72	1.11	0.18370	0.00167	0.15759	0.00337	0.44395	0.00512	11.24444	0.17461	88	2368	23	2687	15
u717A	23-1	176	260	1.48	16.96	0.18406	0.00518	0.20272	0.01165	0.33102	0.00397	8.40051	0.26848	69	1843	19	2690	47
u717A	33-1	200	88	0.44	1.31	0.18556	0.00138	0.12652	0.00277	0.47550	0.00518	12.16554	0.16978	93	2508	23	2703	12
u717A	8-1	71	23	0.32	0.00	0.20981	0.00116	0.08484	0.00104	0.56815	0.00726	16.43547	0.23894	100	2900	30	2904	9
u717A	27-1	180	75	0.42	0.14	0.27092	0.00094	0.11124	0.00119	0.66383	0.00714	24.79654	0.29000	99	3282	28	3311	5

EMD-10, Dacite breccia, Black Flag Formation, Eight Mile Dam (Hole EMD2)

u723	11-1	10226	4775	0.47	3.13	0.12834	0.00027	0.16112	0.00061	0.44963	0.00521	7.98735	0.09605	115	2394	23	2082	4
u723D	13-1	413	86	0.21	54.39	0.14536	0.01614	-0.03434	0.03657	0.09945	0.00202	1.99321	0.22990	27	611	12	2292	192
u723D	21-1	179	144	0.80	2.63	0.17121	0.00278	0.26416	0.00620	0.29219	0.00405	6.89729	0.15600	64	1652	20	2569	27
u723D	20-1	434	144	0.33	29.38	0.17264	0.00694	0.13285	0.01563	0.16843	0.00238	4.00924	0.17730	39	1003	13	2583	67
u723D	18-1	225	175	0.78	15.27	0.17436	0.00570	0.28506	0.01296	0.27061	0.00383	6.50584	0.24206	59	1544	19	2600	54
u723D	16-1	124	150	1.21	16.52	0.17508	0.00659	0.37946	0.01517	0.35006	0.00533	8.45048	0.35761	74	1935	25	2607	63
u723D	23-1	159	78	0.49	13.21	0.17674	0.00507	0.14796	0.01134	0.36097	0.00516	8.79657	0.29579	76	1987	24	2623	48
u723D	20-2	314	84	0.27	20.40	0.17923	0.00535	0.10104	0.01196	0.31183	0.00427	7.70622	0.26471	66	1750	21	2646	50
u723D	8-1	108	88	0.82	16.44	0.17934	0.00610	0.19269	0.01371	0.47005	0.00709	11.62310	0.45185	94	2484	31	2647	56
u723D	7-1	84	79	0.94	0.61	0.18013	0.00166	0.25386	0.00348	0.49755	0.00726	12.35691	0.22505	98	2603	31	2654	15
u723D	6-1	134	194	1.44	0.01	0.18021	0.00099	0.38977	0.00238	0.51618	0.00699	12.82560	0.19549	101	2683	30	2655	9
u723D	2-1	176	106	0.60	1.38	0.18057	0.00157	0.16568	0.00325	0.42676	0.00552	10.62507	0.17488	86	2291	25	2658	14
u723D	22-1	60	57	0.94	0.02	0.18113	0.00152	0.25524	0.00296	0.50607	0.00790	12.63859	0.23543	99	2640	34	2663	14
u723D	3-1	169	97	0.57	0.17	0.18124	0.00097	0.16155	0.00164	0.49840	0.00654	12.45452	0.18422	98	2607	28	2664	9
u723	12-1	80	48	0.59	2.51	0.18118	0.00294	0.15744	0.00627	0.46871	0.00699	11.70893	0.27354	93	2478	31	2664	27
u723	10-1	51	54	1.06	0.06	0.18149	0.00171	0.30026	0.00361	0.53291	0.00843	13.33548	0.25866	103	2754	35	2667	16
u723D	17-1	82	87	1.07	0.03	0.18157	0.00133	0.28890	0.00274	0.51245	0.00756	12.82899	0.22145	100	2667	32	2667	12
u723D	24-1	141	84	0.59	0.88	0.18158	0.00154	0.16785	0.00308	0.48295	0.00657	12.09101	0.20441	95	2540	29	2667	14
u723D	14-1	68	83	1.21	0.04	0.18168	0.00142	0.33007	0.00311	0.50912	0.00779	12.75369	0.22989	99	2653	33	2668	13
u723D	4-1	92	132	1.44	0.03	0.18208	0.00118	0.39737	0.00284	0.50709	0.00708	12.73019	0.20525	99	2644	30	2672	11
u723D	5-1	114	95	0.83	0.49	0.18209	0.00138	0.23680	0.00279	0.47814	0.00657	12.00425	0.19793	94	2519	29	2672	13
u723D	1-1	143	77	0.54	0.85	0.18235	0.00144	0.15611	0.00283	0.46725	0.00625	11.74735	0.19201	92	2471	27	2674	13
u723	9-1	54	53	0.99	0.08	0.18238	0.00182	0.26866	0.00373	0.50373	0.00801	12.66722	0.25063	98	2630	34	2675	16
u723D	19-1	74	28	0.38	2.99	0.18420	0.00358	0.12905	0.00765	0.38606	0.00595	9.80464	0.25761	78	2105	28	2691	32
u723D	15-2	105	58	0.55	0.26	0.18426	0.00135	0.15395	0.00235	0.49234	0.00692	12.50836	0.20809	96	2581	30	2692	12
u723D	15-1	49	13	0.27	0.08	0.18592	0.00184	0.07185	0.00275	0.51472	0.00867	13.19476	0.27150	99	2677	37	2706	16

BF-1, Sandstone, Killaloe Formation, Buldania

u720A	27-1	147	153	1.04	0.56	0.05884	0.00265	0.31568	0.00727	0.10761	0.00152	0.87308	0.04262	117	659	9	561	98
u720A	61-1	554	916	1.65	1.68	0.17892	0.00169	0.20010	0.00359	0.17365	0.00181	4.28381	0.06392	39	1032	10	2643	16
u720A	24-1	59	28	0.47	0.07	0.17951	0.00172	0.12192	0.00251	0.50456	0.00824	12.48836	0.24893	99	2633	35	2648	16
u720A	35-1	133	61	0.46	0.06	0.17950	0.00116	0.11977	0.00169	0.52908	0.00675	13.09472	0.19661	103	2738	28	2648	11
u720A	17-1	52	57	1.10	0.43	0.17965	0.00248	0.17587	0.00480	0.48682	0.00826	12.05832	0.27941	96	2557	36	2650	23
u720A	33-1	50	29	0.59	0.07	0.17973	0.00182	0.14620	0.00268	0.52449	0.00897	12.99735	0.27200	103	2718	38	2650	17

u720A	5-1	68	18	0.27	0.29	0.18000	0.00202	0.06591	0.00327	0.51105	0.00796	12.68306	0.25738	100	2661	34	2653	19
u720A	26-1	127	103	0.81	0.22	0.17999	0.00143	0.10768	0.00232	0.49290	0.00636	12.23217	0.19537	97	2583	27	2653	13
u720A	55-1	76	43	0.56	0.23	0.18002	0.00169	0.14208	0.00277	0.50436	0.00755	12.51849	0.23320	99	2632	32	2653	16
u720A	6-1	115	72	0.63	0.19	0.18007	0.00139	0.16476	0.00244	0.50395	0.00666	12.51202	0.20187	99	2631	29	2654	13
u720A	9-1	72	33	0.45	0.07	0.18009	0.00167	0.11836	0.00254	0.48349	0.00748	12.00523	0.22793	96	2542	32	2654	15
u720A	12-1	59	47	0.80	0.27	0.18027	0.00210	0.17530	0.00386	0.50438	0.00807	12.53649	0.26256	99	2633	35	2655	19
u720A	25-2	132	86	0.65	0.06	0.18026	0.00134	0.16724	0.00212	0.51291	0.00662	12.74800	0.19991	101	2669	28	2655	12
u720A	41-1	98	42	0.43	0.11	0.18059	0.00145	0.11513	0.00115	0.48998	0.00686	12.20009	0.20703	97	2571	28	2658	13
u720A	49-1	142	91	0.64	0.04	0.18061	0.00115	0.17455	0.00192	0.51210	0.00651	12.75253	0.19023	100	2666	30	2658	11
u720A	20-1	91	114	1.26	1.10	0.18092	0.00237	0.17064	0.00482	0.40942	0.00579	10.21272	0.20892	83	2212	28	2661	22
u720A	16-1	43	15	0.35	0.15	0.18115	0.00234	0.10006	0.00907	0.50492	0.00997	12.61155	0.29497	99	2635	26	2663	21
u720A	46-1	55	41	0.74	0.32	0.18124	0.00216	0.13181	0.00364	0.49019	0.00856	12.24963	0.27367	97	2571	39	2664	20
u720A	1-1	87	46	0.53	0.00	0.18144	0.00138	0.14578	0.00185	0.49812	0.00716	12.46144	0.21293	98	2606	37	2666	13
u720A	37-1	92	49	0.53	0.00	0.18141	0.00132	0.14824	0.00179	0.51449	0.00723	12.86896	0.21394	100	2676	31	2666	12
u720A	36-1	109	227	2.08	0.33	0.18154	0.00189	0.13257	0.00345	0.36528	0.00498	9.14319	0.16616	75	2007	31	2667	17
u720A	2-1	58	26	0.45	0.04	0.18160	0.00178	0.11808	0.00255	0.52841	0.00865	13.23091	0.26608	103	2735	36	2668	16
u720A	14-1	61	34	0.56	0.25	0.18169	0.00208	0.13694	0.00369	0.51429	0.00814	12.83392	0.26635	100	2675	35	2668	19
u720A	31-1	48	21	0.44	0.21	0.18166	0.00212	0.11605	0.00327	0.50449	0.00876	12.63604	0.27931	99	2633	38	2668	19
u720A	34-1	55	31	0.56	0.26	0.18196	0.00199	0.15321	0.00338	0.51706	0.00866	12.97253	0.27408	101	2687	37	2671	18
u720A	23-1	84	43	0.51	0.03	0.18228	0.00153	0.10553	0.00217	0.47613	0.00687	11.96639	0.21046	94	2510	30	2674	14
u720A	22-1	148	293	1.98	1.09	0.18243	0.00225	0.15023	0.00448	0.40693	0.00573	10.23592	0.20324	82	2201	26	2675	20
u720A	32-1	81	44	0.54	0.03	0.18250	0.00165	0.14485	0.00275	0.50931	0.00755	12.81573	0.23457	99	2654	32	2676	15
u720A	63-1	656	2631	4.01	2.66	0.18255	0.00213	0.17453	0.00456	0.14973	0.00159	3.76880	0.06291	34	899	9	2676	19
u720A	7-1	56	45	0.82	0.00	0.18262	0.00168	0.19116	0.00260	0.51013	0.00843	12.84471	0.25518	99	2657	36	2677	15
u720A	59-1	401	1671	4.16	0.94	0.18268	0.00133	0.14275	0.00261	0.30888	0.00330	7.78000	0.10617	65	1735	16	2678	12
u720A	38-1	55	24	0.43	0.09	0.18282	0.00207	0.11276	0.00355	0.48136	0.00763	12.13059	0.25114	95	2533	33	2678	19
u720A	66-1	67	67	1.02	0.07	0.18277	0.00210	0.12009	0.00355	0.50390	0.00844	12.70178	0.27118	98	2631	36	2679	19
u720A	45-1	126	128	1.02	0.11	0.18291	0.00140	0.16723	0.00240	0.46020	0.00604	11.60575	0.18562	91	2440	27	2679	13
u720A	10-1	64	23	0.35	0.11	0.18297	0.00175	0.09499	0.00241	0.51246	0.00791	12.92818	0.24747	100	2667	34	2680	16
u720A	47-1	191	232	1.22	0.28	0.18329	0.00120	0.18689	0.00248	0.45149	0.00536	11.41042	0.16241	90	2402	24	2683	11
u720A	43-1	99	47	0.48	0.05	0.18349	0.00137	0.13026	0.00220	0.52172	0.00715	13.19939	0.21663	101	2706	30	2685	12
u720A	25-1	121	77	0.64	0.00	0.18361	0.00112	0.16850	0.00205	0.50664	0.00659	12.82586	0.19293	98	2642	28	2686	10
u720A	42-1	56	48	0.86	0.08	0.18366	0.00177	0.22566	0.00323	0.53551	0.00898	13.56109	0.27621	103	2765	38	2686	16
u720A	57-1	167	182	1.09	0.20	0.18386	0.00123	0.17973	0.00260	0.47132	0.00578	11.93513	0.17525	93	2489	25	2686	11
u720A	13-1	66	30	0.45	0.04	0.18398	0.00196	0.11780	0.00220	0.52028	0.00803	13.18908	0.26126	100	2700	34	2688	18
u720A	28-1	96	110	1.15	0.49	0.18429	0.00196	0.11312	0.00330	0.48092	0.00663	12.19967	0.22522	94	2531	29	2689	18
u720A	4-1	165	878	5.33	1.20	0.18436	0.00229	0.18254	0.00360	0.26796	0.00342	6.80870	0.12606	57	1530	17	2692	20
u720A	15-1	222	1203	5.42	1.34	0.18442	0.00210	0.16485	0.00459	0.18413	0.00222	4.68065	0.08862	40	1090	12	2693	18
u720A	53-1	80	113	1.41	0.01	0.18444	0.00175	0.20303	0.00243	0.42060	0.00624	10.69461	0.19868	84	2263	28	2693	16
u720A	54-1	368	2239	6.09	0.70	0.18460	0.00133	0.15383	0.00175	0.29013	0.00314	7.37826	0.10141	61	1642	16	2695	12
u720A	44-1	81	152	1.86	0.55	0.18473	0.00232	0.16043	0.00437	0.39323	0.00595	10.00888	0.20852	79	2138	28	2695	21
u720A	3-1	126	177	1.40	0.10	0.18482	0.00142	0.13318	0.00232	0.45249	0.00590	11.52550	0.18378	89	2406	26	2696	13
u720A	58-1	58	84	1.46	0.22	0.18493	0.00283	0.11845	0.00513	0.41304	0.00699	10.52566	0.25474	83	2229	32	2697	25
u720A	48-1	71	85	1.20	0.65	0.18510	0.00231	0.19493	0.00460	0.49644	0.00760	12.65836	0.26490	96	2598	33	2698	21
u720A	8-1	109	390	3.57	0.39	0.18555	0.00188	0.12985	0.00337	0.37314	0.00513	9.52291	0.17241	76	2044	24	2699	17
u720A	11-1	178	378	2.12	0.30	0.18567	0.00147	0.16268	0.00188	0.35984	0.00444	9.20598	0.14240	73	1981	21	2703	13
u720A	50-1	339	912	2.69	5.24	0.18611	0.00287	0.21000	0.00635	0.29153	0.00328	7.46344	0.15104	61	1649	16	2704	26
u720A	40-1	102	95	0.94	1.34	0.18622	0.00155	0.17857	0.00270	0.44842	0.00619	11.50712	0.19540	88	2388	28	2708	14
u720A	29-1	71	66	0.94	0.01	0.18623	0.00188	0.11123	0.00155	0.45446	0.00697	11.66877	0.27879	89	2415	31	2709	27
u720A	60-1	471	2727	5.79	1.46	0.18664	0.00177	0.23705	0.00307	0.18586	0.00199	4.77262	0.07236	41	1099	11	2709	16
u720A	19-1	48	41	0.85	0.46	0.18672	0.00338	0.18067	0.00378	0.42830	0.00752	11.02190	0.29469	85	2298	34	2713	30
u720A	39-1	500	963	1.92	8.39	0.18719	0.00259	0.23205	0.00688	0.23806	0.00259	6.12893	0.13360	51	1377	14	2713	29
u720A	18-1	259	1675	6.48	0.91	0.18860	0.00167	0.22469	0.00345	0.29411	0.00338	7.59111	0.11700	61	1662	17	2718	15
u720A	56-1	406	1954	4.81	3.23	0.19099	0.00236	0.16770	0.00506	0.25327	0.00278	6.58601	0.11617	53	1455	14	2730	21
u720A	30-1	523	1127	2.16	9.30	0.19099	0.00371	0.24695	0.00830	0.20940	0.00233	5.51418	0.12981	45	1226	12	2751	32
u720A	62-1	739	1925	2.60	11.44	0.19395	0.00404	0.27174	0.00910	0.15608	0.00172	4.17376	0.10329	34	935	10	2776	34

Table 1A (continued).

Mount	Analysis	U/ppm	Th/ppm	Th/U	f206%	207/206	±	208/206	±	206/238	±	207/235	±	%conc	Age (206/238)	±	Age (207/206)	±
u720A	52-1	238	366	1.54	0.19	0.20202	0.00101	0.14705	0.00157	0.50721	0.00574	14.12778	0.18255	93	2645	25	2843	8
u720A	51-1	47	35	0.75	0.34	0.20436	0.00268	0.17224	0.00501	0.54510	0.00955	15.35923	0.35560	98	2805	40	2861	21

EMD-2, Sandstone, Black Flag Formation, Eight Mile Dam (Hole EMD2)

Mount	Analysis	U/ppm	Th/ppm	Th/U	f206%	207/206	±	208/206	±	206/238	±	207/235	±	%conc	Age (206/238)	±	Age (207/206)	±
ul673c	1-1	358	541	1.51	0.07	0.12880	0.00061	0.43223	0.00203	0.36137	0.00412	6.41784	0.08271	96	1989	20	2082	8
ul673A	51-1	8	2	0.22	5.59	0.17117	0.01504	0.01391	0.03276	0.38664	0.01290	9.12483	0.89252	82	2107	60	2569	147
ul673A	54-1	61	37	0.60	1.43	0.17762	0.00246	0.17132	0.00528	0.50324	0.00742	12.32450	0.26456	100	2628	32	2631	23
ul673A	55-1	56	65	1.15	0.08	0.17786	0.00177	0.30865	0.00420	0.52209	0.00783	12.80330	0.24329	103	2708	33	2633	17
ul673B	2-1	180	160	0.89	4.83	0.17807	0.00238	0.20761	0.00533	0.38881	0.00478	9.54598	0.18358	80	2117	22	2635	22
ul673B	25-1	36	15	0.42	0.24	0.17842	0.00219	0.10847	0.00397	0.52414	0.00879	12.89399	0.28332	103	2717	37	2638	20
ul673A	47-1	126	65	0.51	1.68	0.17926	0.00180	0.13174	0.00376	0.44333	0.00567	10.95739	0.18877	89	2366	25	2646	17
ul673B	13-1	80	38	0.47	0.17	0.17935	0.00143	0.12884	0.00262	0.51246	0.00702	12.67237	0.21136	101	2667	30	2647	13
ul673A	34-1	98	62	0.63	0.40	0.17945	0.00132	0.16623	0.00261	0.51895	0.00678	12.84011	0.20274	102	2695	29	2648	12
ul673A	44-1	67	32	0.48	0.12	0.18000	0.00149	0.12870	0.00264	0.51932	0.00737	12.88863	0.22311	102	2696	31	2653	14
ul673A	16-1	140	116	0.83	0.15	0.18039	0.00098	0.22321	0.00203	0.50515	0.00631	12.56393	0.17869	99	2636	27	2656	9
ul673B	19-1	82	28	0.34	0.16	0.18032	0.00145	0.08987	0.00249	0.50818	0.00702	12.63421	0.21262	100	2649	30	2656	13
ul673B	13-2	42	23	0.56	0.39	0.18043	0.00240	0.13697	0.00479	0.51340	0.00819	12.77215	0.28118	101	2671	35	2657	22
ul673B	1-1	154	113	0.73	2.93	0.18056	0.00190	0.20196	0.00420	0.46737	0.00580	11.63548	0.20056	93	2472	25	2658	17
ul673B	18-1	145	83	0.57	0.11	0.18062	0.00099	0.15168	0.00181	0.49827	0.00619	12.40877	0.17610	98	2606	27	2659	9
ul673B	6-1	142	107	0.75	0.21	0.18081	0.00102	0.20184	0.00207	0.50433	0.00627	12.57277	0.17957	99	2632	27	2660	9
ul673A	43-1	70	34	0.49	1.27	0.18075	0.00211	0.13723	0.00435	0.48574	0.00694	12.10551	0.23677	96	2552	30	2660	19
ul673A	50-1	70	51	0.72	0.20	0.18099	0.00156	0.19287	0.00314	0.49826	0.00709	12.43403	0.21797	98	2606	30	2662	14
ul673B	4-1	187	129	0.69	2.46	0.18114	0.00183	0.17039	0.00396	0.31417	0.00379	7.84673	0.13079	66	1761	19	2663	17
ul673A	41-2	29	35	1.21	0.00	0.18131	0.00188	0.33685	0.00475	0.51380	0.00923	12.84427	0.28044	100	2673	39	2665	17
ul673A	38-1	116	42	0.37	0.00	0.18160	0.00095	0.10220	0.00125	0.50681	0.00645	12.68986	0.18201	99	2643	28	2668	9
ul673A	36-1	98	107	1.09	0.14	0.18199	0.00116	0.29636	0.00270	0.51388	0.00674	12.89452	0.19709	100	2673	29	2671	11
ul673A	49-1	126	59	0.47	0.68	0.18220	0.00138	0.12994	0.00267	0.48303	0.00611	12.13428	0.18854	95	2540	27	2673	13
ul673B	22-1	116	121	1.05	0.08	0.18242	0.00106	0.28249	0.00242	0.52609	0.00675	13.23216	0.19497	102	2725	29	2675	10
ul673A	41-1	145	107	0.74	0.36	0.18244	0.00110	0.20338	0.00226	0.51156	0.00631	12.86802	0.18540	100	2663	27	2675	10
ul673A	42-1	81	64	0.79	0.35	0.18241	0.00154	0.18658	0.00314	0.49415	0.00675	12.42788	0.21056	97	2589	29	2675	14
ul673B	20-1	86	89	1.04	0.12	0.18264	0.00130	0.28468	0.00297	0.50698	0.00686	12.76682	0.20510	99	2644	29	2677	12
ul673A	53-1	101	89	0.87	0.25	0.18280	0.00137	0.25699	0.00303	0.48715	0.00638	12.27820	0.19511	96	2558	28	2678	12
ul673A	40-1	94	84	0.89	0.06	0.18358	0.00131	0.24507	0.00282	0.51587	0.00682	13.05784	0.20598	100	2682	29	2685	12
ul673B	9-1	151	80	0.53	0.81	0.18384	0.00135	0.14650	0.00271	0.44610	0.00553	11.30799	0.17143	88	2378	25	2688	12
ul673	A28-1	168	145	0.86	0.78	0.18394	0.00117	0.23567	0.00254	0.51725	0.00629	13.11795	0.18872	100	2688	27	2689	10
ul673A	46-1	66	24	0.36	0.75	0.18403	0.00208	0.09428	0.00403	0.51106	0.00736	12.96730	0.25140	99	2661	31	2689	19
ul673A	39-1	68	58	0.85	0.06	0.18454	0.00155	0.23336	0.00327	0.50991	0.00724	12.97438	0.22538	99	2656	31	2694	14
ul673B	10-1	195	93	0.47	3.38	0.18547	0.00203	0.13783	0.00437	0.36553	0.00447	9.34753	0.16249	74	2008	21	2702	18
ul673B	11-1	65	28	0.43	0.05	0.18578	0.00150	0.11423	0.00254	0.51252	0.00729	13.12805	0.22580	99	2667	31	2705	13
ul673	A27-1	119	53	0.44	0.08	0.18716	0.00101	0.12201	0.00161	0.54489	0.00694	14.06131	0.20303	103	2804	29	2717	9
ul673	A29-1	106	128	1.20	2.82	0.18753	0.00218	0.32293	0.00506	0.52294	0.00681	13.52165	0.25015	100	2712	29	2721	19
ul673A	52-1	40	27	0.69	0.50	0.18766	0.00254	0.18800	0.00528	0.52443	0.00857	13.56952	0.30515	100	2718	36	2722	22
ul673A	37-1	67	42	0.63	0.09	0.18787	0.00141	0.16897	0.00256	0.52493	0.00747	13.59763	0.22980	100	2720	32	2724	12
ul673B	5-1	11	4	0.36	0.00	0.18861	0.00326	0.10805	0.00431	0.50098	0.01270	13.02797	0.42218	96	2618	55	2730	28
ul673B	15-1	26	8	0.29	0.07	0.18894	0.00327	0.08120	0.00622	0.52647	0.00979	13.71528	0.36970	100	2727	41	2733	29
ul673B	23-1	91	101	1.11	0.28	0.18891	0.00161	0.29690	0.00312	0.53963	0.00715	14.05582	0.22312	102	2782	30	2733	12
ul673B	7-1	65	80	1.23	0.00	0.18915	0.00127	0.33006	0.00309	0.53060	0.00762	13.83753	0.22962	100	2744	32	2735	11
ul673A	48-1	98	78	0.80	0.03	0.18970	0.00118	0.21754	0.00236	0.53302	0.00704	13.94182	0.21308	101	2754	30	2740	10
ul673A	31-1	82	69	0.85	0.10	0.19072	0.00139	0.23472	0.00291	0.52454	0.00718	13.79343	0.22479	99	2718	30	2748	12
ul673c	2-1	1051	176	0.17	0.28	0.19508	0.00040	0.05486	0.00047	0.43446	0.00472	11.68596	0.13223	83	2326	21	2785	3
ul673B	14-1	157	68	0.43	0.67	0.28099	0.00120	0.11596	0.00192	0.62502	0.00767	24.21488	0.32689	93	3130	30	3368	7

BF-6, Sandy matrix of andesite-clast conglomerate, White Flag Formation, White Flag Lake

Mount	Analysis	U/ppm	Th/ppm	Th/U	f206%	207/206	±	208/206	±	206/238	±	207/235	±	%conc	Age (206/238)	±	Age (207/206)	±
u749c	42-1	116	43	0.37	0.33	0.06211	0.00190	0.10902	0.00434	0.12139	0.00134	1.03952	0.03512	109	739	8	678	65
u749c	10-1	234	166	0.71	0.04	0.07099	0.00091	0.20875	0.00233	0.16991	0.00155	1.66303	0.02761	106	1012	9	957	26

u749c	35-1	77	49	0.63	0.15	0.19213	0.00133	0.17925	0.00244	0.51562	0.00562	13.65929	0.18599	97	2681	24	2761	11
u749c	30-1	778	793	1.02	0.04	0.19433	0.00034	0.27191	0.00065	0.50524	0.00397	13.53724	0.11192	95	2636	17	2779	3
u749c	12-1	112	61	0.55	0.18	0.19627	0.00100	0.14673	0.00161	0.53912	0.00536	14.58947	0.17099	99	2780	22	2795	8
u749c	43-1	99	69	0.69	0.17	0.19636	0.00105	0.18277	0.00182	0.54517	0.00555	14.75984	0.17838	100	2805	23	2796	9
u749c	21-1	154	117	0.76	0.04	0.19672	0.00078	0.20921	0.00132	0.54022	0.00503	14.65298	0.15477	99	2784	21	2799	6
u749c	36-1	109	72	0.66	0.23	0.19676	0.00112	0.17986	0.00203	0.52309	0.00525	14.19086	0.17240	97	2712	22	2800	9
u749c	20-1	83	71	0.86	0.00	0.19724	0.00097	0.23150	0.00161	0.53890	0.00573	14.65531	0.17977	99	2779	24	2803	8
u749c	22-1	166	108	0.65	0.18	0.19737	0.00085	0.17589	0.00149	0.53736	0.00492	14.62310	0.15494	99	2772	21	2805	7
u749c	14-1	93	61	0.66	0.09	0.19752	0.00116	0.17634	0.00205	0.55123	0.00572	15.01252	0.18825	101	2830	24	2806	10
u749c	48-1	128	78	0.61	0.05	0.19757	0.00089	0.16387	0.00141	0.54546	0.00527	14.85894	0.16583	99	2806	22	2806	7
u749c	37-1	156	102	0.66	2.42	0.19766	0.00165	0.17340	0.00347	0.49603	0.00470	13.51832	0.18070	93	2597	20	2807	14
u749c	31-1	136	94	0.69	0.04	0.19771	0.00087	0.18493	0.00147	0.54675	0.00519	14.90463	0.16309	100	2812	22	2807	7
u749c	33-1	77	50	0.64	0.13	0.19767	0.00127	0.17127	0.00223	0.54488	0.00594	14.85075	0.19796	100	2804	25	2807	11
u749c	5-1	65	37	0.56	0.11	0.19766	0.00143	0.15072	0.00251	0.55542	0.00625	15.13685	0.21363	101	2848	26	2807	12
u749c	39-1	162	130	0.80	0.06	0.19793	0.00083	0.21468	0.00148	0.54080	0.00502	14.75865	0.15719	99	2787	21	2809	7
u749c	46-1	113	70	0.62	0.16	0.19786	0.00105	0.16669	0.00181	0.54644	0.00539	14.90742	0.17546	100	2810	22	2809	9
u749c	6-1	113	67	0.59	0.20	0.19803	0.00108	0.15461	0.00188	0.55165	0.00545	15.06264	0.17851	101	2832	23	2810	9
u749c	20-2	86	57	0.66	0.22	0.19801	0.00124	0.17325	0.00221	0.55105	0.00587	15.04424	0.19560	100	2830	24	2810	10
u749c	49-1	99	45	0.45	0.50	0.19806	0.00133	0.13648	0.00239	0.48098	0.00495	13.13448	0.17040	90	2532	22	2810	11
u749c	11-2	174	126	0.72	0.15	0.19820	0.00111	0.19066	0.00195	0.54772	0.00524	14.96811	0.17444	100	2816	22	2811	9
u749c	28-1	179	102	0.57	2.26	0.19827	0.00164	0.20343	0.00349	0.38488	0.00355	10.52187	0.13838	75	2099	17	2812	14
u749c	3-1	130	84	0.65	0.10	0.19827	0.00091	0.17041	0.00153	0.54907	0.00527	15.01036	0.16734	100	2821	22	2812	8
u749c	15-1	150	100	0.67	0.06	0.19830	0.00080	0.17977	0.00129	0.55244	0.00517	15.10453	0.16075	101	2835	21	2812	7
u749c	38-1	126	91	0.72	0.18	0.19849	0.00098	0.19725	0.00176	0.53667	0.00516	14.68762	0.16663	98	2770	22	2814	8
u749c	19-1	90	68	0.76	0.15	0.19862	0.00114	0.20172	0.00203	0.54416	0.00570	14.90183	0.18692	99	2801	24	2815	9
u749c	45-1	227	161	0.71	0.04	0.19859	0.00066	0.19381	0.00107	0.53190	0.00464	14.56438	0.14130	98	2749	20	2815	5
u749c	7-1	151	106	0.70	0.14	0.19875	0.00084	0.18714	0.00140	0.54376	0.00509	14.90053	0.15989	99	2799	21	2816	7
u749c	40-1	146	94	0.64	0.07	0.19913	0.00080	0.17279	0.00125	0.54299	0.00510	14.90856	0.15894	99	2796	21	2819	7
u749c	50-1	93	51	0.55	0.15	0.19910	0.00119	0.14541	0.00206	0.54108	0.00563	14.85350	0.18771	99	2788	24	2819	10
u749c	29-1	125	83	0.67	0.06	0.19918	0.00092	0.17724	0.00152	0.54324	0.00528	14.91908	0.16789	99	2797	22	2820	8
u749c	17-1	93	63	0.68	0.08	0.19919	0.00111	0.18462	0.00193	0.54722	0.00567	15.02882	0.18594	100	2814	24	2820	9
u749c	34-1	54	47	0.88	0.11	0.19938	0.00155	0.23319	0.00297	0.55929	0.00675	15.37512	0.23301	102	2864	28	2821	13
u749c	44-1	123	85	0.69	0.12	0.19953	0.00095	0.18660	0.00162	0.54286	0.00552	14.93492	0.16786	99	2795	22	2822	8
u749c	47-1	207	162	0.78	0.06	0.19957	0.00068	0.21532	0.00115	0.54785	0.00486	15.07546	0.14911	100	2816	21	2823	6
u749c	8-1	121	77	0.64	0.10	0.19974	0.00080	0.17180	0.00125	0.54544	0.00531	15.02154	0.17084	99	2806	22	2824	8
u749c	9-1	112	76	0.69	0.04	0.19987	0.00096	0.18121	0.00160	0.54845	0.00543	15.11436	0.17509	100	2819	23	2825	8
u749c	2-1	120	88	0.73	0.10	0.19999	0.00092	0.19666	0.00156	0.55651	0.00548	15.34537	0.17460	101	2852	23	2826	8
u749c	11-1	174	142	0.81	0.05	0.20085	0.00109	0.21705	0.00198	0.51853	0.00491	14.35999	0.16513	95	2693	21	2833	9
u749c	41-1	89	71	0.79	0.11	0.20095	0.00116	0.21024	0.00208	0.54564	0.00569	15.11786	0.18947	99	2807	24	2834	9
u749c	32-1	95	71	0.75	0.45	0.30180	0.00137	0.15808	0.00202	0.69261	0.00713	28.82150	0.33881	98	3393	27	3480	7
KU-1, Sandstone bed in conglomerate, Kurrawang Formation, Kurrawang Syncline																		
u712C	36-1	144	100	0.70	0.37	0.17729	0.00140	0.14919	0.00253	0.43121	0.00525	10.54070	0.16141	88	2311	24	2628	13
u712C	52-1	47	24	0.50	0.00	0.17774	0.00179	0.13478	0.00232	0.51931	0.00892	12.72627	0.26664	102	2696	38	2632	17
u712C	53-1	62	32	0.52	0.13	0.17973	0.00203	0.13351	0.00357	0.51563	0.00811	12.77823	0.26181	101	2681	35	2650	19
u712C	49-1	124	135	1.09	2.46	0.18016	0.00303	0.14816	0.00644	0.35069	0.00475	8.71122	0.19921	73	1938	23	2654	28
u712C	8-1	69	31	0.46	0.08	0.18034	0.00155	0.11732	0.00240	0.48437	0.00702	12.04383	0.21390	96	2546	30	2656	14
u712C	51-1	141	143	1.01	1.44	0.18058	0.00205	0.20959	0.00434	0.45701	0.00582	11.37844	0.20590	91	2426	26	2658	19
u712C	42-1	108	58	0.54	3.68	0.18112	0.00328	0.12155	0.00704	0.49412	0.00675	12.33984	0.29628	97	2588	29	2663	30
u712C	1-1	84	49	0.58	0.13	0.18147	0.00189	0.15807	0.00329	0.50792	0.00704	12.70845	0.21453	99	2648	26	2666	17
u712C	25-1	36	24	0.67	0.29	0.18159	0.00257	0.18119	0.00480	0.50608	0.00924	12.67073	0.30990	99	2640	40	2667	23
u712C	39-1	142	95	0.67	0.00	0.18212	0.00100	0.18463	0.00153	0.50546	0.00626	12.69233	0.19982	99	2637	27	2672	9
u712C	32-1	103	84	0.81	0.02	0.18215	0.00126	0.20563	0.00213	0.49367	0.00665	12.39842	0.19698	97	2587	29	2673	11
u712C	50-1	151	138	0.91	2.49	0.18233	0.00235	0.17930	0.00505	0.46239	0.00574	11.62446	0.22078	92	2450	25	2674	21
u712C	17-1	87	85	0.97	0.26	0.18238	0.00203	0.13940	0.00364	0.37324	0.00534	9.38590	0.17815	76	2045	25	2675	18
u712C	5-1	82	38	0.47	0.11	0.18294	0.00143	0.12276	0.00222	0.52712	0.00734	13.29609	0.22334	99	2729	31	2680	13
u712C	17-1	598	1465	2.45	25.96	0.18342	0.00681	0.37243	0.01561	0.12764	0.00157	3.22786	0.13087	29	774	9	2684	61
u712C	12-1	98	77	0.79	0.00	0.18348	0.00114	0.21210	0.00188	0.51722	0.00685	13.08452	0.20045	100	2687	29	2685	10

Table 1A (continued).

Mount	Analysis	U/ppm	Th/ppm	Th/U	f206%	207/206	±	208/206	±	206/238	±	207/235	±	%conc	Age (206/238)	±	Age (207/206)	±
u712C	24-1	170	178	1.05	0.62	0.18403	0.00136	0.28094	0.00289	0.50259	0.00599	12.75295	0.18853	98	2625	26	2690	12
u712C	10-1	91	80	0.88	0.08	0.18457	0.00149	0.10974	0.00227	0.44627	0.00604	11.35688	0.18840	88	2379	27	2694	13
u712C	27-1	111	85	0.77	0.08	0.18485	0.00140	0.14836	0.00232	0.44428	0.00575	11.32356	0.17865	88	2370	26	2697	12
u712C	47-1	150	68	0.45	0.00	0.18490	0.00099	0.12397	0.00120	0.52209	0.00645	13.31016	0.18736	100	2708	27	2697	9
u712C	4-1	213	408	1.92	1.86	0.18499	0.00202	0.22886	0.00435	0.29937	0.00348	7.63602	0.12912	63	1688	17	2698	18
u712C	26-1	79	52	0.66	0.07	0.18502	0.00174	0.17035	0.00307	0.41139	0.00588	10.49478	0.18953	82	2221	27	2698	16
u712C	44-1	77	108	1.39	0.26	0.18599	0.00177	0.37428	0.00407	0.53897	0.00783	13.82118	0.25340	103	2779	33	2707	16
u712C	43-1	98	148	1.50	0.11	0.18640	0.00160	0.42499	0.00391	0.46283	0.00633	11.89487	0.20269	90	2452	28	2711	14
u712C	7-1	71	104	1.47	0.03	0.18709	0.00172	0.40457	0.00409	0.51056	0.00776	13.17045	0.24631	98	2659	33	2717	15
u712C	41-1	52	54	1.03	0.00	0.18709	0.00169	0.28166	0.00325	0.51688	0.00847	13.33310	0.26208	99	2686	36	2717	15
u712C	15-1	303	1160	3.82	6.22	0.18772	0.00375	0.18998	0.00828	0.17142	0.00197	4.43691	0.10768	37	1020	11	2722	33
u712C	3-1	77	130	1.69	0.64	0.18793	0.00246	0.41701	0.00584	0.45438	0.00583	11.77386	0.22902	89	2415	26	2724	22
u712C	22-1	78	81	1.04	0.00	0.18786	0.00132	0.27773	0.00252	0.53710	0.00761	13.91215	0.23070	102	2771	32	2724	12
u712C	46-1	72	53	0.74	0.15	0.18789	0.00173	0.19708	0.00310	0.53155	0.00792	13.77059	0.25416	101	2748	33	2724	15
u712C	14-1	139	166	1.19	0.00	0.18813	0.00099	0.32615	0.00209	0.51602	0.00628	13.38561	0.18546	98	2682	27	2726	9
u712C	31-1	52	60	1.17	0.06	0.18817	0.00194	0.31636	0.00409	0.53499	0.00880	13.88044	0.28422	101	2762	37	2726	17
u712C	9-1	78	52	0.66	0.40	0.18822	0.00215	0.16128	0.00403	0.35611	0.00507	9.24165	0.17875	72	1964	24	2727	19
u712C	48-1	145	137	0.95	0.03	0.18843	0.00115	0.25256	0.00216	0.51867	0.00653	13.47520	0.19736	99	2694	28	2728	10
u712C	28-1	106	172	1.63	0.04	0.18850	0.00139	0.44529	0.00344	0.51641	0.00682	13.42195	0.21334	98	2684	29	2729	12
u712C	18-1	73	88	1.21	0.81	0.18864	0.00236	0.30957	0.00517	0.49229	0.00713	12.80460	0.25982	95	2581	31	2730	21
u712C	45-1	95	63	0.66	0.00	0.18876	0.00127	0.18431	0.00188	0.52068	0.00716	13.55119	0.21719	99	2702	30	2731	11
u712C	34-1	86	132	1.53	0.00	0.18888	0.00130	0.41364	0.00317	0.53725	0.00760	13.99162	0.23081	101	2772	32	2732	11
u712C	23-1	60	77	1.29	0.03	0.18899	0.00167	0.35760	0.00373	0.51227	0.00796	13.34874	0.25113	98	2666	34	2733	15
u712C	40-1	203	252	1.24	2.96	0.18897	0.00232	0.29879	0.00517	0.43490	0.00519	11.33121	0.20567	85	2328	23	2733	20
u712C	6-1	71	94	1.34	0.13	0.18917	0.00184	0.34995	0.00410	0.49935	0.00720	13.02438	0.23926	95	2611	31	2735	16
u712C	16-1	95	155	1.64	0.00	0.18942	0.00122	0.44361	0.00311	0.51776	0.00696	13.52238	0.21120	98	2690	30	2737	11
u712C	21-1	46	43	0.94	0.43	0.18967	0.00237	0.24680	0.00475	0.51981	0.00872	13.59389	0.30086	99	2698	37	2739	21
u712C	19-1	55	64	1.17	0.15	0.18974	0.00189	0.31600	0.00400	0.53424	0.00859	13.97614	0.27890	101	2759	36	2740	16
u712C	38-1	145	191	1.31	0.00	0.18983	0.00111	0.29956	0.00225	0.50055	0.00616	13.10148	0.18700	95	2616	26	2741	10
u712C	37-1	52	59	1.12	0.05	0.19008	0.00219	0.30243	0.00462	0.52460	0.00854	13.74855	0.29006	99	2719	36	2743	19
u712C	2-1	72	69	0.96	0.12	0.19051	0.00193	0.26146	0.00375	0.53172	0.00689	13.96718	0.24347	100	2749	29	2747	17
u712C	13-1	558	1287	2.30	2.91	0.19085	0.00227	0.28938	0.00505	0.12836	0.00138	3.37774	0.05749	28	778	8	2750	20
u712C	11-1	78	50	0.64	0.00	0.20008	0.00133	0.17530	0.00185	0.55297	0.00772	15.25481	0.24710	100	2837	32	2827	11
u712C	20-1	25	10	0.42	0.27	0.20080	0.00339	0.10966	0.00568	0.56373	0.01199	15.60724	0.44872	102	2882	49	2833	28
u712C	29-1	133	52	0.39	1.34	0.20552	0.00220	0.10783	0.00427	0.40048	0.00513	11.34854	0.20071	76	2171	24	2871	17
u712C	35-1	144	184	1.28	0.02	0.31275	0.00131	0.33491	0.00203	0.70356	0.00873	30.33857	0.41210	97	3434	33	3535	6
u712C	30-1	36	43	1.19	0.01	0.31515	0.00266	0.31569	0.00400	0.72244	0.01345	31.39222	0.67124	99	3505	50	3546	13
KU-2, Sandstone bed in conglomerate, Penny Dam Conglomerate, Penny Dam																		
u725b	27-1	171	88	0.52	1.85	0.17193	0.00171	0.10641	0.00354	0.37228	0.00968	8.82488	0.25557	79	2040	45	2576	17
u725b	53-1	38	11	0.30	0.02	0.17992	0.00184	0.08052	0.00288	0.51003	0.01429	12.65283	0.39240	100	2657	61	2652	17
u725B	3-1	83	38	0.45	0.21	0.18040	0.00128	0.12138	0.00222	0.49839	0.01316	12.39699	0.34932	98	2607	57	2657	12
u725b	25-1	103	54	0.53	0.03	0.18045	0.00110	0.13957	0.00181	0.48919	0.01287	12.17129	0.33740	97	2567	56	2657	10
u725b	39-1	72	36	0.50	0.26	0.18067	0.00137	0.12866	0.00245	0.52725	0.01396	13.13418	0.37345	103	2730	59	2659	13
u725b	53-2	46	14	0.30	0.00	0.18067	0.00138	0.08387	0.00189	0.50189	0.01379	12.50265	0.36791	99	2622	59	2659	13
u725b	37-2	94	60	0.64	0.05	0.18073	0.00140	0.17604	0.00263	0.48170	0.01281	12.00326	0.34326	95	2535	56	2660	13
u725b	59-1	111	73	0.65	0.10	0.18073	0.00099	0.17658	0.00168	0.50793	0.01330	12.65714	0.34698	100	2648	57	2660	9
u725B	1-1	55	31	0.56	0.16	0.18085	0.00148	0.14750	0.00255	0.50121	0.01352	12.49770	0.36422	98	2619	58	2661	14
u725B	55-1	96	48	0.50	0.00	0.18090	0.00091	0.13454	0.00117	0.50856	0.01357	12.68501	0.34729	100	2650	57	2661	8
u725B	6-1	85	50	0.59	0.33	0.18098	0.00128	0.14914	0.00230	0.48330	0.01273	12.05976	0.33906	95	2542	55	2662	12
u725B	4-1	109	57	0.52	0.45	0.18116	0.00129	0.13387	0.00239	0.41611	0.01087	10.39345	0.29018	84	2243	49	2663	12
u725b	21-1	154	73	0.47	0.29	0.18110	0.00096	0.12662	0.00159	0.43818	0.01137	10.94153	0.29676	88	2342	51	2663	9

u725b	43-1	134	54	0.40	0.26	0.18109	0.00106	0.09710	0.00180	0.45812	0.01189	11.43891	0.31231	91	2431	53	2663	10
u725b	61-1	158	62	0.39	0.16	0.18114	0.00084	0.10535	0.00124	0.49474	0.01281	12.35620	0.33197	97	2591	55	2663	8
u725b	66-1	86	40	0.47	0.14	0.18113	0.00115	0.12543	0.00180	0.52060	0.01378	13.00164	0.36371	101	2702	58	2663	10
u725b	19-1	62	29	0.47	0.15	0.18117	0.00161	0.12661	0.00291	0.49576	0.01332	12.38399	0.36347	97	2596	57	2664	15
u725b	46-1	109	48	0.44	0.32	0.18126	0.00108	0.11961	0.00182	0.47692	0.01244	11.91889	0.32748	94	2514	54	2664	10
u725b	14-1	85	40	0.47	0.01	0.18135	0.00108	0.13244	0.00174	0.51185	0.01344	12.79873	0.35387	100	2665	57	2665	10
u725b	51-1	109	70	0.64	0.08	0.18147	0.00099	0.16765	0.00169	0.51129	0.01333	12.79312	0.34918	100	2662	57	2666	9
u725b	42-1	123	79	0.64	0.14	0.18153	0.00090	0.17317	0.00133	0.51844	0.01347	12.97609	0.35095	101	2693	57	2667	8
u725b	65-1	93	47	0.51	0.21	0.18157	0.00124	0.12737	0.00151	0.48525	0.01279	12.14818	0.34059	96	2550	55	2667	11
u725b	5-1	152	70	0.46	0.73	0.18162	0.00121	0.10404	0.00212	0.42388	0.01097	10.61479	0.29229	85	2278	50	2668	11
u725b	17-1	82	36	0.44	0.04	0.18168	0.00105	0.11715	0.00229	0.51400	0.01354	12.87536	0.35621	100	2674	58	2668	10
u725b	38-1	85	44	0.51	0.47	0.18160	0.00169	0.13258	0.00152	0.45480	0.01213	11.38760	0.33408	91	2417	54	2668	15
u725b	2-1	104	43	0.42	0.20	0.18182	0.00125	0.10572	0.00324	0.43529	0.01138	10.91267	0.30377	87	2330	51	2670	11
u725B	12-1	91	44	0.48	0.21	0.18187	0.00107	0.12082	0.00219	0.51109	0.01338	12.81598	0.35277	100	2661	57	2670	10
u725B	50-1	72	36	0.49	0.22	0.18191	0.00141	0.12609	0.00172	0.50799	0.01350	12.74152	0.36438	99	2648	58	2670	13
u725B	7-1	103	63	0.61	0.12	0.18198	0.00105	0.16581	0.00251	0.49320	0.01291	12.37489	0.34021	97	2584	56	2671	10
u725b	34-1	102	54	0.53	0.09	0.18198	0.00119	0.13975	0.00180	0.46719	0.01237	11.72277	0.32875	93	2471	54	2671	11
u725b	48-1	86	50	0.58	0.04	0.18201	0.00109	0.16254	0.00197	0.52560	0.01383	13.19059	0.36525	102	2723	58	2671	10
u725b	52-1	155	64	0.41	0.35	0.18195	0.00111	0.09781	0.00182	0.44204	0.01145	11.08966	0.30309	88	2360	51	2671	10
u725b	23-1	136	44	0.33	0.25	0.18212	0.00129	0.10281	0.00199	0.32992	0.00861	8.28461	0.23107	69	1838	42	2672	12
u725b	45-1	80	33	0.42	0.08	0.18206	0.00124	0.11306	0.00217	0.51102	0.01353	12.82742	0.36118	100	2661	58	2672	11
u725b	58-1	155	75	0.48	0.11	0.18211	0.00141	0.12884	0.00203	0.48906	0.01267	12.28011	0.33073	96	2567	55	2672	13
u725B	13-1	119	100	0.84	0.02	0.18219	0.00087	0.23227	0.00141	0.52545	0.01354	13.19958	0.35496	102	2722	58	2673	8
u725b	20-1	76	54	0.72	0.00	0.18225	0.00103	0.19498	0.00163	0.49000	0.01363	12.31300	0.34326	96	2571	58	2673	9
u725b	26-1	85	50	0.58	0.12	0.18231	0.00126	0.15195	0.00216	0.47550	0.01303	11.95278	0.33826	94	2508	56	2674	11
u725B	30-1	119	72	0.60	0.10	0.18233	0.00105	0.16790	0.00196	0.46597	0.01222	11.71460	0.32401	92	2466	54	2674	10
u725b	47-1	115	47	0.41	0.16	0.18229	0.00100	0.10827	0.00161	0.50934	0.01326	12.80198	0.34909	99	2654	57	2674	9
u725b	60-1	44	18	0.42	0.00	0.18234	0.00134	0.11691	0.00158	0.49626	0.01366	12.47635	0.36638	97	2598	59	2674	12
u725b	57-1	92	32	0.34	0.32	0.18246	0.00142	0.09158	0.00238	0.44716	0.01190	11.24933	0.32209	89	2383	53	2675	13
u725b	31-1	70	39	0.55	0.00	0.18238	0.00102	0.15236	0.00139	0.52550	0.01392	13.21460	0.36677	102	2722	59	2675	9
u725b	45-2	140	112	0.80	0.12	0.18279	0.00098	0.21525	0.00182	0.46587	0.01214	11.74110	0.31991	92	2465	53	2678	9
u725b	22-1	90	43	0.48	0.28	0.18287	0.00127	0.12792	0.00224	0.49309	0.01296	12.43247	0.34830	96	2584	56	2679	12
u725b	49-1	74	37	0.50	0.11	0.18294	0.00137	0.13466	0.00228	0.48744	0.01305	12.29494	0.35253	96	2560	57	2680	11
u725B	28-1	124	62	0.50	0.55	0.18311	0.00125	0.12452	0.00230	0.42025	0.01093	10.61014	0.29383	84	2262	50	2681	11
u725b	8-1	121	64	0.53	0.54	0.18311	0.00121	0.12700	0.00225	0.43564	0.01130	10.99887	0.30313	87	2331	51	2681	10
u725b	10-1	93	44	0.47	0.06	0.18318	0.00104	0.12832	0.00163	0.50730	0.01329	12.81298	0.35224	99	2645	57	2682	10
u725b	16-1	103	54	0.53	0.01	0.18325	0.00103	0.14146	0.00163	0.52105	0.01368	13.16484	0.36224	101	2704	58	2682	10
u725b	64-1	104	75	0.72	0.84	0.18344	0.00141	0.18820	0.00283	0.44733	0.01168	11.31452	0.31808	89	2383	52	2684	12
u725b	9-1	127	46	0.36	0.04	0.18344	0.00090	0.09786	0.00126	0.52128	0.01357	13.18472	0.35703	101	2705	57	2684	8
u725b	60-1	173	117	0.68	0.71	0.18368	0.00130	0.16578	0.00253	0.37689	0.00980	9.54492	0.26520	77	2062	46	2686	12
u725b	32-1	158	119	0.76	0.66	0.18369	0.00128	0.18869	0.00254	0.40185	0.01046	10.17787	0.28283	81	2178	48	2686	12
u725b	33-1	113	61	0.54	0.29	0.18360	0.00109	0.14189	0.00193	0.48759	0.01270	12.34312	0.33853	95	2560	55	2686	10
u725b	41-1	118	54	0.46	0.08	0.18361	0.00103	0.12507	0.00167	0.48457	0.01265	12.26768	0.33585	95	2547	55	2686	10
u725b	44-1	65	34	0.53	0.00	0.18389	0.00114	0.14613	0.00152	0.50940	0.01370	12.91571	0.36605	99	2654	58	2688	10
u725B	63-1	69	39	0.57	0.00	0.18416	0.00100	0.15347	0.00136	0.54643	0.01447	13.87502	0.38404	104	2810	60	2691	9
u725b	11-1	63	26	0.41	0.03	0.18434	0.00137	0.10654	0.00223	0.51835	0.01382	13.17498	0.37620	100	2692	59	2692	12
u725b	15-1	62	25	0.41	0.39	0.18451	0.00164	0.10066	0.00295	0.47472	0.01268	12.07726	0.35263	93	2504	55	2694	15
u725b	40-1	265	58	0.22	0.57	0.18455	0.00115	0.09604	0.00266	0.23179	0.00596	5.89801	0.16037	50	1344	31	2694	10
u725b	56-1	177	102	0.58	0.96	0.18477	0.00156	0.17006	0.00316	0.30042	0.00780	7.65369	0.21659	63	1693	39	2696	14
u725b	62-1	68	36	0.53	0.00	0.18499	0.00133	0.14513	0.00212	0.48133	0.01295	12.27711	0.35253	94	2533	56	2698	12
u725b	24-1	118	58	0.49	0.09	0.18493	0.00109	0.12976	0.00172	0.47326	0.01246	12.06736	0.33403	93	2498	54	2698	10
u725b	35-1	194	110	0.57	0.85	0.18497	0.00145	0.14885	0.00282	0.28745	0.00748	7.33089	0.20595	60	1629	37	2698	13
u725b	36-1	119	48	0.40	0.25	0.18507	0.00117	0.10390	0.00192	0.45878	0.01204	11.70671	0.32493	90	2434	53	2699	10
u725b	29-1	105	67	0.64	0.31	0.18520	0.00139	0.17106	0.00259	0.47016	0.01244	12.00547	0.34083	92	2484	55	2700	12
u725b	37-1	113	44	0.39	0.37	0.18517	0.00130	0.11045	0.00227	0.40668	0.01067	10.38316	0.29054	81	2200	49	2700	12
u725b	18-1	84	55	0.66	0.47	0.18593	0.00156	0.16850	0.00303	0.44033	0.01160	11.28854	0.32305	87	2352	52	2707	14

Table 1A (*continued*).

Mount	Analysis	U/ppm	Th/ppm	Th/U	f206%	207/206	±	208/206	±	206/238	±	207/235	±	%conc	Age (206/238)	±	Age (207/206)	±
KU-3, Sandstone, Mt Belches Sandstone, Mt Belches																		
u720C	8-1	81	66	0.81	0.14	0.17827	0.00175	0.22021	0.00342	0.51237	0.01017	12.59370	0.29233	101	2667	43	2637	16
u720C	3-1	83	50	0.60	0.20	0.17899	0.00175	0.16103	0.00313	0.51183	0.01015	12.63135	0.29269	101	2664	43	2644	16
u720C	24-1	158	87	0.55	0.26	0.17902	0.00134	0.14837	0.00237	0.48912	0.00886	12.07307	0.24691	97	2567	38	2644	12
u720C	1-1	22	13	0.57	0.09	0.17914	0.00387	0.16227	0.00750	0.53241	0.01465	13.15015	0.48707	104	2752	62	2645	36
u720C	18-1	100	99	0.99	0.12	0.18000	0.00136	0.27238	0.00136	0.50616	0.00970	12.56206	0.26959	100	2640	42	2653	13
u720C	41-1	71	46	0.65	0.52	0.18006	0.00217	0.16770	0.00413	0.48493	0.00993	12.03909	0.30111	96	2549	43	2653	20
u720C	45-1	90	65	0.72	0.28	0.18021	0.00178	0.19592	0.00340	0.51225	0.01004	12.72823	0.29297	100	2666	43	2655	16
u720C	30-1	45	105	2.35	0.08	0.18034	0.00178	0.63584	0.00728	0.56142	0.01316	13.95959	0.39669	108	2872	54	2656	22
u720C	48-1	102	89	0.87	0.07	0.18030	0.00140	0.23925	0.00266	0.52215	0.01006	12.98016	0.28102	102	2708	43	2656	13
u720C	11-1	124	114	0.92	0.11	0.18049	0.00137	0.26282	0.00279	0.51877	0.00965	12.91000	0.27039	101	2694	41	2657	13
u720C	37-1	137	158	1.16	0.01	0.18047	0.00124	0.31535	0.00266	0.50667	0.00936	12.60776	0.25848	99	2642	40	2657	11
u720C	61-1	200	120	0.60	0.22	0.18048	0.00108	0.16674	0.00185	0.49287	0.00875	12.26445	0.23838	97	2583	38	2657	10
u720C	15-1	64	30	0.47	0.38	0.18054	0.00238	0.12823	0.00447	0.49144	0.01019	12.23314	0.31695	97	2577	44	2658	22
u720C	13-1	199	162	0.82	0.22	0.18089	0.00105	0.22832	0.00202	0.49647	0.00876	12.38223	0.23855	98	2599	38	2661	10
u720C	14-1	85	66	0.78	0.00	0.18084	0.00136	0.21197	0.00227	0.51724	0.01013	12.89701	0.28165	101	2687	43	2661	13
u720C	2-1	91	36	0.40	0.05	0.18114	0.00160	0.10931	0.00254	0.51722	0.01007	12.91804	0.28885	101	2687	43	2663	15
u720C	19-1	101	103	1.03	0.04	0.18119	0.00151	0.27938	0.00313	0.50764	0.00971	12.68185	0.27648	99	2647	42	2664	14
u720C	16-1	153	229	1.50	0.06	0.18145	0.00112	0.40734	0.00273	0.50717	0.00922	12.68841	0.25249	99	2645	39	2666	10
u720C	20-1	29	12	0.40	0.00	0.18148	0.00231	0.11494	0.00271	0.51628	0.01294	12.91863	0.38123	101	2683	55	2666	21
u720C	38-1	100	91	0.91	0.07	0.18146	0.00166	0.25092	0.00333	0.49699	0.00958	12.43408	0.27784	98	2601	41	2666	15
u720C	60-1	155	117	0.75	0.68	0.18141	0.00175	0.19391	0.00352	0.44230	0.00817	11.06342	0.24191	89	2361	37	2666	16
u720C	59-1	54	61	1.13	0.22	0.18158	0.00217	0.25041	0.00431	0.49061	0.01061	12.28319	0.31902	96	2573	46	2667	20
u720C	12-1	194	113	0.58	0.01	0.18165	0.00096	0.15744	0.00147	0.51195	0.00906	12.82195	0.24465	100	2665	39	2668	9
u720C	43-1	219	161	0.73	0.80	0.18178	0.00156	0.18476	0.00313	0.36042	0.00638	9.03347	0.18630	74	1984	30	2669	14
u720C	57-1	18	15	0.84	1.06	0.18182	0.00578	0.22041	0.01229	0.52003	0.01564	13.03658	0.60502	101	2699	66	2670	53
u720C	31-1	64	60	0.95	0.01	0.18232	0.00196	0.25644	0.00393	0.52606	0.01108	13.22442	0.32797	102	2725	47	2674	18
u720C	28-1	60	29	0.48	0.26	0.18247	0.00125	0.10002	0.00218	0.44626	0.00820	11.22749	0.22889	89	2379	37	2675	11
u720C	49-1	62	28	0.45	0.07	0.18260	0.00147	0.12401	0.00240	0.51266	0.00976	12.90702	0.27857	100	2668	42	2677	13
u720C	39-1	137	150	1.10	0.05	0.18286	0.00120	0.30222	0.00249	0.50909	0.00942	12.83584	0.26163	99	2653	40	2679	11
u720C	17-1	41	16	0.38	0.03	0.18302	0.00276	0.11364	0.00493	0.52191	0.01189	13.16988	0.37983	101	2707	50	2680	25
u720C	40-1	110	61	0.55	0.34	0.18307	0.00154	0.14261	0.00265	0.48716	0.00925	12.29711	0.26695	95	2558	40	2681	14
u720C	42-1	127	102	0.81	0.38	0.18309	0.00193	0.14621	0.00366	0.38971	0.00727	9.83808	0.22177	79	2121	34	2681	17
u720C	9-1	244	169	0.69	0.37	0.18315	0.00119	0.17973	0.00221	0.37027	0.00650	9.35041	0.18190	76	2031	31	2682	11
u720C	21-1	64	43	0.67	0.14	0.18336	0.00199	0.17540	0.00360	0.51367	0.01063	12.98621	0.31873	100	2672	45	2683	18
u720C	53-1	157	21	0.14	0.21	0.18347	0.00134	0.01163	0.00158	0.40335	0.00738	10.20315	0.20944	81	2184	34	2684	12
u720C	54-1	53	53	0.99	0.19	0.18366	0.00201	0.26842	0.00399	0.52501	0.01152	13.29457	0.34159	101	2720	49	2686	18
u720C	5-1	105	94	0.90	0.04	0.18407	0.00135	0.25258	0.00253	0.49218	0.00941	12.49101	0.26612	96	2580	41	2690	12
u720C	28-1	60	29	0.48	0.26	0.18407	0.00217	0.10002	0.00352	0.47827	0.01010	12.13824	0.30857	94	2520	44	2690	19
u720C	49-1	62	28	0.45	0.07	0.18406	0.00237	0.12699	0.00436	0.52372	0.01099	13.29147	0.34487	101	2715	47	2690	21
u720C	23-1	121	41	0.33	0.16	0.18417	0.00146	0.08072	0.00228	0.49175	0.00920	12.48733	0.26494	96	2578	40	2691	13
u720C	36-1	93	87	0.94	0.06	0.18429	0.00152	0.25679	0.00298	0.51204	0.00994	13.01076	0.28648	99	2665	42	2692	14
u720C	33-1	45	22	0.48	0.19	0.18464	0.00246	0.13487	0.00430	0.51926	0.01169	13.21946	0.36413	100	2696	50	2695	22
u720C	52-1	24	31	1.30	0.05	0.18459	0.00500	0.22461	0.01039	0.42482	0.01215	10.81256	0.45164	85	2282	55	2695	45
u720C	6-1	100	18	0.18	0.24	0.18552	0.00161	0.05020	0.00245	0.51571	0.00992	13.19162	0.29129	99	2681	42	2703	14
u720C	4-1	177	45	0.25	0.18	0.18613	0.00130	0.05380	0.00187	0.41072	0.00738	10.54053	0.21138	82	2218	34	2708	11
u720C	56-1	36	12	0.34	0.00	0.18745	0.00244	0.10163	0.00261	0.51333	0.01414	13.26764	0.42309	98	2671	60	2720	21
u720C	51-1	79	34	0.43	0.23	0.19363	0.00203	0.11026	0.00353	0.53348	0.01066	14.24254	0.33761	99	2756	45	2773	17
u720C	46-1	89	81	0.91	0.11	0.19634	0.00177	0.24561	0.00341	0.53104	0.01044	14.37612	0.32511	98	2746	44	2796	15
u720C	47-1	54	34	0.63	0.25	0.19838	0.00252	0.14140	0.00444	0.50318	0.01079	13.76317	0.36137	93	2627	46	2813	21
u720C	44-1	78	48	0.62	0.09	0.19845	0.00186	0.17270	0.00317	0.53535	0.01076	14.64867	0.33990	98	2764	45	2814	15
u720C	25-1	94	71	0.76	0.12	0.20092	0.00158	0.19779	0.00265	0.51587	0.00997	14.29131	0.31078	95	2682	42	2834	13

u720C	27-1	158	38	0.24	0.00	0.20114	0.00105	0.06422	0.00084	0.52856	0.00954	14.65838	0.28448	96	2735	40	2835	8
u720C	22-1	162	116	0.72	0.10	0.20507	0.00116	0.19180	0.00189	0.55324	0.00999	15.64277	0.30621	99	2839	41	2867	9
u720C	34-1	102	147	1.44	0.06	0.21735	0.00148	0.38389	0.00170	0.57982	0.01110	17.37614	0.36665	100	2948	45	2961	11
u720C	29-1	167	92	0.55	0.13	0.21749	0.00117	0.14405	0.00148	0.57373	0.01030	17.20467	0.33340	99	2923	42	2962	9
u720C	35-1	84	53	0.63	0.27	0.21941	0.00194	0.16705	0.00329	0.58348	0.01153	17.65175	0.39956	100	2963	47	2976	14
u720C	58-1	193	27	0.14	0.13	0.22030	0.00111	0.03429	0.00127	0.57460	0.01019	17.45326	0.33209	98	2927	42	2983	8
u720C	10-1	247	277	1.12	0.51	0.22282	0.00122	0.18468	0.00214	0.43410	0.00750	13.33672	0.25015	77	2324	34	3001	9
u720C	7-1	120	60	0.50	0.04	0.22380	0.00130	0.13705	0.00186	0.65895	0.01229	20.33320	0.41096	108	3263	48	3008	9
u720c	26-1	15	7	0.48	0.21	0.22581	0.00536	0.13589	0.00988	0.60236	0.01910	18.75440	0.78597	101	3039	77	3023	38
u720c	55-1	72	34	0.47	0.24	0.22659	0.00202	0.12547	0.00322	0.59146	0.01205	18.47860	0.42940	99	2995	49	3028	14
MB, Sandstone, Merougil Formation base, Lake Lefroy																		
u712B	48-1	111	154	1.39	5.53	0.16129	0.00499	0.10762	0.01097	0.32866	0.00647	7.30880	0.28275	74	1832	31	2469	52
u712B	11-1	123	121	0.98	2.14	0.17474	0.00344	0.21448	0.00746	0.31773	0.00616	7.65534	0.22429	68	1779	30	2604	33
u712B	28-1	285	303	1.06	6.07	0.17564	0.00381	0.16980	0.00842	0.22485	0.00396	5.44503	0.16108	50	1307	21	2612	36
u712B	14-1	177	363	2.05	6.64	0.17599	0.00505	0.21923	0.01128	0.23854	0.00447	5.78833	0.20933	53	1379	23	2615	48
u712B	50-1	155	177	1.14	0.96	0.17691	0.00173	0.32907	0.00391	0.43372	0.00784	10.57915	0.22856	89	2322	35	2624	16
u712B	49-1	116	76	0.66	1.39	0.17710	0.00239	0.16228	0.00494	0.42021	0.00786	10.26064	0.25030	86	2261	36	2626	22
u712B	31-1	89	77	0.87	0.43	0.17752	0.00111	0.19555	0.00363	0.48355	0.00943	11.83585	0.27576	97	2543	41	2630	22
u712B	8-1	177	183	1.04	3.39	0.17793	0.00325	0.19291	0.00710	0.31750	0.00571	7.78931	0.21164	67	1778	28	2634	18
u712B	41-1	97	80	0.82	0.59	0.17800	0.00212	0.16675	0.00409	0.39887	0.00766	9.78912	0.23313	82	2164	35	2634	30
u712B	46-1	128	159	1.24	0.70	0.17825	0.00179	0.32523	0.00401	0.51389	0.00947	12.63013	0.27859	101	2673	41	2637	20
u712B	7-1	137	152	1.11	0.64	0.17859	0.00176	0.26439	0.00371	0.39881	0.00729	9.82041	0.21407	82	2164	34	2640	17
u712B	13-1	98	80	0.82	0.44	0.17859	0.00198	0.21525	0.00397	0.47426	0.00913	11.67800	0.27329	95	2502	40	2640	16
u712B	46-2	174	236	1.36	0.67	0.17861	0.00155	0.36394	0.00357	0.48016	0.00854	11.82460	0.24510	96	2528	37	2640	18
u712B	65-1	105	97	0.92	0.27	0.17902	0.00150	0.24784	0.00298	0.50402	0.00954	12.44077	0.26933	100	2631	41	2644	14
u712B	59-1	101	64	0.64	0.85	0.17910	0.00210	0.16709	0.00418	0.51210	0.00973	12.64568	0.29778	101	2666	41	2645	19
u712B	37-1	210	182	0.86	5.95	0.17939	0.00422	0.18219	0.00934	0.23373	0.00423	5.78102	0.18176	51	1354	22	2647	39
u712B	58-1	183	191	1.04	2.76	0.17953	0.00292	0.13020	0.00620	0.31207	0.00570	7.72493	0.20036	66	1751	28	2649	27
u712B	34-1	137	164	1.20	0.12	0.17998	0.00122	0.31329	0.00261	0.48917	0.00892	12.13885	0.24540	97	2567	39	2653	11
u712B	36-1	75	50	0.67	0.92	0.18005	0.00244	0.17028	0.00490	0.49721	0.00998	12.34320	0.31573	98	2602	43	2653	22
u712B	1-1	156	166	1.07	0.10	0.18027	0.00101	0.27970	0.00209	0.45052	0.00511	11.19785	0.14856	90	2398	23	2655	9
u712B	45-1	169	177	1.05	1.07	0.18021	0.00170	0.28571	0.00372	0.47544	0.00847	11.81373	0.25016	94	2507	37	2655	16
u712B	39-1	179	118	0.66	2.43	0.18038	0.00264	0.19013	0.00568	0.32591	0.00585	8.10573	0.19892	68	1819	28	2656	24
u712B	51-1	198	374	1.89	5.77	0.18036	0.00375	0.43850	0.00876	0.36159	0.00653	8.99229	0.26255	75	1990	31	2656	34
u712B	2-1	117	149	1.27	0.05	0.18051	0.00107	0.32920	0.00234	0.50347	0.00610	12.53078	0.17708	99	2629	26	2658	10
u712B	17-1	73	52	0.71	0.18	0.18057	0.00179	0.18698	0.00320	0.49795	0.01002	12.39753	0.29137	98	2605	43	2658	16
u712B	4-1	127	96	0.75	0.00	0.18066	0.00104	0.20187	0.00182	0.49563	0.00594	12.34546	0.17200	98	2595	26	2659	10
u712B	15-2	132	119	0.90	0.94	0.18080	0.00191	0.23238	0.00402	0.45112	0.00829	11.24560	0.25096	90	2400	37	2660	18
u712B	32-1	122	137	1.13	0.20	0.18073	0.00155	0.23479	0.00303	0.42228	0.00781	10.52301	0.22440	85	2271	35	2660	14
u712B	55-1	220	198	0.90	3.45	0.18077	0.00309	0.20605	0.00677	0.30156	0.00536	7.51636	0.19664	64	1699	27	2660	28
u712B	20-1	132	148	1.12	0.59	0.18103	0.00167	0.26533	0.00351	0.44954	0.00827	11.22072	0.24205	90	2393	37	2662	15
u712B	64-1	176	176	1.00	1.29	0.18097	0.00189	0.26802	0.00410	0.41859	0.00745	10.44441	0.22693	85	2254	34	2662	17
u712B	27-1	100	81	0.81	0.57	0.18122	0.00191	0.18185	0.00374	0.46315	0.00887	11.57274	0.26609	92	2453	39	2664	17
u712B	22-1	148	185	1.25	3.11	0.18152	0.00354	0.26899	0.00786	0.30447	0.00566	7.62052	0.21773	64	1713	28	2667	32
u712B	26-1	214	130	0.61	1.61	0.18156	0.00185	0.09828	0.00373	0.42350	0.00745	10.60177	0.22701	85	2276	34	2667	17
u712B	30-1	71	85	1.19	0.02	0.18149	0.00194	0.32393	0.00427	0.51364	0.01042	12.85312	0.30954	100	2672	44	2667	18
u712B	43-1	80	70	0.87	0.06	0.18158	0.00177	0.23311	0.00346	0.49054	0.00975	12.28119	0.28510	96	2573	42	2667	16
u712B	16-1	177	143	0.81	2.05	0.18162	0.00239	0.15581	0.00502	0.37104	0.00669	9.29171	0.21946	76	2034	31	2668	22
u712B	62-1	126	131	1.04	0.20	0.18165	0.00145	0.28018	0.00300	0.50560	0.00933	12.66293	0.26572	99	2638	40	2668	13
u712B	5-1	64	56	0.87	0.03	0.18174	0.00152	0.20817	0.00271	0.48429	0.00696	12.13543	0.21267	95	2546	30	2669	14
u712B	56-1	86	70	0.81	0.69	0.18203	0.00214	0.21331	0.00434	0.48639	0.00956	12.20738	0.29438	96	2555	41	2671	19
u712B	3-1	111	84	0.75	0.00	0.18214	0.00108	0.20598	0.00187	0.50771	0.00627	12.75002	0.18295	99	2647	27	2672	10
u712B	9-1	122	116	0.95	0.95	0.18208	0.00208	0.21116	0.00432	0.40783	0.00755	10.23832	0.23504	83	2205	35	2672	19
u712B	17-2	73	71	0.98	0.19	0.18222	0.00199	0.23003	0.00387	0.46952	0.00940	11.79638	0.28263	93	2481	41	2673	18
u712B	52-1	125	129	1.03	0.10	0.18216	0.00140	0.28358	0.00289	0.50320	0.00928	12.63837	0.26342	98	2628	40	2673	13
u712B	15-1	125	99	0.79	0.08	0.18228	0.00139	0.21427	0.00260	0.51263	0.00941	12.88399	0.26716	100	2668	40	2674	13
u712B	35-1	83	81	0.98	1.77	0.18229	0.00306	0.23238	0.00659	0.43215	0.00857	10.86180	0.29877	87	2315	39	2674	28

Table 1A (*continued*).

Mount	Analysis	U/ppm	Th/ppm	Th/U	f206%	207/206	±	208/206	±	206/238	±	207/235	±	%conc	Age (206/238)	±	Age (207/206)	±
u712B	44-1	180	148	0.82	2.89	0.18233	0.00267	0.16732	0.00573	0.38851	0.00691	9.76701	0.23848	79	2116	32	2674	24
u712B	18-1	125	54	0.43	0.01	0.18245	0.00140	0.12020	0.00227	0.48269	0.00892	12.14266	0.25338	95	2539	39	2675	13
u712B	10-1	307	310	1.01	8.59	0.18265	0.00461	0.22260	0.01032	0.17153	0.00303	4.31977	0.14074	38	1021	17	2677	42
u712B	37-1	86	64	0.75	0.71	0.18263	0.00234	0.17349	0.00465	0.46408	0.00907	11.68634	0.28840	92	2458	40	2677	21
u712B	54-1	457	368	0.81	8.98	0.18265	0.00481	0.33268	0.01095	0.11523	0.00203	2.90194	0.09705	26	703	12	2677	44
u712B	6-1	138	93	0.67	0.62	0.18271	0.00141	0.16703	0.00270	0.43532	0.00522	10.96677	0.16511	87	2330	23	2678	13
u712B	19-1	113	54	0.48	0.25	0.18277	0.00153	0.12815	0.00260	0.47311	0.00888	11.92262	0.25636	93	2497	39	2678	14
u712B	33-1	159	235	1.48	3.62	0.18271	0.00326	0.27486	0.00727	0.34688	0.00634	8.73873	0.23662	72	1920	30	2678	30
u712B	61-1	63	50	0.78	0.10	0.18275	0.00224	0.21921	0.00443	0.49996	0.01043	12.59803	0.32104	98	2614	45	2678	20
u712B	53-1	102	57	0.56	3.00	0.18290	0.00387	0.11588	0.00822	0.40807	0.00803	10.29094	0.31521	82	2206	37	2679	35
u712B	60-1	178	113	0.64	2.97	0.18301	0.00301	0.18362	0.00650	0.32316	0.00584	8.15458	0.21121	67	1805	28	2680	27
u712B	24-1	127	114	0.90	1.18	0.18324	0.00243	0.24655	0.00519	0.39190	0.00729	9.90149	0.23906	79	2132	34	2682	22
u712B	66-1	207	227	1.10	3.63	0.18323	0.00304	0.17654	0.00660	0.27960	0.00500	7.06360	0.18279	59	1589	25	2682	27
u712B	29-1	128	64	0.50	1.30	0.18326	0.00211	0.12384	0.00420	0.44458	0.00823	11.23357	0.25846	88	2371	37	2683	19
u712B	42-1	150	155	1.03	1.14	0.18341	0.00193	0.29325	0.00422	0.44184	0.00801	11.17326	0.24661	88	2359	36	2684	17
u712B	63-1	131	131	1.00	0.13	0.18341	0.00137	0.27492	0.00277	0.50151	0.00922	12.68284	0.26223	98	2620	40	2684	12
u712B	25-1	186	177	0.95	2.74	0.18362	0.00265	0.15837	0.00609	0.34191	0.00609	8.65629	0.21024	71	1896	29	2686	24
u712B	57-1	164	232	1.41	2.01	0.18414	0.00266	0.20899	0.00570	0.33252	0.00602	8.44244	0.20698	69	1851	29	2691	24
u712B	12-1	156	63	0.40	0.36	0.18494	0.00155	0.05999	0.00262	0.44333	0.00798	11.30484	0.23477	88	2366	36	2698	14
u712B	23-1	134	99	0.74	0.20	0.18529	0.00146	0.19302	0.00262	0.45886	0.00842	11.72276	0.24450	90	2435	37	2701	13
u712B	47-1	207	208	1.00	6.01	0.18600	0.00391	0.20306	0.00865	0.32186	0.00576	8.25431	0.24135	66	1799	28	2707	35
u712B	40-1	131	147	1.12	2.21	0.18682	0.00281	0.26458	0.00617	0.41078	0.00763	10.58127	0.26789	82	2218	35	2714	25
u712B	21-1	778	1036	1.33	11.08	0.18761	0.00564	0.22997	0.01264	0.07002	0.00122	1.81113	0.06627	16	436	7	2721	50
MQ, Sandstone, Merougil Formation, Speedway Quarry, Kambalda																		
u711B	7-1	2778	1885	0.68	0.34	0.13308	0.00021	0.17845	0.00044	0.41044	0.00503	7.53133	0.09452	104	2217	23	2139	3
u711B	37-1	1215	912	0.75	0.03	0.14510	0.00028	0.20454	0.00055	0.39271	0.00488	7.85653	0.10063	93	2135	23	2289	3
u711B	51-1	1669	734	0.44	0.05	0.14622	0.00027	0.11790	0.00041	0.33704	0.00416	6.79516	0.08641	81	1872	20	2302	3
u711B	42-1	559	453	0.81	0.07	0.14822	0.00040	0.21850	0.00080	0.41468	0.00527	8.47426	0.11298	96	2236	24	2325	5
u711B	3-1	931	561	0.60	0.12	0.15299	0.00032	0.13479	0.00054	0.42988	0.00535	9.06809	0.11688	97	2305	24	2380	4
u711B	22-1	238	67	0.28	0.57	0.15705	0.00107	0.07265	0.00192	0.40819	0.00560	8.83913	0.14183	91	2207	26	2424	12
u711B	4-1	999	991	0.99	0.23	0.15760	0.00035	0.26521	0.00076	0.43958	0.00548	9.55229	0.12368	97	2349	25	2430	4
u711B	30-1	487	325	0.67	0.45	0.16179	0.00047	0.17789	0.00073	0.44082	0.00550	9.83388	0.13014	95	2354	25	2474	4
u711B	5-1	748	454	0.61	0.02	0.16179	0.00047	0.17789	0.00073	0.44082	0.00550	9.83388	0.13014	95	2354	25	2474	4
u711B	12-1	182	85	0.47	0.30	0.17823	0.00090	0.13244	0.00154	0.47909	0.00648	11.77364	0.17700	96	2523	28	2637	8
u711B	28-1	1110	990	0.89	0.09	0.16206	0.00032	0.24506	0.00064	0.48602	0.00624	11.94978	0.16080	97	2553	27	2637	5
u711B	35-1	441	179	0.41	0.02	0.17832	0.00049	0.11284	0.00065	0.48602	0.00624	11.94978	0.16080	97	2553	27	2637	5
u711B	23-1	528	221	0.42	0.07	0.16890	0.00046	0.11417	0.00142	0.43715	0.00554	10.18030	0.13519	92	2338	25	2547	5
u711B	29-1	218	102	0.47	0.07	0.16932	0.00089	0.12950	0.00142	0.45507	0.00691	10.62368	0.16256	95	2418	28	2551	9
u711B	3-1	486	305	0.63	0.00	0.17283	0.00048	0.17860	0.00078	0.46168	0.00591	11.00155	0.14773	95	2447	26	2585	5
u711B	22-1	600	451	0.75	0.01	0.17324	0.00042	0.20317	0.00073	0.47138	0.00597	11.25976	0.14863	96	2490	26	2589	4
u711B	19-1	401	45	0.11	0.59	0.17529	0.00079	0.03035	0.00107	0.41646	0.00541	10.06534	0.14365	86	2244	25	2609	8
u711B	30-1	487	325	0.67	0.02	0.17711	0.00046	0.17789	0.00073	0.48694	0.00621	11.89101	0.15859	97	2557	27	2626	4
u711B	12-1	182	85	0.47	0.30	0.17823	0.00090	0.13244	0.00154	0.47909	0.00648	11.77364	0.17700	96	2523	28	2637	8
u711B	35-1	441	179	0.41	0.02	0.17832	0.00049	0.11284	0.00066	0.48602	0.00624	11.94978	0.16080	97	2553	27	2637	5
u711B	62-1	64	25	0.40	0.00	0.17910	0.00124	0.10852	0.00142	0.50418	0.00828	12.45049	0.23176	100	2632	36	2645	12
u711B	33-1	240	201	0.84	0.05	0.17938	0.00067	0.22396	0.00124	0.46880	0.00620	11.59474	0.16459	94	2478	27	2647	6
u711B	6-1	208	81	0.39	0.02	0.17959	0.00073	0.10154	0.00095	0.53304	0.00719	13.19870	0.19234	104	2754	30	2649	7
u711B	32-1	107	83	0.78	0.03	0.17963	0.00107	0.21225	0.00195	0.50491	0.00741	12.50495	0.20647	99	2635	32	2649	10
u711B	16-1	196	122	0.62	0.35	0.17979	0.00094	0.17590	0.00175	0.48155	0.00653	11.93698	0.18047	96	2534	28	2651	9
u711B	38-1	75	36	0.48	0.05	0.18008	0.00133	0.13344	0.00209	0.45664	0.00722	11.33775	0.20735	91	2425	32	2654	12
u711B	44-1	340	105	0.31	0.03	0.18031	0.00063	0.08378	0.00073	0.48102	0.00629	11.95883	0.16669	95	2532	27	2656	6
u711B	1-1	84	85	1.01	0.10	0.18048	0.00135	0.27098	0.00277	0.48927	0.00754	12.17521	0.21844	97	2568	33	2657	12
u711B	59-1	180	125	0.70	0.03	0.18044	0.00085	0.19362	0.00146	0.46954	0.00644	11.68147	0.17573	93	2482	28	2657	8
u711B	65-1	168	92	0.55	0.03	0.18046	0.00090	0.14835	0.00140	0.49754	0.00690	12.37979	0.18964	98	2603	30	2657	8
u711B	9-1	235	282	1.20	0.80	0.18059	0.00134	0.13600	0.00260	0.32333	0.00442	8.05084	0.13148	68	1806	22	2658	12

u711B	58-1	170	46	0.27	0.07	0.18080	0.00087	0.07389	0.00111	0.48234	0.00667	12.02442	0.18266	95	2537	29	2660	8
u711B	31-1	238	141	0.59	0.02	0.18097	0.00067	0.16211	0.00102	0.49586	0.00660	12.37262	0.17627	98	2596	28	2662	6
u711B	11-1	514	229	0.45	0.29	0.18119	0.00057	0.12283	0.00094	0.48119	0.00613	12.02131	0.16206	95	2532	27	2664	5
u711B	56-1	116	32	0.28	0.01	0.18120	0.00107	0.07400	0.00147	0.49994	0.00736	12.49024	0.20644	98	2614	32	2664	10
u711B	61-1	155	70	0.45	0.00	0.18125	0.00083	0.12043	0.00100	0.49675	0.00695	12.41417	0.18936	98	2600	30	2664	8
u711B	8-1	275	195	0.71	0.00	0.18142	0.00063	0.19822	0.00101	0.50071	0.00662	12.52496	0.17628	98	2617	28	2666	6
u711B	25-1	210	73	0.35	0.00	0.18138	0.00068	0.09057	0.00070	0.51253	0.00690	12.81754	0.18481	100	2667	29	2666	6
u711B	26-1	165	116	0.70	0.00	0.18169	0.00078	0.19458	0.00068	0.49689	0.00688	12.44819	0.18682	97	2600	30	2668	7
u711B	46-1	340	91	0.27	0.00	0.18165	0.00054	0.07397	0.00050	0.49350	0.00641	12.35988	0.16909	97	2586	28	2668	5
u711B	64-1	156	48	0.31	0.02	0.18168	0.00090	0.08356	0.00110	0.48782	0.00681	12.21991	0.18798	96	2561	30	2668	8
u711B	34-1	338	169	0.50	0.02	0.18173	0.00058	0.13767	0.00085	0.48811	0.00632	12.23022	0.16795	96	2562	27	2669	5
u711B	21-1	222	120	0.54	0.01	0.18200	0.00071	0.14910	0.00105	0.50356	0.00672	12.63677	0.18135	98	2629	29	2671	6
u711B	52-1	114	67	0.59	0.14	0.18198	0.00105	0.15823	0.00168	0.50705	0.00742	12.72229	0.20858	99	2644	32	2671	10
u711B	27-1	243	139	0.57	0.04	0.18210	0.00067	0.15295	0.00099	0.48758	0.00651	12.24195	0.17478	96	2560	28	2672	6
u711B	36-1	236	158	0.67	0.01	0.18207	0.00076	0.18132	0.00112	0.50367	0.00675	12.64412	0.18358	98	2630	29	2672	6
u711B	57-1	62	19	0.30	0.00	0.18208	0.00133	0.08442	0.00132	0.51370	0.00851	12.89660	0.24412	100	2672	36	2672	12
u711B	2-1	234	81	0.35	0.07	0.18223	0.00071	0.09502	0.00092	0.49587	0.00661	12.45900	0.17865	97	2596	28	2673	6
u711B	41-1	346	133	0.38	0.02	0.18218	0.00069	0.11108	0.00096	0.50142	0.00656	12.59479	0.17697	98	2620	28	2673	6
u711B	54-1	117	62	0.53	0.10	0.18222	0.00115	0.14910	0.00187	0.53147	0.00788	13.35287	0.22474	103	2748	33	2673	10
u711B	13-1	361	140	0.39	0.07	0.18234	0.00056	0.10345	0.00075	0.51715	0.00670	13.00185	0.17774	100	2687	28	2674	5
u711B	24-1	243	156	0.64	0.01	0.18233	0.00072	0.18025	0.00114	0.46868	0.00625	11.78230	0.16919	93	2478	27	2674	6
u711B	40-1	289	126	0.44	0.00	0.18230	0.00064	0.12087	0.00077	0.50461	0.00667	12.68338	0.17861	98	2634	29	2674	6
u711B	66-1	559	236	0.42	0.00	0.18247	0.00044	0.11438	0.00051	0.47826	0.00608	12.03242	0.15927	94	2520	27	2675	4
u711B	48-1	287	94	0.33	0.02	0.18271	0.00063	0.08596	0.00076	0.48149	0.00628	12.12941	0.16850	95	2534	27	2678	6
u711B	43-1	272	110	0.40	0.00	0.18218	0.00069	0.11112	0.00092	0.51191	0.00656	12.90984	0.18216	98	2665	29	2679	5
u711B	50-1	125	75	0.61	0.00	0.18281	0.00092	0.16420	0.00131	0.50281	0.00667	12.67360	0.20147	98	2626	31	2679	8
u711B	17-1	203	150	0.74	0.04	0.18312	0.00075	0.19750	0.00075	0.49127	0.00666	12.40362	0.18165	96	2576	27	2681	7
u711B	53-1	162	81	0.50	0.00	0.18314	0.00086	0.13302	0.00086	0.51404	0.00721	12.98047	0.19899	100	2674	31	2682	8
u711B	18-1	207	50	0.24	0.02	0.18356	0.00076	0.06893	0.00092	0.49325	0.00663	12.48388	0.18159	96	2585	29	2685	8
u711B	45-1	158	49	0.31	0.00	0.18360	0.00083	0.08337	0.00081	0.50964	0.00713	12.90127	0.19654	99	2655	30	2686	7
u711B	39-1	92	46	0.51	0.11	0.18379	0.00140	0.11729	0.00230	0.44988	0.00677	11.40023	0.20159	89	2395	30	2687	7
u711B	55-1	45	14	0.31	0.02	0.18400	0.00185	0.08250	0.00275	0.53486	0.00987	13.56925	0.29953	103	2762	41	2689	13
u711B	60-1	119	41	0.34	0.01	0.18397	0.00101	0.09411	0.00128	0.51835	0.00728	13.14847	0.21727	100	2692	33	2689	17
u711B	10-1	324	219	0.68	0.03	0.18412	0.00060	0.18346	0.00095	0.49192	0.00644	12.48838	0.17338	96	2579	28	2690	5
u711B	49-1	82	112	1.37	0.00	0.18469	0.00112	0.37504	0.00260	0.51889	0.00806	13.21344	0.22948	100	2694	34	2695	10
u711B	63-1	144	48	0.33	0.00	0.18494	0.00086	0.09028	0.00088	0.50660	0.00720	12.91787	0.20017	98	2642	31	2698	8
u711B	47-1	44	32	0.73	0.00	0.18656	0.00153	0.20240	0.00242	0.54443	0.00968	14.00394	0.28690	103	2802	40	2712	13
u711B	15-1	13	4	0.32	0.00	0.18794	0.00272	0.08576	0.00265	0.48370	0.01321	12.53399	0.40702	93	2543	57	2724	24

E178, Sandstone bed in conglomerate, Jones Creek Conglomerate, Jones Creek

u711A	32-1	134	108	0.80	1.17	0.17834	0.00139	0.23754	0.00297	0.45463	0.00542	11.17936	0.16802	92	2416	24	2638	13
u711A	35-1	164	191	1.16	6.61	0.17865	0.00301	0.43104	0.00701	0.38134	0.00459	9.39300	0.20545	79	2083	21	2640	28
u711A	43-1	96	105	1.10	0.23	0.17959	0.00114	0.29668	0.00244	0.52119	0.00613	12.90572	0.18971	102	2704	28	2649	11
u711A	14-1	137	130	0.95	0.04	0.17969	0.00091	0.25970	0.00177	0.52156	0.00637	12.92234	0.17809	102	2706	27	2650	8
u711A	52-1	60	82	1.38	0.03	0.17970	0.00122	0.37291	0.00285	0.51527	0.00712	12.76681	0.20594	101	2679	30	2650	11
u711A	46-1	111	137	1.24	0.00	0.17978	0.00080	0.34216	0.00179	0.51549	0.00629	12.77797	0.17265	101	2680	27	2651	7
u711A	6-1	133	73	0.55	0.18	0.17992	0.00094	0.14846	0.00080	0.50421	0.00615	12.50817	0.17306	99	2632	26	2652	7
u711A	10-1	105	135	1.29	0.21	0.17993	0.00117	0.34703	0.00264	0.49709	0.00631	12.33182	0.18445	98	2601	27	2652	9
u711A	47-1	46	52	1.13	0.00	0.17989	0.00128	0.30424	0.00265	0.49846	0.00728	12.36354	0.21043	98	2607	31	2652	11
u711A	49-1	33	4	0.12	0.15	0.18011	0.00454	0.03283	0.00454	0.48574	0.01127	12.06226	0.43805	96	2552	49	2654	42
u711A	50-1	134	50	0.37	1.18	0.18035	0.00143	0.10035	0.00143	0.45272	0.00537	11.25741	0.16975	91	2407	24	2656	13
u711A	17-1	99	119	1.21	0.05	0.18041	0.00112	0.33065	0.00112	0.51548	0.00664	12.82221	0.19192	101	2680	28	2657	8
u711A	20-1	363	309	0.85	0.05	0.18042	0.00052	0.23331	0.00052	0.49710	0.00544	12.36591	0.14409	98	2601	23	2657	10
u711A	16-1	118	143	1.21	1.38	0.18063	0.00165	0.34097	0.00094	0.51284	0.00641	12.77273	0.20898	100	2669	27	2659	5
u711A	53-1	91	84	0.92	0.00	0.18071	0.00091	0.25040	0.00167	0.53086	0.00671	13.22717	0.18756	103	2745	28	2659	15
u711A	27-1	37	16	0.42	0.61	0.18078	0.00128	0.11269	0.00454	0.48882	0.00785	12.18418	0.26772	96	2566	34	2660	8
u711A	29-1	60	67	1.12	2.52	0.18081	0.00237	0.25860	0.00276	0.43727	0.00612	10.90146	0.23966	88	2338	27	2660	22
u711A	8-1	109	56	0.52	0.02	0.18084	0.00102	0.14230	0.00102	0.50929	0.00641	12.69900	0.18315	100	2654	27	2661	9

129

Table 1A (continued).

Mount	Analysis	U/ppm	Th/ppm	Th/U	f206%	207/206	±	208/206	±	206/238	±	207/235	±	%conc	Age (206/238)	±	Age (207/206)	±
u711A	9-1	122	120	0.99	0.00	0.18093	0.00082	0.27038	0.00158	0.50190	0.00619	12.52072	0.17116	99	2622	27	2661	8
u711A	37-1	44	26	0.59	0.12	0.18102	0.00175	0.15883	0.00323	0.51117	0.00756	12.75839	0.23818	100	2662	32	2662	16
u711A	28-1	36	39	1.06	0.48	0.18129	0.00227	0.30239	0.00494	0.49328	0.00790	12.32989	0.26525	97	2585	34	2665	21
u711A	36-1	170	188	1.11	5.71	0.18138	0.00226	0.32773	0.00514	0.43864	0.00511	10.96949	0.19838	88	2345	23	2665	21
u711A	44-1	243	128	0.52	2.37	0.18130	0.00128	0.15294	0.00271	0.49286	0.00550	12.31994	0.17164	97	2583	24	2665	12
u711A	18-1	68	42	0.62	0.00	0.18146	0.00108	0.17188	0.00159	0.51425	0.00710	12.86669	0.20204	100	2675	30	2666	10
u711A	24-1	185	178	0.96	0.04	0.18167	0.00075	0.26403	0.00149	0.51925	0.00596	13.00686	0.16483	101	2696	25	2668	7
u711A	41-1	127	206	1.62	0.27	0.18164	0.00100	0.44819	0.00251	0.49723	0.00596	12.45250	0.17184	98	2602	26	2668	9
u711A	38-1	103	104	1.01	0.00	0.18182	0.00084	0.27655	0.00162	0.50812	0.00625	12.73823	0.17397	99	2649	27	2670	8
u711A	34-1	11	5	0.50	0.00	0.18195	0.00266	0.13836	0.00345	0.56817	0.01322	14.25411	0.41322	109	2900	54	2671	24
u711A	54-1	93	91	0.97	0.00	0.18202	0.00098	0.26384	0.00191	0.53454	0.00670	13.41490	0.19095	103	2761	28	2671	9
u711A	5-1	130	132	1.01	2.92	0.18212	0.00202	0.29596	0.00452	0.47505	0.00587	11.92902	0.20991	94	2506	26	2672	18
u711A	39-1	175	219	1.25	0.00	0.18220	0.00069	0.33951	0.00153	0.51585	0.00589	12.95876	0.16174	100	2682	25	2673	6
u711A	15-1	202	125	0.62	0.71	0.18234	0.00105	0.17374	0.00206	0.51044	0.00593	12.83319	0.17464	99	2659	25	2674	10
u711A	25-1	87	132	1.53	2.90	0.18269	0.00271	0.49694	0.00650	0.43245	0.00569	10.89310	0.22889	87	2317	26	2677	25
u711A	45-1	167	132	0.79	2.59	0.18266	0.00161	0.22168	0.00350	0.47864	0.00557	12.05491	0.18629	94	2521	24	2677	15
u711A	48-1	219	51	0.23	0.71	0.18259	0.00092	0.06602	0.00165	0.51867	0.00583	13.05818	0.16819	101	2694	25	2677	8
u711A	19-1	104	69	0.66	0.32	0.18280	0.00115	0.18282	0.00211	0.49448	0.00628	12.46340	0.18509	97	2590	27	2678	10
u711A	2-1	120	76	0.63	6.17	0.18290	0.00339	0.22550	0.00756	0.39304	0.00499	9.91147	0.23543	80	2137	23	2679	31
u711A	11-1	246	142	0.58	9.89	0.18340	0.00275	0.19014	0.00614	0.46198	0.00537	11.68228	0.23486	91	2448	24	2684	25
u711A	1-1	62	20	0.32	0.04	0.18357	0.00142	0.08663	0.00215	0.51639	0.00732	13.07040	0.22193	100	2684	31	2685	13
u711A	7-1	43	25	0.59	0.10	0.18356	0.00188	0.15584	0.00338	0.52992	0.00825	13.41192	0.26386	102	2741	35	2685	17
u711A	31-1	95	112	1.18	0.31	0.18365	0.00127	0.32892	0.00282	0.48893	0.00618	12.38045	0.18758	96	2566	27	2686	11
u711A	26-1	132	121	0.91	0.02	0.18428	0.00089	0.24849	0.00169	0.51124	0.00618	12.98999	0.17624	99	2662	26	2692	8
u711A	23-1	170	94	0.55	26.40	0.18493	0.00639	0.25967	0.01448	0.28793	0.00383	7.34151	0.28298	60	1631	19	2698	57
u711A	12-1	211	60	0.28	9.06	0.18522	0.00283	0.10885	0.00623	0.45603	0.00538	11.64603	0.23780	90	2422	24	2700	25
u711A	40-1	50	24	0.48	0.00	0.18637	0.00122	0.12909	0.00149	0.51611	0.00741	13.26207	0.21866	99	2683	31	2710	11
u711A	3-1	245	95	0.39	24.86	0.18853	0.00510	0.21536	0.01150	0.34172	0.00421	8.88256	0.27623	69	1895	20	2729	45
u711A	4-1	23	10	0.42	0.07	0.19711	0.00263	0.11272	0.00452	0.55040	0.01039	14.95836	0.36573	101	2827	43	2802	22
u711A	55-1	114	61	0.54	0.06	0.19758	0.00090	0.14298	0.00131	0.55780	0.00678	15.19616	0.20507	102	2858	28	2806	7
u711A	13-1	238	164	0.69	2.09	0.19761	0.00129	0.19425	0.00272	0.54929	0.00624	14.96640	0.20645	101	2822	26	2807	11
u711A	51-1	85	38	0.45	0.00	0.19810	0.00094	0.11790	0.00150	0.53193	0.00673	14.52890	0.20406	98	2750	28	2811	8
u711A	33-1	191	138	0.72	0.47	0.19848	0.00090	0.20073	0.00169	0.51673	0.00591	14.14085	0.18116	95	2685	25	2814	7
u711A	42-1	148	83	0.56	3.23	0.19847	0.00183	0.16850	0.00390	0.50561	0.00597	13.83639	0.21954	94	2638	26	2814	15
u711A	21-1	166	99	0.60	0.59	0.19902	0.00113	0.17433	0.00212	0.54533	0.00646	14.96393	0.20587	100	2806	27	2818	9
u711A	22-1	131	74	0.56	0.32	0.20167	0.00107	0.15394	0.00187	0.52146	0.00638	14.50005	0.20221	95	2705	27	2840	9
u711A	30-1	150	95	0.64	0.39	0.20246	0.00098	0.18319	0.00176	0.50987	0.00598	14.23294	0.18842	93	2656	26	2846	8

sedimentary recycling of felsic rocks from contemporaneous volcanoplutonic arcs (including older crustal components), and preserved source terranes globally.

Acknowledgements

Funding for research was provided through the Australian Minerals Industry Research Association (AMIRA grant P437 to M.E.B., R.A.C. and B.K.) and an Australian Research Council Collaborative Research Grant (to M.E.B. and R.A.C.). We are grateful to the management and research personnel of industry sponsors (BHP Minerals, Goldfields Exploration, Normandy Exploration, WMC Resources), Paul Sauter and Roger Bateman of Kalgoorlie Consolidated Gold Mines, and the past and present research management of AMIRA (Dave Tucker and Alan Goode) for their support during the project. We would also like to acknowledge Jack Hallberg for his help and enthusiasm throughout the project, and Marian Dahl, Ian Fletcher and Neal McNaughton for their assistance with SHRIMP age dating, particularly during the early stages of the project. B.K. acknowledges Jon Claoué-Long for kindly providing the full list of SHRIMP zircon dates on the Kapai Slate, which stimulated research into the problem of distinguishing provenance ages from post-depositional ages. The manuscript was improved considerably by the reviews of Jim Crowley, Ken Eriksson and Rob Rainbird.

References

Ahmat, A.L., 1993. Mafic/ultramafic rocks of the Gindalbie Terrane: a review of the Bulong and Carr Boyd Complexes. Aust. Geol. Surv. Org. Rec. 1993 (54), 23–27.

Barley, M.E., Eisenlohr, B., Groves, D.I., Perring, C.S., Vearncombe, J.R., 1989. Late Archaean convergent margin tectonics and gold mineralization: a new look at the Norseman-Wiluna Belt. Geology 17, 826–829.

Barley, M.E., Groves, D.I., 1987. Hydrothermal alteration of Archaean supracrustal sequences in the central Norseman-Wiluna Belt, Western Australia: a brief review. In: Geol. Dept. & Univ. Extension, Univ. West. Aust. Publ. 11, 51–66.

Barley, M.E., Krapez, B., Groves, D.I., Kerrich, R., 1998a. The Late Archaean bonanza: metallogenic and environmental consequences of the interaction between mantle plumes, lithospheric tectonics and global cyclicity. Precambrian Res. 91, 65–90.

Barley, M.E., Krapez, B., Brown, S.J.A., Hand, J., Cas, R.A.F., 1998b. Mineralised volcanic and sedimentary successions in the Eastern Goldfields Province, Western Australia. In: Aust. Miner. Ind. Res. Assoc. Final Report, 437.

Bavington, O.A., 1981. The nature of sulfidic metasediments at Kambalda and their broad relationships with associated ultramafic rocks and nickel ores. Econ. Geol. 76, 1606–1628.

Binns, R.A., Gunthorpe, R.J., Groves, D.I., 1976. Metamorphic patterns and development of greenstone belts in the eastern Yilgarn Block. In: Windley, B.F. (Ed.), The Early History of the Earth. Wiley, London, pp. 303–313.

Busby, C., Smith, D., Morris, W., Fackler-Adams, B., 1998. Evolutionary model for convergent margins facing large ocean basins: Mesozoic Baja California, Mexico. Geology 26, 227–230.

Campbell, I.H., Hill, R.I., 1988. A two stage model for the formation of the granite–greenstone terrains of the Kalgoorlie-Norseman area, Western Australia. Earth Planet. Sci. Lett. 90, 11–25.

Claoué-Long, J.C., Compston, W., Cowden, A., 1988. The age of the Kambalda greenstones resolved by ion-microprobe: implications for Archaean dating methods. Earth Planet. Sci. Lett. 89, 239–259.

Curray, J.R., Moore, D.G., Lawver, L.A., Emmel, F.J., Raitt, R.W., Henry, M., Kieckhefer, R., 1978. Tectonics of the Andaman Sea and Burma. Am. Assoc. Pet. Geol. Mem. 29, 189–198.

Dickinson, W.R., Suczek, C.A., 1979. Plate tectonics and sandstone compositions. Am. Assoc. Pet. Geol. Bull. 63, 2164–2182.

Folk, R.L., 1968. Petrology of Sedimentary Rocks. Hemphills, Austin, TX.

Foster, J.G., Lambert, D.D., Frick, L.R., Maas, R., 1996. Re–Os isotopic evidence for genesis of Archaean nickel ores from uncontaminated komatiites. Nature 382, 703–706.

Froude, D.O., Ireland, T.R., Kinny, P.D., Wiiliams, I.S., Compston, W., Williams, I.R., Myers, J.S., 1983. Ion microprobe identification of 4,100–4,200 Myr-old terrestrial zircons. Nature 304, 616–618.

Hallberg, J.A., 1970. The petrology and geochemistry of metamorphosed basic volcanic and related rocks between Coolgardie and Norseman, Western Australia. Ph.D. thesis, Univ. West. Aust., Nedlands.

Hallberg, J.A., 1985. Geology and Mineral Deposits of the Leonora-Laverton Area Northeastern Yilgarn Block Western Australia. Hesperian Press, Carlisle.

Hammond, R.L., Nisbet, B.W., 1992. Towards a structural and tectonic framework for the central Norseman-Wiluna greenstone belt, Western Australia. In: Geol. Dept. & Univ. Extension, Univ. West. Aust. Publ. 22, 39–49.

Hand, J., 1998. The sedimentological and stratigraphic evolution of the Archaean Black Flag Beds, Kalgoorlie, Western Australia: implications for regional stratigraphy and basin setting within the Kalgoorlie Terrane. Ph.D. thesis (unpublished), Monash University, Melbourne.

Hartmann, L.A., Takehara, L., Leite, J.A.D., McNaughton, N.J., Vasconcellos, M.A.Z., 1997. Fracture sealing in zircon as evaluated by electron microprobe analysis and back-scattered electron imaging. Chem. Geol. 141, 67–72.

Hill, R.I., Campbell, I.H., Chappell, B.W., 1992. Crustal growth, crustal reworking, and granite genesis in the southeastern Yilgarn Block, Western Australia. In: Geol. Dept. & Univ. Extension, Univ. West. Aust. Publ. 22, 203–212.

Ingersoll, R.V., Bullard, T.F., Ford, R.L., Grimm, J.P., Pickle, J.D., Sares, S.W., 1984. The effect of grain size on detrital modes: a test of the Gazzi–Dickinson point-counting method. J. Sediment. Petrol. 54, 103–116.

Ingersoll, R.V., Graham, S.A., Dickinson, W.R., 1995. Remnant ocean basins. In: Busby, C.J., Ingersoll, R.V. (Eds.), Tectonics of Sedimentary Basins. Blackwell Science, Oxford, pp. 363–391.

Kent, A.J.R., Hagemann, S.G., 1996. Constraints on the timing of lode-gold mineralisation in the Wiluna greenstone belt, Yilgarn Craton, Western Australia. Aust. J. Earth Sci. 443, 573–588.

Kinny, P.D., Wijbrans, J.R., Froude, D.O., Williams, I.S., Compston, W., 1990. Age constraints on the geological evolution of the Narryer Gneiss Complex, Western Australia. Aust. J. Earth Sci. 37, 51–69.

Krapez, B., 1996. Sequence-stratigraphic concepts applied to the identification of basin-filling rhythms in Precambrian successions. Aust. J. Earth Sci. 43, 355–380.

Krapez, B., 1997. Sequence-stratigraphic concepts applied to the identification of depositional basins and global tectonic cycles. Aust. J. Earth Sci. 44, 1–36.

Krapez, B., Brown, S., Hand, J., 1997. Stratigraphic signatures of depositional basins in Archaean volcanosedimentary successions of the Eastern Goldfields Province. Aust. Geol. Surv. Org. Rec. 1997 (41), 33–38.

Krynine, P.D., 1942. Differential sedimentation and its products during one complete geosynclinal cycle. Ann. Prim. Congr. Panamericano de Ingen. de Minas Geolog. (Chile), Geologia 1st Pt 2, 537–561.

Marston, R.J., Travis, G.A., 1976. Stratigraphic implications of heterogeneous deformation in the Jones Creek Conglomerate (Archaean), Kathleen Valley, Western Australia. Geol. Soc. Aust. J. 23, 141–156.

Martin, D.McB., Li, Z.X., Nemchin, A.A., Powell, C.McA., 1998. A pre-2.2. Ga age for giant hematite ores of the Hamersley Province, Australia. Econ. Geol. 93, 1084–1090.

Morris, P.A., Witt, W.K., 1997. Geochemistry and tectonic setting of two contrasting Archaean felsic volcanic associations in the Eastern Goldfields, Western Australia. Precambrian Res. 83, 83–107.

Myers, J.S., 1997. Preface: Archaean geology of the Eastern Goldfields of Western Australia – regional overview. Precambrian Res. 83, 1–10.

Nance, W.B., Taylor, S.R., 1977. Rare earth element patterns and crustal evolution. II. Archean sedimentary rocks from Kalgoorlie Australia. Geochim. Cosmochim. Acta 41, 225–231.

Nelson, D.R., 1995. Compilation of SHRIMP U–Pb zircon geochronology data, 1994. In: Geol. Surv. West. Aust. Rec. 1995/3.

Nelson, D.R., 1996. Compilation of SHRIMP U–Pb zircon geochronology data, 1995. In: Geol. Surv. West. Aust. Rec. 1996/5.

Nelson, D.R., 1997. Evolution of the Archaean granite–greenstone terranes of the Eastern Goldfields, Western Australia: SHRIMP U–Pb zircon constraints. Precambrian Res. 83, 57–81.

Nilsen, T.H., McLaughlin, R.J., 1985. Comparison of tectonic framework and depositional patterns of the Hornelen strike-slip basin of Norway and the Ridge and Little Sulphur Creek strike-slip basins of California. Soc. Econ. Paleontol. Mineral. Spec. Publ. 37, 79–103.

Pettijohn, F.J., Potter, P.E., Siever, R., 1972. Sand and Sandstone. Springer, New York.

Pidgeon, R.T., 1986. The correlation of acid volcanics in the Archean of Western Australia. In: West. Aust. Miner. Pet. Inst. Rep. 27.

Pidgeon, R.T., 1992. Recrystallization of oscillatory zoned zircon: some geochronological and petrological implications. Contrib. Mineral. Petrol. 110, 463–472.

Pidgeon, R.T., Hallberg, J.A., 2000. Age relationships in supracrustal sequences of the northern part of the Murchison Terrane, Archaean Yilgarn Craton, Western Australia: a combined field and zircon U–Pb study. Aust. J. Earth Sci. 47, 153–168.

Pidgeon, R.T., Wilde, S.A., 1990. The distribution of 3.0 Ga and 2.7 Ga volcanic episodes in the Yilgarn Craton of Western Australia. Precambrian Res. 48, 309–325.

Savage, M.D., Barley, M.E., McNaughton, N.J., 1996. SHRIMP U–Pb dating of 2.95 to 3.0 Ga intermediate to silicic rocks in the southern Yellowdine Terrane, Western Australia. Geol. Soc. Aust. Abstr. 41, 376.

Smith, J.B., Barley, M.E., Groves, D.I., Krapez, B., McNaughton, N.J., Bickle, M.J., Chapman, H.J., 1998. The Sholl Shear Zone, West Pilbara: evidence for a domain boundary structure from integrated tectonostratigraphic analyses, SHRIMP U–Pb dating and isotopic and geochemical data of granitoids. Precambrian Res. 88, 143–171.

Smithies, R.H., Hickman, A.H., Nelson, D.R., 1999. New constraints on the evolution of the Mallina Basin, and their bearing on relationships between the constrasting eastern and western granite–greenstone terranes of the Archaean Pilbara Craton, Western Australia. Precambrian Res. 94, 11–28.

Swager, C.P., 1997. Tectono-stratigraphy of late Archaean greenstone terranes in the southern Eastern Goldfields, Western Australia. Precambrian Res. 83, 11–42.

Swager, C.P., Witt, W.K., Griffin, T.J., Ahmat, A.L., Hunter, W.M., McGoldrick, P.J., Wyche, S., 1992. Late Archaean granite–greenstones of the Kalgoorlie Terrane, Yilgarn Craton, Western Australia. In: Geol. Dept. & Univ. Extension, Univ. West. Aust. Publ. 22, 107–122.

Taylor, S.R., McLennan, S.M., 1985. The Continental Crust: its Composition and Evolution. Blackwell Science, Oxford.

Valloni, R., Maynard, J.B., 1981. Compositional suites of terrigenous deep-sea sands of the present continental margins. Sedimentology 31, 353–364.

Wayne, D.M., Sinha, A.K., Hewitt, D.A., 1992. Differential response of zircon U–Pb isotopic systematics to metamorphism across a lithologic boundary: an example from the Hope Valley Shear Zone, southeastern Massachusetts, USA. Contrib. Mineral. Petrol. 109, 408–420.

TECTONOPHYSICS

www.elsevier.com/locate/tecto

Nd isotopic evidence for Early to Late Archean (3.4–2.7 Ga) crustal growth in the Western Superior Province (Ontario, Canada)

Philippe Henry *, Ross K. Stevenson, Youcef Larbi, Clément Gariépy

GEOTOP — Dépt des Sciences de la Terre — UQAM, CP 8888, Montréal (PQ), Canada H3C 3P8

Received 22 January 1999; accepted for publication 16 August 1999

Abstract

New Sm–Nd analyses of Archean metavolcanic, plutonic, gneissic and metasedimentary rocks from Sachigo and Berens River subprovinces (North Caribou terrane) and Winnipeg River and Wabigoon subprovinces in conjunction with previously published data are used to trace the crustal evolution of the Western Superior Province (WSP) from ca. 3.4 to 2.7 Ga. The extensive WSP geochronological database, which contains 3.17–2.55 Ga plutonic rocks, allows the calculation of precise $\epsilon_{Nd}{}^t$ values and the identification of the main periods of new crustal addition at 3.4 ± 0.1, 3.02–2.92 and 2.76–2.69 Ga. Mass-balance calculations indicate that the Western Superior Province represents a collage of Archean crusts comprising 7%, 44% and 48 wt% of crust created at 3.4, 3.0 and 2.7 Ga, respectively. Although these calculations depend on the isotopic signatures of the mantle and those of the average crust at different times, several important observations can be made: (1) the importance of pre-2.7 Ga terranes recycled in the 2.7 Ga orogen; (2) the existence of time intervals of 150–300 Ma dominated by intra-crustal processes resulting in a large range of crystallization ages (U–Pb zircon ages) that do not reflect new additions from the mantle; (3) much of the new crustal additions at 2.7 Ga were contaminated by crustal material subducted into the source region (TTG suites) or were contaminated during intrusion into the crust (sanukitoids). © 2000 Elsevier Science B.V. All rights reserved.

Keywords: Archean; crust; mantle; neodymium; recycling

1. Introduction

The Archean was a major period of continental crust formation during which very different geological assemblages were assembled and stabilized to form large cratonic areas. In the Canadian Shield, this is exemplified by the formation of the Superior and the Slave Provinces that now lie at the core of the North American continent.

The Western Superior Province (WSP, Fig. 1) has been divided into a series of elongated, broadly east–west-trending subprovinces (Sp), based upon similarities in lithologic assemblages, structural traits and metamorphic grades (e.g. Card and Ciesielski, 1986). The alternating pattern of the subprovinces and their characteristic elongated nature suggest that the craton was formed by accretion of crustal segments in much the same way as Phanerozoic orogens formed (e.g. Langford and Morin, 1975; Blackburn et al., 1985; Corfu

---

* Corresponding author. Present address: Dépt. des Géosciences, UFC-UFR Sciences, 16 route de Gray, 25030 Besançon Cedex, France.
Tel.: +33-3-81-66-65-55; fax: +33-3-81-66-65-58.
E-mail address: philippe.henry@univ-fcomte.fr (P. Henry)

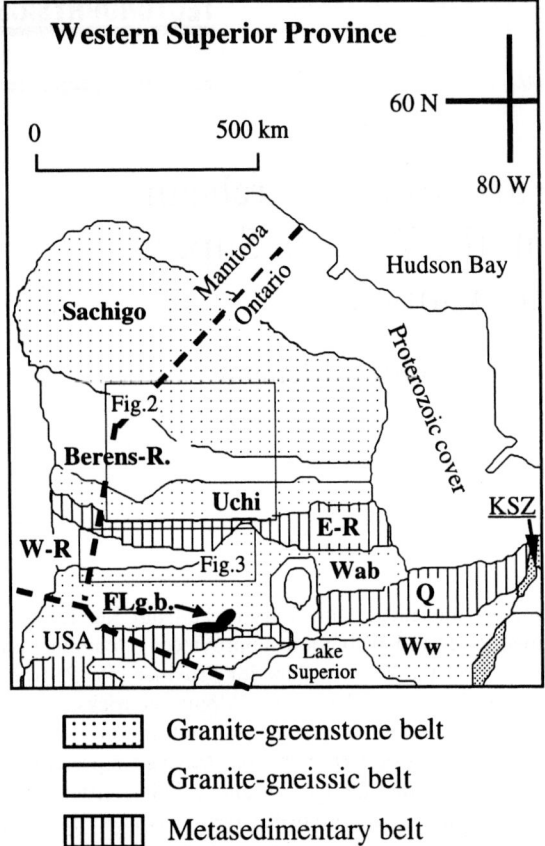

Fig. 1. Simplified map of the Western Superior Province (Stevenson, 1995 after Card and Ciesielski, 1986) showing the various subprovinces: Sachigo, Berens River, Uchi, English River (E-R), Winnipeg River (W-R), Wabigoon (Wab), Quetico (Q) and Wawa (Ww) Sp. The map also shows the location of the Finlayson Lake greenstone belt (FL g.b.) where 3 Ga metasedimentary rocks have been sampled (I–IV, Table 1) and the positions of Figs. 2 and 3. KSZ designs the Kapuskasing Structural Zone.

chronology of geological events, and permits us to calculate precise initial Nd isotopic compositions (ϵ^t_{Nd} values) and to monitor the evolution of these compositions as a function of the ages of crystallization or deposition.

In this study, we attempt to identify the reservoirs involved in the formation of the WSP based on 210 whole rock Sm–Nd analyses of mafic to felsic metavolcanic rocks, plutonic, gneissic and metasedimentary rocks from all subprovinces in the WSP (Noble, 1989; Stevenson, 1995; Henry et al., 1998; Larbi et al., 1999 and this study). This data set reveals the existence of three major periods of crustal growth from a homogeneous Archean depleted mantle at 3.4 ± 0.1, 3.02–2.92 and 2.76–2.69 Ga, and leads to an estimation of the relative abundances of these three crustal end-members in the WSP as a whole and in the individual subprovinces.

2. Results

2.1. Analytical techniques and data treatments

Nd isotope analyses were obtained at the Université du Québec à Montréal following the techniques outlined in Henry et al. (1998). The analyses were performed on a VG-Sector 54 mass spectrometer for which replicate analyses of the La Jolla Nd standard yielded a $^{143}Nd/^{144}Nd$ ratio of 0.511849 ± 12 (2σ on 21 analyses).

The Sm–Nd data presented in the following sections are recalculated as Nd-depleted mantle model ages (also called time of Nd crustal residence or crust separation ages from a depleted mantle) in order to simplify the presentation of the data. To calculate Nd model ages, we need to estimate the Nd isotopic composition of the Archean depleted mantle. Based on previous studies, the Nd isotopic composition of the 2.67–3.02 Ga depleted mantle is now well established at $+3.0\pm0.5$ (e.g. Machado et al., 1986; Shirey and Hanson, 1986; Tilton and Kwon, 1990; Stevenson, 1995; Henry et al., 1998) by data from mafic volcanic rocks sampled in all Sp in the WSP. Furthermore, the evolution in time of the depleted mantle can be deduced from models that are

et al., 1985; Ludden et al., 1986), i.e. in a convergent plate setting with Terrane accretion and concomitant to late-collisional magmatism.

Isotopic studies of the Archean shields provide an opportunity to study the early evolution of the Earth by comparing the evolution of depleted mantle-like sources and enriched, mantle or crustal, reservoirs. In the WSP, which contains rocks dated between 3.17 and 2.55 Ga (e.g. Corfu, 1996; Davis, 1996), the large database of U–Pb ages from zircons provides constraints for the constructed by fitting a line or curve through the isotopic compositions of mantle-derived basalts from different periods through geologic history. Models have been constructed using quadratic (e.g. De Paolo, 1981) and linear (e.g. Jacobsen, 1988) regressions as well as correlations between Hf and Nd isotopic data (Vervoort et al., 1996). The Hf–Nd correlation model closely approximates the Nd isotopic compositions for basaltic samples from this study and our previous work (Henry et al., 1998) and varies from ϵ_{Nd}^t between +1 and +1.5 at 3.6 Ga (Vervoort et al., 1996) to $\epsilon_{Nd}^t = +3.0 \pm 0.5$ at 2.7 Ga, values that agree well with the De Paolo (1981) and Jacobsen (1988) models and which correspond to 0.5131 ($\epsilon_{Nd} = +9$) and 0.2136 for the $^{143}Nd/^{144}Nd$ and $^{147}Sm/^{144}Nd$ ratios, respectively, for the present-day MORB's source depleted mantle. However, Nd model ages (T_{DM}) are highly dependent on the $^{147}Sm/^{144}Nd$ ratios of the rocks and these ratios may change as a result of chemical fractionation between Sm and Nd during extensive single-stage or multi-stage magmatic or sedimentary processes. In most cases, this results in T_{DM} ages being minimum estimates of crustal extraction ages. In an effort to present realistic model ages for the purposes of discussion, we chose to calculate the Nd model ages by a two-stage calculation (T_{2DM} after Jacobsen, 1988) where a sample's $^{147}Sm/^{144}Nd$ ratio is used to correct the isotopic composition from the present to the age of crystallization or deposition, and thereafter, an average crustal $^{147}Sm/^{144}Nd$ ratio is used to intersect with the depleted mantle curve. The aim of this latter method is to compensate for intracrustal fractionation effects. Following Allègre and Rousseau (1984), we used a $^{147}Sm/^{144}Nd$ ratio of 0.11 to model crustal evolution, which represents an average between values used in numerous previous studies (0.10 and 0.12). Moreover, this value agrees with the average $^{147}Sm/^{144}Nd$ ratio of metasedimentary rocks when the average is weighted for the Nd concentrations:

$([Nd]_i(^{147}Sm/^{144}Nd)_i)/([Nd]_i).$

This equation produces average $^{147}Sm/^{144}Nd$ ratios of 0.112 and 0.106 for metasedimentary rocks in Table 1 (samples #15–20 and I–IV) and for seven analyses of Quetico sediments reported in Henry et al. (1998), respectively.

2.2. Sm–Nd data

This paper presents 75 new analyses (Table 1) of various rocks from the Sachigo–Berens River (the North Caribou Terrane) and Uchi Sp (Fig. 2) and the Winnipeg River and Wabigoon Sp (Fig. 3) including four metasedimentary rocks from the ca. 3 Ga Finlayson Lake greenstone belt (Fig. 1). The following sections briefly summarize the main results for each Sp.

2.2.1. Sachigo and Berens River Sp (North Caribou Terrane)

New data from this region are shown in Fig. 4 along with Sm–Nd data from the North Spirit Lake greenstone belt (Stevenson, 1995). The Nd isotopic compositions of the ca. 2.7 Ga plutonic rocks from Berens River Sp range between those of a sanukitoid (Stormer Lake, sample # J with a ϵ_{Nd}^t value of +1.8), which has the most juvenile isotopic compositions reflecting its origin from a depleted mantle (e.g. Shirey and Hanson, 1986), and a pegmatite (Margot Lake, # F with a ϵ_{Nd}^t value of −1.4), which is associated with leucogranites and has a Nd isotopic composition consistent with an origin by crustal anatexis of earlier crust ($T_{2DM} = 3.03$ Ga), such as the early Sachigo TTG suites. A comparison with data from the Sachigo Sp (Stevenson, 1995) shows that early crustal material was isolated from the depleted mantle at 2.92–3.02 Ga and subsequently evolved (arrows on Fig. 4) to negative $\epsilon_{Nd}^{2.7Ga}$ values. In contrast, the positive $\epsilon_{Nd}^{2.7Ga}$ values of ca. 2.7 Ga TTG suite and other granitoid rocks from the Berens River Sp, corresponding to an average T_{2DM} age of 2.86 Ga, do not appear to result from re-melting of older crust because of a lack of crust of appropriate age. Thus, the Nd isotopic compositions of these 2.7 Ga middle suites likely suggest mixtures of juvenile material having ϵ_{Nd}^t values close to that of the depleted mantle (+3) with older crustal reservoirs. An appropriate candidate would be the 2.92–3.02 Ga crust of the Sachigo Sp (Stevenson, 1995). Overall, the data strongly suggest that the

Table 1
Sm–Nd data

Locations[a]	N°	Sample descriptions	Age (Ga)[b]	Sm[c]	Nd[c]	$^{147}Sm/^{144}Nd$	$^{143}Nd/^{144}Nd$	2σ	ϵ_{Nd}^{t} [d]	T_{2DM}^{e}
North Caribou Terrane (Sachigo and Berens River Sp)										
A	C 93-58	Granite, Varveclay Lake batholith	2.700	4.31	28.31	0.0920	0.510720	27	−1.0	3.01
B	C 93-53	Bi-tonalite, Cochrane River	2.708 (Corfu and Stone, 1999)	2.28	15.50	0.0888	0.510700	13	−0.2	2.95
C	C 93-61	Bi-tonalite, Namiwan Lake	2.736 (Corfu and Stone, 1999)	1.29	12.62	0.0616	0.510186	9	−0.3	2.98
D	C 93-44	Granite, Deer Lake batholith	2.736 (Corfu and Stone, 1999)	1.14	6.51	0.1056	0.510995	13	0.0	2.96
North Spirit greenstone belt										
E	C 93-46	Gneissic tonalite, Pakwan stock	2.726 (Corfu and Stone, 1999)	2.08	12.52	0.1003	0.510884	10	−0.4	2.98
F	C 93-47	Pegmatite, Margot Lake	2.697 (Corfu and Stone, 1999)	3.33	13.60	0.1479	0.511700	13	−1.4	3.03
G	C 93-43	Granite, Sparling Lake	2.712 (Corfu and Stone, 1999)	1.65	12.14	0.0819	0.510650	12	1.3	2.85
H	C 93-49	Granite, Throat River	2.697 (Corfu and Stone, 1999)	6.63	42.86	0.0935	0.510826	11	0.5	2.90
I	C 93-41a	Gneissic bi-tonalite, Berens Lake	2.860 (Corfu and Stone, 1999)	1.22	8.81	0.0836	0.510571	15	1.3	2.97
J	C 93-39	Qz-monzonite, Stormer Lake stock	2.700 (Corfu and Stone, 1999)	7.94	52.60	0.0912	0.510851	9	1.8	2.80
K	C 93-66	Granite, Job Lake	2.703 (Corfu and Stone, 1999)	3.19	25.87	0.0746	0.510524	9	1.3	2.84
L	C 93-65	Bi-tonalite, Bloodvein River	2.705 (Corfu and Stone, 1999)	0.56	4.42	0.0762	0.510533	14	0.9	2.87
M	C 87-23	Granodiorite, Williams Lake batholith	2.701 (Corfu and Stott, 1993a)	3.52	24.08	0.0884	0.510800	14	1.8	2.80
Uchi subprovince										
Birch Uchi greenstone belt										
N	C 93-51	Qz-monzonite, Okanse Lake	2.700	7.17	40.60	0.1067	0.511133	6	1.9	2.79
Red Lake greenstone belt										
a	C 93-63	Bi-tonalite, Douglas Lake	2.734 (Corfu and Stone, 1999)	3.21	18.37	0.1057	0.510997	14	0.0	2.96
b	RL-94-14	Felsic metavolc. rock, Pipestone Bay	2.925 (Corfu and Wallace, 1986)	1.61	10.02	0.0969	0.510764	7	1.1	3.04
c	RL 94-17	Felsic tuff, Pipestone Bay	2.940 (Corfu and Wallace, 1986)	1.55	8.92	0.1049	0.510894	12	0.8	3.08
d	RL-94-3	Granodiorite, Killala-Baird batholith	2.704 (Corfu and Andrews, 1987)	1.86	10.19	0.1102	0.511117	10	0.4	2.90
e	RL-94-23	Granodiorite, Mac Kenzie stock	2.720 (Corfu and Andrews, 1987)	2.79	17.86	0.0945	0.510845	11	0.8	2.89
f	RL-94-24a	Granite, Mac Kenzie stock	2.720 (Corfu and Andrews, 1987)	4.41	30.52	0.0873	0.510740	7	1.3	2.86
g	RL-94-22	Rhyolitic tuff, Cochenour	2.893 (Corfu and Wallace, 1986)	0.36	1.29	0.1688	0.512209	10	2.0	2.94
h	RL-94-13a	Felsic dyke, Rahill Bay	2.757 (Corfu and Wallace, 1986)	0.42	1.56	0.1621	0.512045	13	0.7	2.93
I	RL-94-9a	Ilapilli tuff, Mac Neely Bay	2.748 (Corfu and Wallace, 1986)	4.72	19.94	0.1430	0.511728	9	1.2	2.88
j	RL-94-6	Granodiorite, Dome Stock batholith	2.718 (Corfu and Wallace, 1986)	5.82	35.63	0.0987	0.510963	11	1.6	2.83
k	RL 94-25c	Rhyolitic tuff, Heyson sequence	2.739 (Corfu and Wallace, 1986)	13.49	69.10	0.1180	0.511365	7	2.9	2.75
Meen–Dempster greenstone belt										
O	C 86-29	Granodiorite, Stoughton Creek Pluton	2.732 (Corfu and Stott, 1993b)	1.98	11.25	0.1062	0.511100	13	1.8	2.82
P	C 84-6	Tonalite, Kawashe Lake stock	2.722 (Corfu and Stott, 1993b)	2.38	13.70	0.1050	0.511006	10	0.3	2.93
Q	C 86-30	Tonalite, Dobie Lake Pluton	2.750 (Corfu and Stott, 1993b)	3.08	20.47	0.0910	0.510817	10	1.9	2.83
Lake Saint-Joseph greenstone belt										
R	C 86-22	Heterolithic tuff-breccia, Lake St. Joseph	2.713 (Corfu and Stott, 1993b)	7.54	31.82	0.1431	0.511687	11	0.2	2.93
S	C 86-32	Granodiorite, Osnaburgh Pluton	2.694 (Corfu and Stott, 1993a)	2.90	18.26	0.0959	0.510829	12	−0.4	2.95
Pickle Lake greenstone belt										
T	C 87-12	Felsic metavolcanic rock, Pickle Crow	2.860 (Corfu and Stott, 1993b)	2.64	14.45	0.1103	0.511127	14	2.4	2.89
U	C 87-18	Granodiorite, Pickle Lake stock	2.740 (Corfu and Stott, 1993a)	4.77	33.54	0.0859	0.510716	15	1.6	2.85

Sample	ID	Description	Age (Ga)[b]	Sm[c]	Nd[c]	$^{147}Sm/^{144}Nd$	$^{143}Nd/^{144}Nd$	±	ε_{Nd}[d]	T_{2DM}[e]
V	C 87-16	Granodiorite, Second Loon Pluton	2.711 (Corfu and Stott, 1993b)	2.75	16.83	0.0989	0.511019	12	2.6	2.75
W	C 86-20	Tonalitic gneiss, Seach-Achapi bath.	2.821 (Corfu and Stott, 1993b)	0.75	2.78	0.1635	0.512040	13	0.4	3.01
Fort Hope greenstone belt										
X	C 84-23	Crystal tuff, Miminiska Lake	2.715 (Corfu and Stott, 1993b)	1.86	11.36	0.0989	0.511022	15	2.7	2.75
Winnipeg River subprovince										
1a	K-94-1a	Tonalitic gneiss, Tannis Lake	3.051 (Davis et al., 1988)	0.37	1.87	0.1198	0.510966	25	−2.4	3.40
1b	K-94-1b	Tonalitic gneiss, Tannis Lake	3.051 (Davis et al., 1988)	4.01	29.50	0.0822	0.510161	8	−3.3	3.47
2	K-94-5	Granodiorite, Pelican Pouch Lake	2.700	3.48	23.20	0.0905	0.510738	13	−0.2	2.94
3	K-94-7a	Tonalitic gneiss, Kenora	2.875 (Corfu, 1988)	0.53	3.03	0.1063	0.510991	15	1.4	2.98
4	K-94-8	Granodiorite, Kenora	2.709 (Corfu, 1988)	1.93	11.16	0.1045	0.510853	11	−2.7	3.14
5	K-94-13a	Gneiss, Kenora (road 666)	2.830 (Corfu, 1988)	1.47	10.39	0.0853	0.510511	6	−0.9	3.11
6	K-94-12	Granite, Black Sturgeon Lake	2.760 (Corfu, 1988)	1.65	10.28	0.0969	0.510909	11	1.7	2.85
7	K-94-10	Granodiorite, Redditt	2.690 (Corfu, 1988)	4.38	29.73	0.0890	0.510700	9	−0.5	2.96
8	RL 94-30	Tonalite, Daniels Lake	2.840 (Corfu, 1988)	3.96	30.57	0.0782	0.510372	10	−0.9	3.11
9	RL 94-26	Tonalitic gneiss, Cedar Lake	3.170 (Corfu, 1988)	9.24	58.52	0.0954	0.510543	12	0.5	3.29
10a	RL-94-27a	Granodiorite, Cedar Lake	2.760 (Corfu, 1988)	1.61	10.36	0.0937	0.510667	6	−1.9	3.12
10b	RL-94-27b	Granite, Cedar Lake	2.760 (Corfu, 1988)	1.94	9.17	0.1282	0.511266	13	−2.4	3.16
Wabigoon subprovince										
11	K-94-2	Amphibolite, south of Ingolf	2.702 (Davis et al., 1988)	2.63	10.05	0.2015	0.512876	9	3.0	2.71
12	K-94-4	Granite, High Lake stock (g)	2.727 (Davis and Smith, 1991)	1.77	10.67	0.1001	0.510971	10	1.4	2.85
13a	K-94-14B	Granite, south of Silver Lake	2.700	6.17	41.15	0.0906	0.510784	6	0.7	2.88
13b	K-94-14c	Felsic tuff	2.700	2.21	13.13	0.1018	0.511017	8	1.4	2.82
14	K-94-17a	Granite, Hawk Lake stock	2.740	0.42	2.21	0.1183	0.511272	11	1.0	2.89
Minnitaki Lake greenstone belt										
15	95 TB 37	Siltstone, Sandybeach Lake	2.700	2.54	17.12	0.0897	0.510930	12	3.9	2.65
Vermillion Lake greenstone belt										
16	95 TB 40	Black shale, Hudson (road 664)	2.700	4.98	34.06	0.0883	0.510846	7	2.7	2.73
17	95 TB 42	Sandstone, Sioux Lookout	2.700	5.26	30.65	0.1038	0.510909	15	−1.5	3.04
18a	95 TB 44a	Conglomerate-matrice, Sioux Lookout	2.700	2.37	15.96	0.0899	0.510548	12	−3.7	3.20
18b	95 TB 44b	Granitic pebble in 44a	2.700	2.63	11.10	0.1430	0.511635	17	−0.9	3.00
19	95 TB 13b	Gneiss, Sturgeon Lake area	2.700	1.37	7.49	0.1104	0.510867	18	−4.6	3.27
Savant Lake greenstone belt										
20a	95 TB 26a	Sandstone, Wiggle Creek	2.700	3.96	21.02	0.1139	0.511063	14	−2.0	3.08
20b	95 TB 26b	Siltstone, Wiggle Creek	2.700	3.96	23.52	0.1016	0.510922	15	−0.4	2.96
Finlayson Lake greenstone belt										
I	N-1	Metasedimentary rock	3.000	4.62	14.32	0.1949	0.512693	10	1.8	3.05
II	DF-14	Metasedimentary rock	3.000	2.01	10.68	0.1135	0.511080	9	1.8	3.05
III	2-34 MH	Metasedimentary rock	3.000	3.01	14.18	0.1282	0.511349	9	1.3	3.09
IV	DP-O2	Metasedimentary rock	3.000	1.78	5.91	0.1824	0.512480	15	2.5	3.00

[a] Letters and numbers are used in text and to simplify the sample locations on Figs. 1–3.
[b] Ages are based on field relationships (in italics) or refer to previous works.
[c] Sm and Nd concentrations in ppm.
[d] ε_{Nd} values calculated with present-day CHUR $^{143}Nd/^{144}Nd$ and $^{147}Sm/^{144}Nd$ ratios of 0.512638 and 0.1967, respectively.
[e] T_{2DM} is a two-stage Nd model age (Jacobsen, 1988) in Ga (see text).

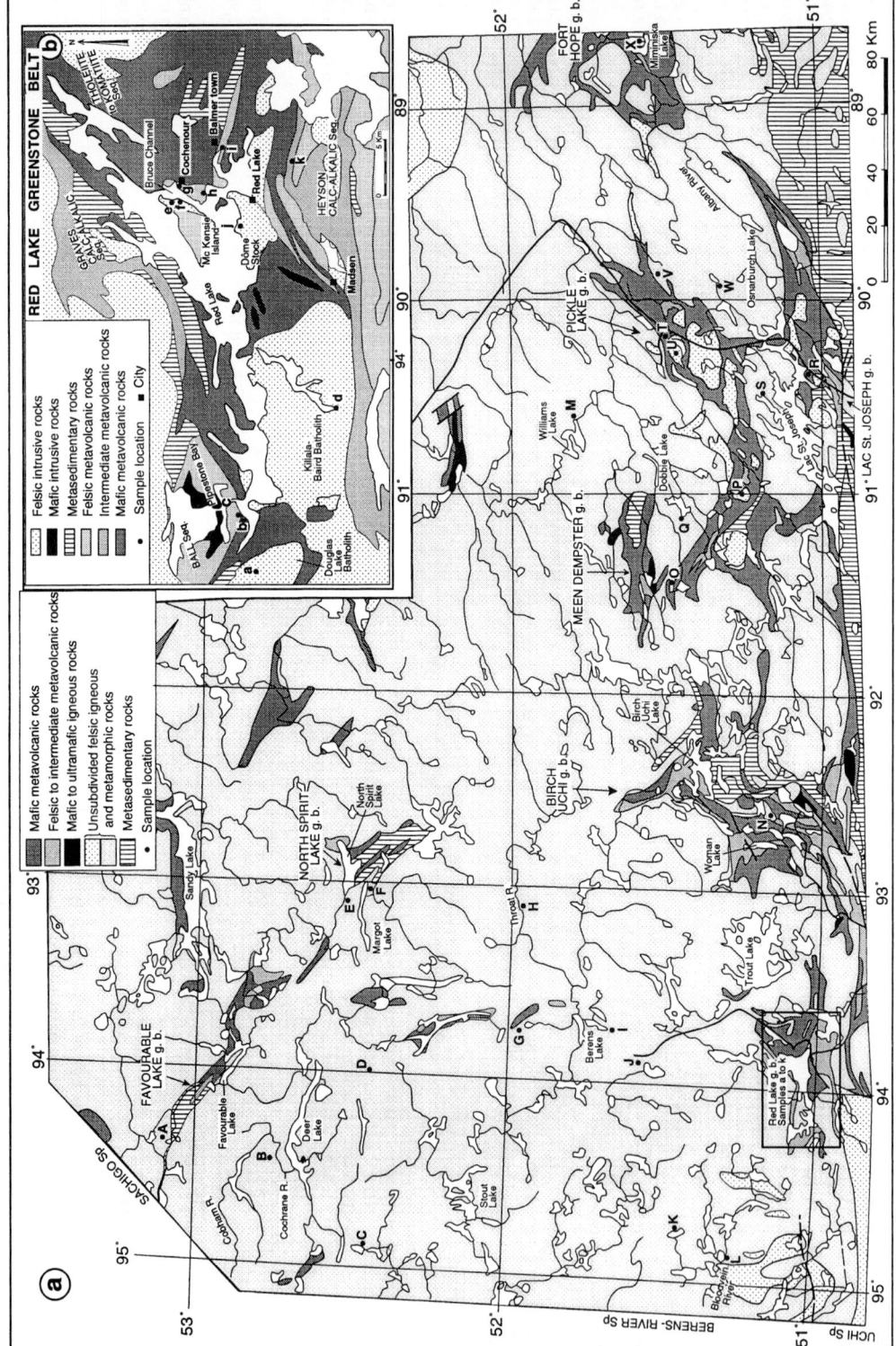

Fig. 2. Locations of samples from the North Caribou Terrane (Sachigo and Berens River Sp) and Uchi Sp. (a) Simplified map (after O.G.S. map #2440, Ontario Geological Survey and Holweg, 1980) showing the locations of the samples from Sachigo and Berens River Sp. (# A–M, Table 1) and Uchi Sp (# N to X, Table 1). (b) Detailed map of the Red Lake greenstone belt (after Corfu and Wallace, 1986) with the locations of the samples # a to k (Table 1).

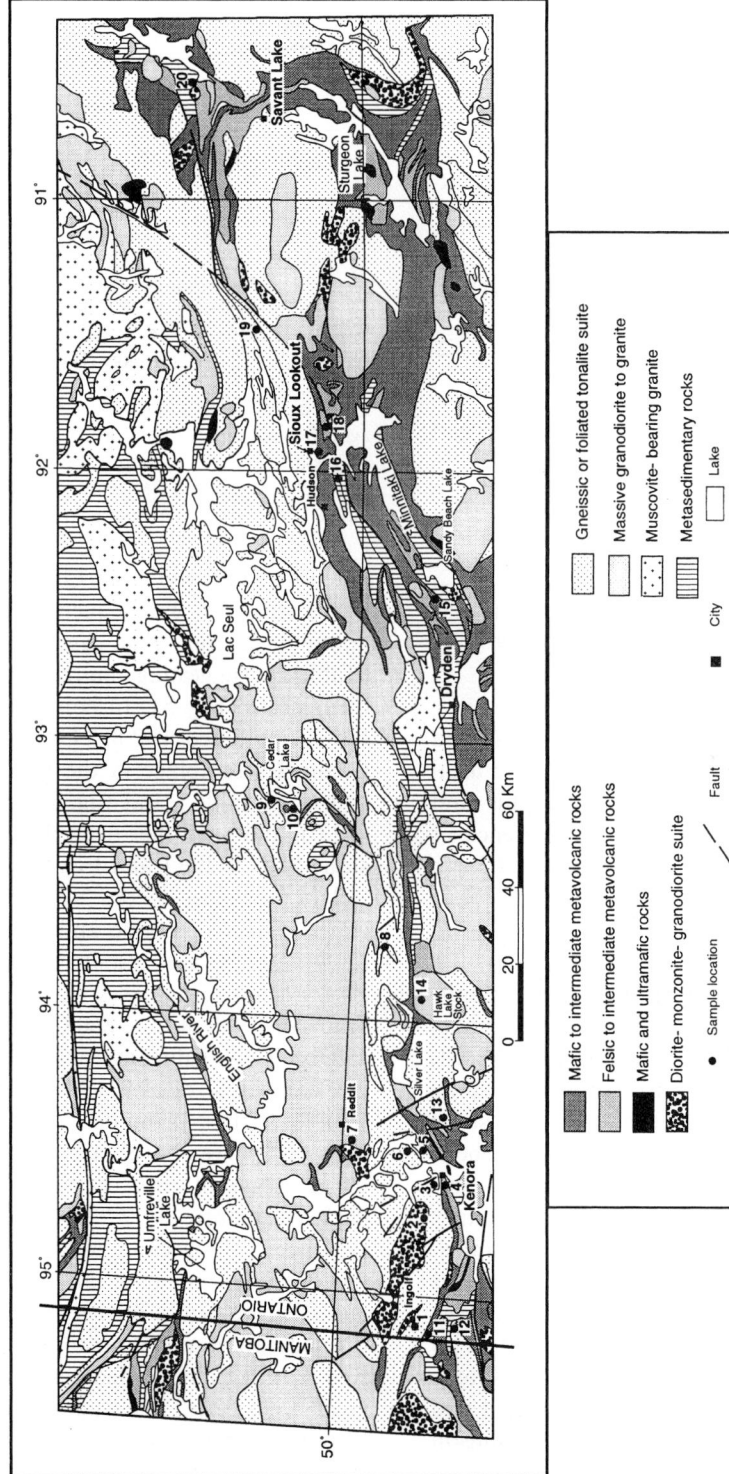

Fig. 3. Locations of samples # 1–20 (Table 1) from Winnipeg River Sp and Wabigoon Sp (after O.G.S. 1991, map #2542).

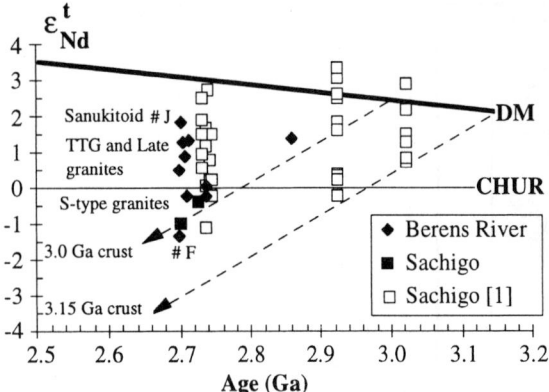
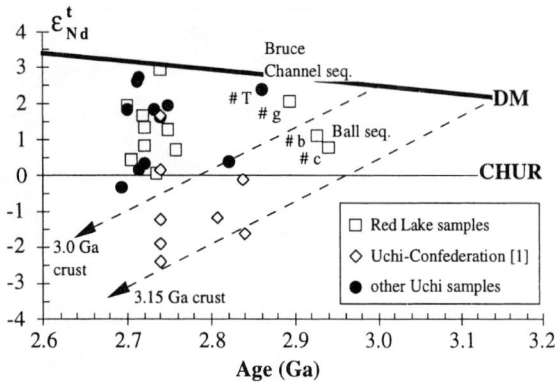

Fig. 4. ϵ_{Nd}^{t} values as a function of crystallization or depositional ages showing the Nd compositions of the samples from the North Caribou Terrane (Sachigo and Berens River Sp). The ϵ_{Nd}^{t} values can be interpreted as mixtures between 2.92–3.02 Ga crust and 2.76–2.70 Ga juvenile inputs from the depleted mantle. DM = depleted mantle evolution curve (see text); CHUR = Chondritic Uniform Reservoir. The arrows represent the isotopic evolution of felsic crust isolated from the depleted mantle at 3.0 and 3.15 Ga and evolving with a $^{147}Sm/^{144}Nd$ ratio of 0.11. [1] Stevenson (1995).

Fig. 5. ϵ_{Nd}^{t} values as a function of crystallization or depositional ages showing Sm–Nd data of the samples from Uchi Sp. The ϵ_{Nd}^{t} values between +3 and −3 indicate additions from both the depleted mantle (DM) and from the recycled crust. [1] Noble (1989), other symbols as in Fig. 4.

ca. 2.7 Ga orogeny both created new crust and recycled old crust. The importance of crustal recycling in the Berens River metamorphic and plutonic rocks is consistent with the presence of inherited zircons in the ca. 2.7 Ga rocks (Corfu and Stone, 1999).

2.2.2. Uchi Subprovince

Samples from the Uchi Sp encompass granitoids, felsic flows and tuffs from the Red Lake, Confederation Lake, Meen-Dempster, Lake St. Joseph, Pickle Lake and Fort Hope greenstone belts. Data for the Uchi-Confederation greenstone belt are from Noble (1989). The Uchi Sp has been further subdivided into five lithotectonic assemblages, each comprising a volcano-sedimentary sequence: the Balmer, the Ball, the Bruce Channel, the Confederation and the Woman assemblages (Stott and Corfu, 1991). The Balmer assemblage comprises basalts and komatiites deposited in an oceanic plateau setting with isotopic compositions indicative of the presence of both a depleted and an enriched mantle (Tomlinson, 1996; Tomlinson et al., 1998).

Two felsic flows (# b and c, Table 1 and Fig. 5, dated at 2.925 and 2.940 Ga, Corfu and Wallace, 1986) from the Ball sequence have ϵ_{Nd}^{t} values of about +1 and T_{2DM} ages of about 3.05 Ga (Table 1), suggesting the recycling of ca. 3 Ga or older crust (Fig. 5). In contrast, two other felsic flows, one dated at 2.893 Ga and from the Bruce Channel assemblage in the Red Lake greenstone belt (# g, Table 1 and Fig. 5, with a ϵ_{Nd}^{t} value of +2.0), and one other from the Pickle Lake greenstone (# T, Table 1 and Fig. 5, with a ϵ_{Nd}^{t} value of +2.4) plot close to the Nd isotopic composition of the depleted mantle and thus have T_{2DM} ages (2.94 and 2.89 Ga, respectively) close to their U–Pb crystallization ages (2.893 Ga, Corfu and Wallace, 1986 and 2.860 Ga, Corfu and Stott, 1993b). These values are similar to the T_{2DM} crustal creation ages of 2.92–3.02 Ga determined for samples in the North Spirit greenstone belt (Fig. 4, Sachigo Sp, see above), and together, they confirm the creation of new crustal segments from the depleted mantle at 2.92–3.02 Ga. Furthermore, evidence for recycling of 2.9–3.0 Ga crust from the older assemblages can be seen in granitoids and felsic volcanics of the younger, 2.76–2.69 Ga, Confederation and Woman assemblages from the Red Lake, Uchi-Confederation and Pickle Lake greenstone belts. The majority of ϵ_{Nd}^{t} values fall between depleted mantle (+3) and ca. 3.0 Ga crustal values (−1) associated with an average T_{2DM} age of 2.87 Ga, which results from such mixtures. Thus, volcanic and granitoid rocks of this era represent both new crustal growth from the depleted mantle as well as recycling of older crust, as discussed above for the North Caribou terrane.

2.2.3. Winnipeg River Subprovince

Samples from this subprovince were chosen to characterize regions where U–Pb zircon dating had identified some of the oldest crust in the Superior Province. The Tannis Lake (# 1) and Cedar Lake (# 9) gneisses yielded U–Pb zircon ages of 3.05 Ga (Davis et al., 1988) and 3.17 Ga (Corfu, 1988), respectively. Moreover, their Nd isotopic compositions (ϵ_{Nd}^t values between 0.5 and −3.3, Table 1 and Fig. 6) suggest the presence of older crustal material whose ages can be approximated by the T_{2DM} ages between 3.5 and 3.3 Ga. These T_{2DM} ages are consistent with radiogenic $^{207}Pb/^{204}Pb$ ratios (15.1–14.7) determined for K-feldspars in this region (Corfu, 1988) which could be interpreted as inherited Pb from a crust older than 3.2 Ga. The Daniels Lake gneiss (# 8), dated at 2.84 Ga (Corfu, 1988), has an ϵ_{Nd}^t value of −0.9 and a T_{2DM} of 3.11 Ga that appears to reflect remelting

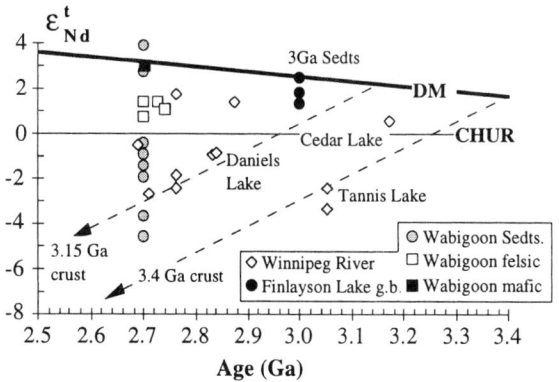

Fig. 6. ϵ_{Nd}^t values as a function of crystallization or depositional ages showing Sm–Nd data of the samples from Winnipeg River and Wabigoon Sp. The Winnipeg River Sp contains some of the oldest rocks found in the WSP. The ϵ_{Nd}^t values of the 3.17 Ga Cedar Lake and the 3.05 Ga Tannis Lake gneisses imply that the earliest crust was separated from the depleted mantle at 3.4±0.1 Ga. The data also suggest mixtures between crustal precursors and younger (at 3.0 then 2.7 Ga) inputs from the depleted mantle. Such mixtures yield mixed T_{2DM} ages illustrated here by the evolution of a hypothetical 3.15 Ga crust that is unknown in the WSP.

of crust of at least 3.0 Ga in age. Other tonalitic gneisses and granitoids from the Kenora region have ϵ_{Nd}^t values between −3 and +2 and T_{2DM} ages between 2.85 and 3.16 Ga. These results indicate that the Winnipeg River Sp received new crustal additions during the ca. 2.7 Ga orogeny and that some of the older (oldest) crust was recycled during this period.

2.2.4. Wabigoon Subprovince

The Wabigoon Sp is generally subdivided into the Western, Central and Eastern regions. The Central Wabigoon contains supracrustal sequences and granitoids of ca. 3.0 Ga (e.g. Davis and Jackson, 1988), whereas the Eastern and Western Wabigoon contain supracrustal sequences of 2.73 Ga (Davis and Jackson, 1988; Blackburn et al., 1991). Sedimentary rocks (amphibolite facies) from the Finlayson Lake greenstone belt in the Central Wabigoon (# I–IV, Table 1) have ϵ_{Nd}^{3Ga} values ranging from +1.3 to +2.5 (Fig. 6) in agreement with the ϵ_{Nd}^{3Ga} values of three gneisses and a felsic tuff (between +3.2 and +2.4, Henry et al., 1998) from the 3 Ga old Lumby Lake greenstone belt. The isotopic compositions of both the gneisses and the sedimentary rocks from the 3.0 Ga supracrustal sequences plot close to juvenile depleted mantle values at 3.0 Ga. This suggests that the Central Wabigoon was largely composed of new crust at 3.0 Ga. Samples from the 2.73 Ga Eastern Wabigoon region are derived from the Kenora, Dryden, Sioux Lookout and Sturgeon Lake areas. The rock samples vary from mafic and felsic metavolcanic to sedimentary and granitoids. The mafic (# 11 with ϵ_{Nd}^t of +3.0) and felsic (# 13 with ϵ_{Nd}^t of +3.4) metavolcanic samples are derived from the Lake of the Woods greenstone belt and have a depleted mantle isotopic composition at 2.7 Ga. The ϵ_{Nd}^t values of granites and granodiorites from this area (# 12, 13a, 14, Table 1) range from +0.7 to +1.4, with T_{2DM} ages between 2.80 and 2.91 Ga, once again reflecting either melting of a heterogeneous source (juvenile and older crustal mixtures) or contamination of juvenile magmas on emplacement. Sedimentary rocks (and a granitic clast from a conglomerate) from the Sioux Lookout (# 15–18) Sturgeon Lake (# 19) and Savant Lake (# 20) areas have variable

$\epsilon_{Nd}^{2.7Ga}$ values between −4.6 and + 3.9. The negative values from the Sioux Lookout area reflect the fault bounded nature of the basin where the north side of the basin is bounded by the Winnipeg River Sp and likely received detritus eroded from the 3.0 Ga or older gneisses. Thus sediments in the Wabigoon reflect mixtures of older sources such as the Winnipeg River or Central Wabigoon with juvenile material derived from the erosion of 2.73 Ga volcanic units as seen in the shales (# 16, $\epsilon_{Nd}^{2.7Ga} = +2.7$) and siltstones (# 15, $\epsilon_{Nd}^{2.7Ga} = +3.9$). The crustal trends drawn on Fig. 6, which correspond to T_{2DM} ages of 3.15 and 3.40 Ga, demonstrate that the metasedimentary rocks from the Wabigoon Sp record a complex evolution of the subprovince from 2.7, 3.0 and ca. 3.4 Ga crust. The complex nature of the crustal evolution in the southern part of the WSP was also discussed by Beakhouse and McNutt (1991). They suggested that Na$_2$O-rich, ca. 2.7 Ga old plutons with juvenile isotopic compositions in the Wabigoon Sp were produced by partial melting of amphibolites, whereas, more K$_2$O-rich granodioritic plutons with crustal isotopic compositions in the Winnipeg River Sp were formed by partial melting of older crustal reservoirs. In addition, U–Pb ages of detrital zircons in the Quetico metasedimentary belt (e.g. Davis et al., 1990; Davis, 1996, 1998) demonstrate that the ages of the eroded materials vary between ca. 2.7 and 3.0 Ga. These studies, along with mixing trends for the Sm–Nd and Pb–Pb isotopic systems in the Wabigoon, Quetico and Wawa Sp (Henry et al., 1998) imply that the southern part of the WSP can be described as a collage of Archean crusts that reflect variable ages and mixtures of depleted mantle and older recycled crust.

3. Discussion

3.1. Crustal growth versus crustal recycling

The Sm–Nd data presented above are shown in an ϵ_{Nd}^t value versus age of crystallization or deposition diagram (Fig. 7) together with Nd data from Noble (1989), Stevenson (1995), Henry et al. (1998) and Larbi et al. (1999) in order to illustrate

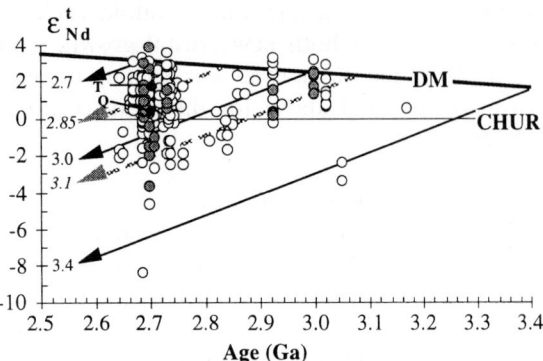

Fig. 7. ϵ_{Nd}^t values as a function of crystallization or depositional ages. The ϵ_{Nd}^t values range from depleted mantle values (DM) to values indicative of older crustal precursors. The >3.05 Ga gneisses of the Winnipeg River Sp argue for the existence of an Early Archean crust for which a Nd model age of 3.4 Ga can be suggested. The crustal evolution is modeled with a ^{147}Sm/^{144}Nd ratio of 0.11 leading to mixed Nd model ages of 3.1 and 2.85 Ga (gray arrows). These ages, without geological significance, are produced by the mixture between 3.4 and 3.0 Ga crusts, then by 2.7 Ga crust and older crustal materials, respectively. The letters Q and T are 2.69 Ga metasedimentary rocks from Quetico and Wawa Sp, respectively and represent possible average compositions for the WSP crust (see text). White circles: igneous rocks, gray circles: metasedimentary rocks and black circles: averages of Quetico and Wawa metasedimentary rocks.

the evolution of the WSP. This plot of 210 Nd isotopic compositions allows us to identify the periods in which crustal growth predominated over crustal recycling. Four periods are identified:

(1) 2.67 to 2.76 Ga: $-8 < \epsilon_{Nd}^t < +4$. The range of values in this period is interpreted as the result of mixing between depleted mantle-like sources and crustal end-members. Such mixtures reflect the construction of new crust for which the average Nd isotopic composition will depend on the relative amounts of juvenile terranes and older crustal end-members, respectively. The ϵ_{Nd}^t value of the depleted mantle (DM on Fig. 7) is constrained by the higher ϵ_{Nd}^t values (of about +3) as in the Hf–Nd correlation (Vervoort et al., 1996) while the ϵ_{Nd}^t values of the recycled older crustal components are estimated from average ϵ_{Nd}^t values of older crustal rocks projected to 2.67–2.76 Ga.

(2) 2.76 to 2.92 Ga: the ϵ_{Nd}^t values decrease with time. This period is characterized by a paucity of volcanic activity and thus relatively less mantle

input compared to the preceding and following periods. The ϵ_{Nd}^t values of the ca. 2.89 Ga volcanic units of the Uchi Sp indicate the greatest mantle affinity; +2.0 to +2.4 for felsic volcanic rocks in the Red Lake (# g) and the Pickle Lake greenstone belts (# T), respectively. Thereafter, however, the ϵ_{Nd}^t values decrease towards −1 at 2.83 Ga (# 5, a gneiss from Kenora area) and −2 at 2.76 Ga (# 10a and b, the Cedar Lake granites intruding the oldest tonalite # 9 dated at 3.17 Ga in the Winnipeg River Sp, Table 1). Fig. 7 indicates that if the ϵ_{Nd}^t values of the 2.76 and 2.92 Ga rocks are projected to 2.7 Ga, they would overlap with the majority of the negative ϵ_{Nd}^t values of the younger 2.67–2.76 Ga rocks and the ϵ_{Nd}^t values of older 2.92–3.02 Ga rocks. These data strongly suggest that the latter part of this period was characterized predominantly by crustal recycling rather than crustal growth. This would imply that the majority of U–Pb ages between 2.76–2.92 Ga in the WSP, and especially the U–Pb ages of detrital zircons (e.g. Davis et al., 1990; Davis, 1996, 1998), are the product of intra-crustal processes. The average composition of the crustal end-member recycled during the ca. 2.7 Ga events can be estimated from the average $\epsilon_{Nd}^{2.7Ga}$ values and T_{2DM} ages of the 2.76–2.92 Ga group and yield −2±1 and 3.1±0.1 Ga, respectively (Figs. 7 and 8).

(3) 2.92 to 3.02 Ga: 0 < ϵ_{Nd}^t < +3. This range, like that discussed above, is interpreted to record mixtures of juvenile and older crustal end-members. The ϵ_{Nd}^t values obtained from mafic rocks from the Lumby Lake greenstone belt in Central Wabigoon (Henry et al., 1998), Red Lake (Uchi Sp) and North Spirit Lake (Sachigo Sp, Stevenson, 1995) greenstone belts suggest that ϵ_{Nd}^t values between +2 and +3 are representative of the depleted mantle. Other 2.92–3.02 Ga plutonic rocks having lower ϵ_{Nd}^t values argue for a mixture between ca. 3.0 Ga juvenile and older crustal end-members as suggested by T_{2DM} ages slightly older than the ages of crystallization (Fig. 8). These model ages represent a mixture of 2.92 to 3.02 Ga material from the depleted mantle with crust from the oldest terranes identifiable in the WSP (see below).

(4) > 3.05 Ga: evidence for Early Archean crusts. The two > 3.05 Ga tonalite gneisses from

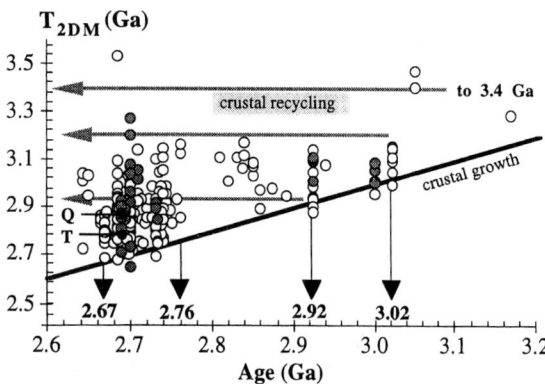

Fig. 8. Nd model ages versus crystallization or depositional ages diagram. The Nd model ages correspond to the T_{2DM} defined by Jacobsen (1988). The black line, defined by T_{2DM} = ages corresponds to the formation and isolation of continental crust from the depleted mantle without crustal recycling. In contrast, crustal recycling would produce a horizontal trend as indicated by the gray arrows. This figure suggests that the main periods of crustal formation (isolation of crust from the depleted mantle) were 2.67–2.76, 2.92–3.02, and 3.4 Ga. The rocks dated between 2.76 and 2.92 Ga have the same model ages that the 2.92–3.02 Ga rocks, showing the importance of intra-crustal processes during the 2.76–2.92 Ga interval without significant input from the mantle to modify the average composition of the WSP. Symbols as Fig. 7.

the Tannis Lake area yield $\epsilon_{Nd}^{3.05Ga}$ values of −2.4 and −3.3, and the 3.17 Ga tonalite gneiss from Cedar Lake has an $\epsilon_{Nd}^{3.17Ga}$ value of 0.5. In addition with the contribution of recycled crust recorded by Nd isotopic compositions of the 2.92–3.02 Ga plutons, these values suggest the participation of Early Archean crust in the formation of the WSP. Evidence for the presence of Early Archean crust within the WSP had also been demonstrated by 3.3 Ga U–Pb ages from detrital zircons (Davis et al., 1990; Davis, 1996; Davis, 1998). The T_{2DM} ages of the three oldest rocks from Winnipeg River Sp suggest an average age of 3.4±0.1 Ga for the oldest crust in the WSP.

3.2. Quantification of the crustal growth in WSP

To build a general model of crustal growth in the WSP, we need to estimate the ages when crust was created from the mantle and the ϵ_{Nd}^t values of the depleted mantle and crustal end-members at these different ages.

The isotopic composition of the depleted mantle has been treated in the previous sections. The ages when crust was isolated from the depleted mantle are defined by the ages when Archean ϵ_{Nd}^t values plot on the mantle evolution curve (Fig. 7), in other words when the T_{2DM} ages and the true ages are identical (Fig. 8). This classical interpretation of the Sm–Nd isotopic system implies that the main periods of crust formation were 2.67–2.76 Ga; 2.92–3.02 Ga. A third and older period of crust formation is implied by the T_{2DM} ages of 3.4 ± 0.1 Ga calculated from the Sm–Nd compositions of the oldest gneisses of the Winnipeg River Sp (Cedar Lake and Tannis Lake gneisses, # 9 and # 1a,b in Table 1, respectively) but which have younger crystallization ages of 3.17 and 3.05 Ga, respectively. The crustal recycling trend shown in Figs. 7 and 8 suggests that any new additions from the mantle between these periods were minor. For example, the local volcanic activity such as that found in the Uchi Sp (where enriched mantle was invoked, e.g. Tomlinson et al., 1998) did not produce significant amounts of new crust to modify the composition of the WSP.

Thus, our calculations are based on the creation of an initial crust at 3.4 ± 0.1 Ga followed by a second period of new crustal growth, between 3.02 and 2.92 Ga, which is recorded by samples from the North Spirit (Sachigo Sp), Red Lake (Uchi Sp) and Lumby Lake and Finlayson Lake (Central Wabigoon region) greenstone belts. To simplify the modelling of the evolution of the WSP, we define this period as a single crustal growth event at 3.0 Ga. During this period, mixing occurs between the 3.0 Ga additions from the depleted mantle and the 3.4 Ga crust. This simplification is supported by the average crustal recycling trend defined by the 2.92–2.76 Ga rocks, which indicates an origin from a 3.0 Ga end-member, which itself is a mixture of new and old crusts (Figs. 7 and 8). In fact, the crustal recycling trend observed from 2.92 to 2.76 Ga intersects the Nd composition of the 3 Ga detrital sedimentary rocks of the Finlayson Lake greenstone belt (average $\epsilon_{Nd}^{3Ga} = +1.7$). On this basis, the mixing parameters for the 3.0 Ga event are: (1) the depleted mantle having a ϵ_{Nd}^{3Ga} value about of $+2.5$, which is defined both by the higher ϵ_{Nd}^t values of metavolcanic rocks from ca. 3 Ga supracrustal sequences and by the Hf–Nd correlation (Vervoort et al., 1996) and (2) a crustal end-member having an average ϵ_{Nd}^{3Ga} value of -3 corresponding to the value at 3.0 Ga of the 3.4 Ga crust with a $^{147}Sm/^{144}Nd$ ratio of 0.11.

Finally, most of the rocks dated from 2.76 to 2.69 Ga have ϵ_{Nd}^t values between $+3$ to -2, a range of values that can be produced by similar mixtures of new crustal material derived from the 2.7 Ga depleted mantle and an older hybrid crust that is an average of the preceding crusts. The rare exceptions, having lower ϵ_{Nd}^t values, are plutonic rocks emplaced in or near the Winnipeg River Sp (e.g. Wabigoon–Winnipeg River Sp boundary zone), or near older terranes such as the ca. 3 Ga supracrsutal sequence within the Central Wabigoon region. For these rocks, the ϵ_{Nd}^t values suggest a higher participation of pre-3 Ga crust. The average ϵ_{Nd}^t values of the crust, created by the mixture between new and old terranes at 3.0 and 2.7 Ga, respectively, can be estimated from the ϵ_{Nd}^t values of detrital metasedimentary rocks (e.g. Allègre and Rousseau, 1984; Taylor and McLennan, 1985), although care must be taken to ensure that the sediments are well-mixed and thus representative of an average continental crust. The Nd concentrations and $^{147}Sm/^{144}Nd$ ratios can be used to discriminate between metasedimentary rocks that are representative of the average eroded continental crust and those that could be considered as first-cycle sediments from local sources and are generally biased towards more mafic compositions (e.g. McLennan and Hemming, 1992). On this basis, and following Henry et al. (1998), we have used the ϵ_{Nd}^t values of Quetico and Timiskaming-type (Wawa Sp) metasedimentary rocks (Q and T on Figs. 7 and 8, respectively) having, on average, $[Nd]=20$ ppm and $^{147}Sm/^{144}Nd=0.11$ to estimate the average ϵ_{Nd}^t values of the southern Wabigoon and Quetico Sp ($\epsilon_{Nd}^{2.7Ga} = +0.6$) and the Wawa Sp ($\epsilon_{Nd}^{2.7Ga} = +1.8$), respectively. The average ϵ_{Nd}^t value of the crustal end-member can be estimated by the crustal recycling trend recorded by 2.92–2.76 Ga rocks or from the $\epsilon_{Nd}^{2.7Ga}$ value of the Finlayson Lake sediments having on average a $^{147}Sm/^{144}Nd$ ratio of about 0.11. Both methods suggest that the 3 Ga crust formed with an average ϵ_{Nd}^{3Ga} value of $+1.7$ and evolved to a $\epsilon_{Nd}^{2.7Ga}$ value of -2 ± 0.5 with $^{147}Sm/^{144}Nd$ ratios between 0.10 and 0.12.

Having characterized the period of crustal growth versus crustal recycling and having defined the average Nd composition of the crust in each period, we propose to calculate the relative contribution of juvenile versus older crust by each period within the WSP. To quantify the crustal growth, we must account for the difference in Nd concentrations found in juvenile crustal sections, essentially represented by tholeiitic-calc-alkaline volcanic rocks and TTG suites, and older granitoid plutons that represent more differenciated material enriched in incompatible elements such as LREE. Another factor is the fact that new crust in the form of high-standing arcs could be subjected to greater erosion in comparison with older terranes. To account for these biases, we assigned Nd concentrations of 15 and 30 ppm for new and old crustal materials, respectively, and, following the example of Allègre and Rousseau (1984), the relative differences in erosion rates of juvenile versus mature arcs/terranes are accounted for in the mass-balance calculation by use of an erosion coefficient, K, of 2–4.

The relative amounts of new and old terranes can be calculated from the following mass-balance equation:

$$\epsilon_{Nd_m} = ((f_{old}\epsilon_{Nd_{old}}[Nd]_{old}) + K(f_{new}\epsilon_{Nd_{new}}[Nd]_{new}))$$
$$/((f_{old}[Nd]_{old}) + K(f_{new}[Nd]_{new}))$$

where ϵ_{Nd_m}, $\epsilon_{Nd_{new}}$, and $\epsilon_{Nd_{old}}$ represent the ϵ_{Nd} values for the mixture (m) of juvenile (new) and recycled (old) crustal sections; $f_{new} = M_{new}/(M_{new} + M_{old})$ and $f_{new} + f_{old} = 1$, M being the mass of each new or old crust; [Nd] is the Nd concentrations and K the erosion coefficient (see above).

The results of the calculations are shown in Fig. 9 for $K=2$, which implies that the two endmembers, old crust or new crust, contribute in a same way to the total Nd budget of the 2.7 crust. Using the average Sm–Nd composition of the Quetico metasedimentary rocks (Henry et al., 1998) as an average for 2.7 Ga WSP continental crust and a K value of 2, the calculation yields an Archean crust composed of 48%, 44% and 8 wt%

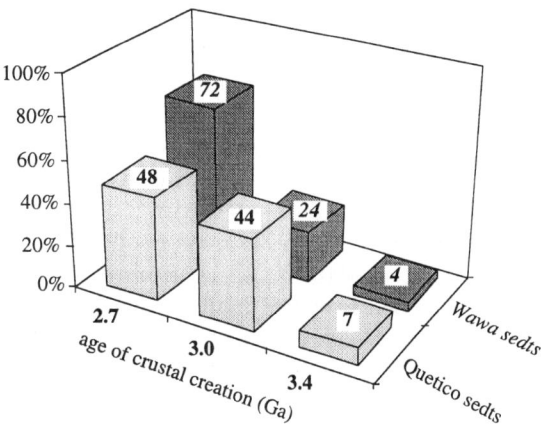

Fig. 9. Relative amounts of crust created at 3.4, 3.0 and 2.7 Ga. Based on mass balance calculations discussed in the text, we propose that the WSP is formed, on average, by 7%, 44% and 48 wt% of crust isolated from the depleted mantle at 3.4, 3.0 and 2.7 Ga, respectively. These results are obtained by considering the average composition of the WSP, as represented by the composition of the Quetico metasedimentary rocks (Henry et al., 1998). The use of Timiskaming-type metasedimentary rocks from the Wawa granite-greenstone belt leads to greater quantities of juvenile crust because these sediments are often of a more local, and thus juvenile, origin.

of crust created at 2.7, 3.0 and 3.4 Ga, respectively. A larger erosion coefficient (e.g. $K=4$) results in a smaller amount of juvenile crust with a crustal composition of 33%, 48% and 19 wt% created at 2.7, 3.0 and 3.4 Ga, respectively. Use of the Wawa Sp sediments (Timiskaming-type sedimentary rocks, Henry et al., 1998) to represent a more juvenile average crust yields 72–55, 24–32, and 4–13% for $K=2$–4. The difference in the modeled values underlines the dependency of the calculations on the choice of average crustal (sedimentary sequence) end-member and the K value. We suggest that the calculations based on the Quetico sediments with $K=2$ are more representative of the WSP because:

(1) The average $\epsilon_{Nd}^{2.7Ga}$ values of rocks from Sachigo ($+0.5$), Berens River ($+0.1$), Uchi ($+0.9$), English River ($+0.3$) and Wabigoon ($+0.4$) Sp agree within error with the Quetico sediments ($+0.6$). Only the average of $\epsilon_{Nd}^{2.7Ga}$ values of rocks from Winnipeg River (-3.1) and Wawa ($+2.1$) Sp are significantly different. These differences could reflect a real difference in the crusts of

these two subprovinces or they could be produced by a sampling bias in favor of the older gneisses in the Winnipeg River Sp and in favor of more mafic compositions within the Wawa Sp.

(2) The average Nd isotopic composition of shales and greywackes from the Quetico Sp (average $\epsilon_{Nd}^{2.7Ga}$ value of $+0.6$ with T_{2DM} ages of 2.88 Ga, Henry et al., 1998) are in good agreement with Nd isotope data for shales from other parts of the Superior Province such as the Abitibi Sp having Nd model ages of 2.85 Ga (Dia et al. 1990).

(3) The relative amounts of 3.4, 3.0 and 2.7 Ga crust, calculated with Quetico compositions and an erosion coefficient $K=2$, agree with the frequency of U–Pb ages of detrital zircons from sedimentary sequences of the WSP (Davis et al., 1990; Davis, 1996, 1998) if grouped into similar time intervals of <2.76, 2.76–3.02 and >3.02 Ga.

3.3. Cratonic evolution

In comparison with other Precambrian terranes such as the Kaapvaal, Slave, Yilgarn, North Atlantic and Birimian cratons, the Superior Province is unique for the amount of late Archean crust that it contains and is one of the largest Archean cratons. It has traditionally been regarded as a largely late Archean Craton of 2.7–2.8 Ga in age. This is largely due to the early concentration on data from the Abitibi subprovince which, although it is the most economically important region of the Superior Province, is bereft of the older multi-cyclic greenstone belt sequences that characterize the western part of the Superior Province (e.g. Corfu and Davis, 1992).

The Kaapvaal Craton, for example, certainly has a longer, and perhaps more varied, history than the Superior Craton, dating from at least 3.6 Ga (e.g. de Witt et al., 1992; Jahn and Condie, 1995). Sedimentary units from this craton have also been the subject of extensive geochemical and isotopic studies aimed at placing constraints on the crustal protolith of the craton and its evolution (Wronkiewicz and Condie, 1989; Stevenson and Patchett, 1990; Jahn and Condie, 1995). The overall conclusion of these studies is that there is little evidence for crust/rocks older than 3.6 Ga, which is about the maximum age of igneous units found within the Kaapvaal Province (e.g. Armstrong et al., 1991). In contrast, a number of Nd model ages from our data for the Superior Craton exceed the ages of the oldest igneous units within the craton by up to 200 Ma (3.4 vs. 3.2 Ga). Nd isotope studies of sediments from the Abitibi subprovince lead to a similar conclusion (Dia et al., 1990; Feng et al., 1993), and detrital zircons as old as 3.2–3.5 Ga have also been identified in sedimentary rocks from the Superior Province (Gariépy et al., 1984; Davis, 1996). These data suggest that the sedimentary basins of the Superior Province received material(s) from external sources in the form of sediments that were likely derived through the erosion of pre-existing older cratons. Possible candidates could be the North Atlantic craton (Bridgwater et al., 1973) or gneisses of the Minnesota River Valley (Morey and Van Schmus, 1988). An alternative explanation may be that the >3.2 Ga crust resides at depth or has not been identified yet. Nevertheless, the signature of this older protolith has crept into the isotopic compositions of both the sediments and the pre-tectonic volcanic and tonalitic units throughout the Superior Province through crustal recycling such as subduction of sediments into the tonalite source areas beneath island arcs (Henry et al., 1998).

This exterior sedimentary influence suggests that the terranes of the Superior Province did not form in an open ocean basin far from continental contaminants, as has been suggested for the Proterozoic Birimian Terranes of West Africa (Abouchami et al., 1990; Boher et al., 1992). Rather, the Superior Province seems to represent a collage of arc terranes (Williams, 1990; Percival et al., 1994) which encompass small earlier continental slivers (for example, Lumby Lake gneisses within Wabigoon Sp., Cedar Lake gneisses and others within Winnipeg River Sp.) as well as sediments that received material from distal and more ancient crusts. An environnment similar to that of southeast Asia and Indonesia is perhaps more likely.

The early Birimian terranes of West Africa are thought to reflect the geochemical characteristic of oceanic plateaux created by mantle plume magmatism (Abouchami et al., 1990) and have been held as a model for the formation of the Superior

Province. Terranes of such nature have also been identified in the Superior Province (Desrocher et al., 1993; Tomlinson, 1996), but they represent only a small percentage of the overall terranes. This does not, however, necessarily diminish their importance as possible catalysts in the formation of the initial subduction zones that formed the Superior crust. The juncture of ordinary oceanic crust with thick piles of mafic to ultramafic volcanic rocks of the plateaux could have served as a point of lithospheric weakness leading to the initiation of subduction.

4. Conclusions

Sm–Nd compositions from 2.65–3.17 Ga volcanic, plutonic, gneissic and metasedimentary rocks provide a general overview of the formation of the WSP crust. Nd isotopic data suggest the existence of three main events, 3.4, 3.0 and 2.7 Ga, where crust is created and isolated from the depleted mantle. Granitoids that formed between these ages are essentially the products of recycled crustal precursors. Finally, this model of crustal growth in the WSP is in agreement with previous studies (e.g. Dia et al., 1990; Davis, 1996) and proposes that the WSP is formed by 48%, 44% and 7 wt% of crusts created at 2.69–2.76, 2.92–3.02 and 3.4 ± 0.1 Ga, respectively.

The Sm–Nd data discussed above and shown in conventional ϵ_{Nd}^{t} values versus age diagrams, allow the following observations with regard to the debate concerning the processes of crustal growth in the Archean (by example, see Albarède, 1998 and De Smet et al., 1998):

(1) The concept of episodic crustal growth in the Archean must be considered. The data from the WSP documents two periods of important crustal growth, each lasting up to 190 Ma (3.02–2.92 and 2.76–2.67 Ga) which are separated by a 160 Ma period dominated by crustal recycling. Local volcanism does exist during the 'crustal period', but on the whole, the amount of new crustal addition does not modify the average composition of the WSP crust.

(2) The crustal isotopic signatures produced during the 2.92–2.76 Ga period imply that the pre-2.76 Ga crust was thick and/or hot enough to initiate crustal partial melting. The heat for partial melting could be derived through tectonic thickening of the crust or episodic mantle plume volcanism.

(3) With the exception of a few plutons emplaced into the greenstone belts that exhibit depleted mantle-like isotopic compositions, most of the pre-, syn- and post-tectonic plutonic rocks are contaminated by old crust. Although the crustal contamination of the sanukitoid suites can be shown to have occurred during emplacement in the continental crust (Stevenson et al., 1999), mixing relationships for the more predominant TTG suites strongly suggest that the crustal contamination occurred in the magma source area (Henry et al., 1998). This strongly correlates the creation of new crust and the recycling of early continental crust in Archean subduction zones.

Acknowledgements

The authors thanks F. Corfu (Royal Ontario Museum) and D. Stone (Ontario Geological Survey) who graciously donated samples from the North Caribou Terrane (labeled C 93) and D. King and P. Fralick who provided the four metasedimentary rocks from the Finlayson Lake greenstone belt (# I–IV). Jean Carignan, Christophe Innocent, Jean David and John Ludden are thanked for frank discussions and suggestions that led to improvements in this paper. R. Lapointe and GEOTOP's students are thanked for their help in maintaining the mass spectrometer laboratory at UQAM. Serge André (Univ. Franche-Comté) is thanked for his help in drawing Figs. 2 and 3. A. Dickin and an anonymous reviewer are acknowledged for their constructive comments. This work was funded by LITHOPROBE, NSERC and FCAR GRANTS to Stevenson and Gariépy. LITHOPROBE contribution No. 1058.

References

Abouchami, W., Boher, M., Michard, A., Albarède, F., 1990. A major 2.1 Ga event of mafic magmatism in West Africa:

An early stage of crustal accretion. J. Geophys. Res. 95, 17605–17629.

Albarède, F., 1998. The growth of continental crust. In: Vauchez, A., Meissner, R.O. (Eds.), Continents and their Mantle Roots. Tectonophysics 296, 1–14.

Allègre, C.J., Rousseau, D., 1984. The growth of the continent through geological time studied by Nd isotope analysis of shales. Earth Planet. Sci. Lett. 67, 19–34.

Armstrong, R.A., Compston, W., Retief, E.A., Williams, I.S., Welke, H.J., 1991. Zircon ion microprobe studies bearing on the age and evolution of the Witwatersrand triad. Precambrian Res. 53, 243–266.

Beakhouse, G.P., McNutt, R.H., 1991. Constrasting types of Late Archean plutonic rocks in northwestern Ontario: implications for crustal evolution in the Superior Province. Precambrian Res. 49, 141–165.

Blackburn, C.E., Bond, W.D., Breaks, F.W., Davis, D.W., Edwards, G.R., Poulsen, K.H., Trowell, N.F., Wood, J., 1985. Evolution of Archean volcanic-sedimentary sequences of the western Wabigoon Subprovince and its margins: a review. In: Evolution of Archean Supracrustal Sequences, Geol. Assoc. Can. Spec. Pap. 28, 89–116.

Blackburn, C.E., Johns, G., Ayer, J., Davis, D.W., 1991. Wabigoon Subprovince, Thurston, P.C., Williams, H.R., Sutcliffe, R.H., Stott, G.M. (Eds.), Geology of Ontario Ont. Geol. Surv. Special Vol. 4., 303–381.

Boher, M., Abouchami, W., Michard, A., Albarède, F., Arndt, N., 1992. Crustal growth in West Africa. J. Geophys. Res. 97, 345–369.

Bridgwater, D., Watson, J., Windley, B.F., 1973. The Archean craton of the North Atlantic region. Philos. Trans. R. Soc. London A 273, 493–512.

Card, K.D., Ciesielski, A., 1986. Subdivisions of the Superior Province of the Canadian Shield. Geosci. Can. 13, 5–14.

Corfu, F., 1988. Differential response of U–Pb systems in coexisting accessory minerals, Winnipeg River Subprovince, Canadian Shield: implications for Archean crustal growth and stabilization. Contrib. Mineral. Petrol. 98, 312–325.

Corfu, F., 1996. U–Pb geochronology and evolution of the Uchi, English River and Winnipeg River Subprovinces, Superior Province. In: Harrap, R.M., Helmstaedt, H. (Eds.), Western Superior Transect Second Annual Workshop (Oct. 20–21, 1995). Lithoprobe Report 53., 74–80.

Corfu, F., Andrews, A.J., 1987. Geochronological constraints on the timing of magmatism, deformation, and gold mineralization in the Red Lake greenstone belt, northwestern Ontario. Can. J. Earth Sci. 24, 1302–1320.

Corfu, F., Davis, D.W., 1992. A U–Pb geochronological framework for the western Superior Province, Ontario, Thurston, P.C., Williams, H.R., Sutcliffe, R.H., Stott, G.M. (Eds.), Geology of Ontario, Ont. Geol. Surv., Special Vol. 4., 1335–1346.

Corfu, F., Stone, D., 1999. Age structure and orogenic significance of the Berens River composite batholiths, western Superior Province. Can. J. Earth Sci. 35, 1089–1109.

Corfu, F., Stott, G.M., 1993a. Age and petrogenesis of two Late Archean magmatic suites, northwestern Superior Province, Canada: zircon U–Pb and Lu–Hf isotopic relations. J. Petrol. 34, 817–838.

Corfu, F., Stott, G.M., 1993b. U–Pb geochronology of the Central Üchi Subprovince, Superior Province. Can. J. Earth Sci. 30, 1179–1196.

Corfu, F., Wallace, H., 1986. U–Pb zircon ages for magmatism in the Red Lake greenstone belt, northwestern Ontario. Can. J. Earth Sci. 23, 27–42.

Corfu, F., Krogh, T.E., Ayres, L.D., 1985. U–Pb zircon and sphene geochronology of a composite Archean batholith, Favourable Lake area, northwestern Ontario. Can. J. Earth Sci. 22, 1436–1451.

Davis, D.W., 1996. Provenance and depositional age constraints on Sedimentation in the Western Superior Transect Area from U–Pb Ages of Zircons. In: Harrap, R.M., Helmstaedt, H. (Eds.), 1996 Western Superior Transect Second Annual Workshop (Oct. 20–21, 1995), Lithoprobe Report 53, 18–23.

Davis, D.W., 1998. Speculations on the formation and crustal structure of the Superior province from U–Pb geochronology, Western Superior Lithoprobe Transect Fourth Annual Workshop, Harrap, R.M., Helmstaedt, H.H. (Eds.), Lithoprobe Report 65, 21–28.

Davis, D.W., Jackson, M., 1988. Geochronology of the Lumby Lake greenstone belt: A 3 Ga complex within the Wabigoon subprovince, northwest Ontario. Geol. Soc. Am. Bull. 100, 818–824.

Davis, D.W., Smith, P.M., 1991. Archean gold mineralization in the Wabigoon Subprovince, a product of crustal accretion: Evidence from U–Pb geochronology in the Lake of the Wood area, Superior Province. Can. J. Geol. 99, 337–353.

Davis, D.W., Sutcliffe, R.H., Trowell, N.F., 1988. Geochronological constraints on the tectonic evolution of a Late Archaean greenstone belt, Wabigoon subprovince, northwest Ontario, Canada. Precambrian Res. 39, 171–191.

Davis, D.W., Pezutto, F., Ojakangas, R., 1990. The age and provenance of metasedimentary rocks in the Quetico subprovince, Ontario, from single zircon analyses: Implications for Archean sedimentation and tectonics in the Superior Province. Earth Planet. Sci. Lett. 99, 95–105.

De Paolo, D.J., 1981. Neodymium isotopes in the Colorado Front Range and crust-mantle evolution in the Proterozoic. Nature 291, 193–196.

De Smet, J.H., Van den Berg, A.P., Vlaar, N.J., 1998. Stability and growth of continental shields in mantle convection models including recurrent melt production, Vauchez, A., Meissner, R.O. (Eds.), Tectonophysics 296, 15–29. Special Issue: Continents and their Mantle Roots.

de Wit, M.J., Roering, C., Hart, R.J., Armstrong, R.A., de Ronde, C.E.J., Green, R.W.E., Tredoux, M., Peberdy, E., Hart, R.A., 1992. Formation of an Archean continent. Nature 357, 553–562.

Desrocher, J.-P., Hubert, C., Ludden, J.N., Pilot, P., 1993. Accretion of oceanic plateau fragments in the Abitibi greenstone belt. Can. Geol. 21, 451–454.

Dia, A., Dupré, B., Allègre, C.J., Gariépy, C., 1990. Sm-nd and trace element analysis on Canadian shales: consequences for the continental crust evolution through geological time. Can. J. Earth Sci. 27, 758–766.

Feng, R., Kerrich, R., Maas, R., 1993. Geochemical, oxygen, and neodymium isotope compositions of metasediments from the Abitibi greenstone belt and Pontiac Subprovince, Canada: Evidence for ancient crust and Archean terrane juxtaposition. Geochim. Cosmochim. Acta 57, 641–658.

Gariépy, C., Allègre, C.J., Lajoie, J., 1984. U–Pb systematics in single zircons from the Pontiac sediments, Abitibi greenstone belt. J. Can. Earth Sci. 21, 1296–1304.

Henry, P., Stevenson, R.K., Gariépy, C., 1998. Late Archean mantle composition and crustal growth in the Western Superior Province of Canada: Neodymium and lead isotopic evidence from the Wawa, Quetico and Wabigoon subprovinces. Geochim. Cosmochim. Acta 62, 143–157.

Jacobsen, S.B., 1988. Isotopic constraints on crustal growth and recycling. Earth Planet. Sci. Lett. 90, 315–329.

Jahn, B.M., Condie, K., 1995. Evolution of the Kaapvaal Craton as viewed from geochemical and Sm–Nd analyses of intracratonic pelites. Geochim. Cosmochim. Acta 59, 2239–2258.

Langford, F.F., Morin, J.A., 1975. The development of the Superior Province of northwestern Ontario by merging island arcs. Am. J. Sci. 276, 1023–1034.

Larbi, Y., Stevenson, R., Breaks, F., Machado, N., Gariépy, C., 1999. Age and isotopic composition of late Archean leucogranites: implications for continental collision in the Western Superior Province. J. Can. Earth Sci. 36, 495–510.

Ludden, J., Hubert, C., Gariépy, C., 1986. The tectonic evolution of the Abitibi greenstone belt of Canada. Geol. Mag. 123, 153–166.

Machado, N., Brooks, C., Hart, S.R., 1986. Determination of initial $^{87}Sr/^{86}Sr$ and $^{143}Nd/^{144}Nd$ in primary minerals from mafic and ultramafic rocks: Experimental procedure and implications for the isotopic characteristics of the Archean mantle under the Abitibi greenstone belt, Canada. Geochim. Cosmochim. Acta 50, 2335–2348.

McLennan, S.M., Hemming, S., 1992. Samarium/neodymium elemental and isotopic systematics in sedimentary rocks. Geochim. Cosmochim. Acta 56, 887–898.

Morey, G.B., Van Schmus, W.R., 1988. Correlation of Precambrian rocks of the Lake Superior region, United States. In: US Geological Survey Professional Paper, F1–F31.

Noble, S.R. 1989. Geology geochemistry and isotope geology of the Trout Lake Batholith and the Uchi-Confederation lakes greenstone belt, northwestern Ontario, Canada. Ph.D. thesis, University of Toronto, Toronto, Canada, 288 pp.

Ontario Geological Survey, Holweg, J.D., 1980. In: Ontario Geological Survey Geological Highway Map, Northern Ontario, O.G.S., Map 2440.

Ontario Geological Survey, 1991. In: Bedrock Geology of Ontario, West-central sheet, Ontario Geological Survey, Map 2542, Scale 1: 1 000 000.

Percival, J., Stern, R.A., Skulski, T., Card, K.D., Mortensen, J., Begin, N., 1994. Minto Block, Superior Province, missing link in deciphering assembly of the craton at 2.7 Ga. Geology 22, 839–842.

Shirey, S.B., Hanson, G.N., 1986. Mantle heterogeneity and crustal recycling in Archean granite-greenstone belts: Evidence from neodymium isotopes and trace elements in the Rainy Lake area, Superior Province, Ontario, Canada. Geochim. Cosmochim. Acta 50, 2631–2651.

Stevenson, R.K., 1995. Crust and mantle evolution in the Late Archean: Evidence from a Sm–Nd isotopic study of the North Spirit Lake greenstone belt, northwestern Ontario, Canada. Geol. Soc. Am. Bull. 107 (12), 1458–1467.

Stevenson, R.K., Patchett, P.J., 1990. Implications for the evolution of continental crust from hafnium isotope systematics of Archean detrital zircons. Geochim. Cosmochim. Acta 54, 1683–1697.

Stevenson, R., Henry, P., Gariépy, C., 1999. Assimilation-fractional crystallization origin of Archean Sanukitoid Suites: Western Superior Province, Canada. Precambrian Res. 96, 83–99.

Stott, G.M., Corfu, F., 1991. The Uchi Subprovince. In: Thurston, P.C., Williams, H.R., Sutcliffe, R.H., Stott, G.M. (Eds.), Geology of Ontario, Ont. Geol. Surv., Spec. Vol. 4., 145–236.

Taylor, S.R., McLennan, S.M., 1985. The Continental Crust: its Composition and Evolution. Blackwell Scientific, Oxford. 312 pp.

Tilton, G., Kwon, S.T., 1990. Isotopic evidence for crust-mantle evolution with emphasis on the Canadian Shield. Chem. Geol. 83, 149–163.

Tomlinson, K.Y., 1996. The geochemistry and tectonic setting of early Precambrian greenstone belts, northern Ontario, Canada. Ph.D. thesis, University of Portsmouth, UK, 287 pp.

Tomlinson, K.Y., Stevenson, R.K., Hughes, D.J., Hall, R.P., Thurston, P.C., Henry, P., 1998. The Red Lake greenstone belt, Superior Province: evidence of plume-related magmatism at 3 Ga and evidence of an older enriched source. Precambrian Res. 89, 59–76.

Vervoort, J.D., Patchett, P.J., Gehrels, G.E., Nutman, A.P., 1996. Constraints on early Earth differentiation from hafnium and neodymium isotopes. Nature 379, 624–627.

Williams, H.R., 1990. Subprovince accretion in the south-central Superior Province. Can. J. Earth Sci. 27, 570–581.

Wronkiewicz, D.J., Condie, K.C., 1989. Geochemistry and provenance of sediments from the Pongola Supergroup, South Africa: Evidence for a 3.0 Ga old continental craton. Geochim. Cosmochim. Acta 53, 1537–1549.

Episodic continental growth models: afterthoughts and extensions

Kent C. Condie *

Department of Earth & Environmental Science, New Mexico Institute of Mining & Technology, Socorro, NM 87801, USA

Received 15 January 1999; accepted for publication 16 August 1999

Abstract

Although there are still problems that need to be resolved with episodic models for continental growth, a large number of geochemical and geophysical observations can be explained with these models. The models have in common episodic collapse of subducted slabs through the 660 km seismic discontinuity, mantle convection changing from layered to whole mantle, and each collapse correlates with major episodes of crustal formation. Isotopic age distributions of juvenile continental crust suggest that crustal formation episodes at 2.7, 1.9, and 1.2 Ga may have been very short, each ≤100 My in length. The supercontinent cycle probably operates independently of slab avalanches at the 660 km seismic discontinuity, except for the first supercontinent at 2.7 Ga, which may have formed in response to the first slab avalanche. The production rate of continental crust may increase during slab avalanches in response to increased production rate of oceanic plateaus and of subduction-related crust.

Crustal recycling rate may drop significantly below crustal production rate during slab avalanches due to the formation of supercontinents, which trap juvenile crust. If supercontinents are responsible for the preservation of large amounts of juvenile continental crust, then they play an important role in the growth of continental crust with time. During each of three superevents at 2.7, 1.9, and 1.2 Ga, the mantle may have been well mixed, thus losing geochemical evidence for enriched mantle sources. © 2000 Elsevier Science B.V. All rights reserved.

Keywords: continental growth; isotopic ages; mantle plumes; supercontinents

1. Introduction

Several models have been published in the last few years that propose a relationship between catastrophic slab avalanches in the mantle and the episodic distribution of ages in continental crust (Condie, 1998a). Most of these models have in common episodic collapse of subducted slabs that accumulate at the 660 km seismic discontinuity, changing mantle convection from layered to whole mantle, and each catastrophic collapse correlates with major episodes of crustal formation.

Several important problems have appeared for these models and it is the purpose of this short contribution to discuss these problems and to suggest some solutions. First, let us briefly review the major models that have appeared in the literature in recent years.

2. Review of models

One of the first models for episodic continental growth is that of McCulloch and Bennett (1994).

---

* Tel.: +1-505-835-5531. fax: +1-505-835-6436.
  E-mail address: kcondie@nmt.edu (K.C. Condie)

0040-1951/00/$ - see front matter © 2000 Elsevier Science B.V. All rights reserved.
PII: S0040-1951(00)00061-5

They propose a non-recycling model involving three reservoirs: continental crust, depleted mantle, and primitive mantle. The model is based chiefly on Nd isotopic data but also accommodates Sr and Pb isotopes and incompatible element distributions in each reservoir. It assumes that the volume of depleted mantle increases with time in a stepwise manner, which is linked to major episodes of continental crust formation at 3.6, 2.7, and 1.8 Ga. The isotopic and trace element composition of the upper mantle is buffered by progressive extraction of continental crust and increasing size of the depleted mantle reservoir. The buffered composition of the upper mantle remains depleted in incompatible elements through time.

Stein and Hofmann (1994) were among the first to advocate that episodic instability at the 660 km seismic discontinuity may control the growth of continental crust. They suggested that convection patterns changed in the mantle from layered convection (the normal case), when the growth rates of continental crust were relatively low, to whole-mantle convection when the growth rates were high. Whole-mantle convection occurs in short-lived episodes during which subducted slabs that have accumulated at the 660 km discontinuity catastrophically sink into the lower mantle, in a manner similar to that proposed by Tackley (1997). Stein and Hofmann (1994) show that such a model can account for several geochemical observations, including: (1) ϵ_{Nd} values of the upper mantle are maintained between primitive and depleted mantle values; (2) trace element models for crustal extraction require that the mantle reservoir source is larger than the upper mantle (above the 660 km discontinuity), but less the entire mantle; (3) primordial ^{3}He and radiogenic ^{40}Ar must be present in both the lower and upper mantle; (4) trace element distributions and Pb isotopes require the upper mantle and continental crust to have evolved as an open system. One of the important features of the Stein and Hofmann (1994) model is that during periods of whole-mantle convection, plumes rise from the D" layer above the core and replenish incompatible elements to the upper mantle, which has been depleted by oceanic crust and arc formation.

Based on the same theme of instability at the 660 km discontinuity and using parameterized mantle convection, Davies (1995) proposed catastrophic global magmatic and tectonic events at 1 to 2 Gy spacings. The favored models show layered convection, which becomes unstable and breaks down episodically to whole-mantle convection as in the Stein and Hofmann (1994) model. The Early Archean overturn cycles may occur on time scales of a few hundred million years. During the catastrophic mantle overturns, hot lower mantle material is transferred to the upper mantle and may be responsible for rapid episodic growth of juvenile crust, as well as replenishing the upper mantle with incompatible elements.

Peltier et al. (1997) extended thermal constraints to evaluate the catastrophic mantle models more thoroughly. These investigators quantified the physical processes that control the Rayleigh number at the 660 km discontinuity, which in turns controls the frequency of slab avalanches at this discontinuity. They also suggest a correlation between avalanche events and the supercontinent cycle. Their results imply that slab avalanches occur at a spacing of 400–600 My, and that they are brought about by the growth of an instability in the thermal boundary layer at the 660 km discontinuity. During and after slab avalanches a large mantle downwelling is produced directly above the avalanches, and this downwelling attracts fragments of continental lithosphere, thus leading to the formation of a supercontinent.

Based on the episodic age distribution of continental crust, Condie (1998a) proposed that continents grew episodically, with major periods of growth at 2.7, 1.9, and 1.2 Ga (Fig. 1). Each maximum in continental growth reflects a superevent in the mantle caused by catastrophic slab avalanching at the 660 km discontinuity, and also correlates with supercontinent formation. The total duration of each of the three major superevent cycles is about 800 My. In addition to the three major superevents in the Precambrian (Fig. 1), the superplume events in the late Paleozoic and Mid-Cretaceous (Larson, 1991) may have been caused by minor slab avalanches at the 660 km seismic discontinuity. The intensity of slab avalanches should decrease with time because the seismic discontinuity becomes increasingly permeable to

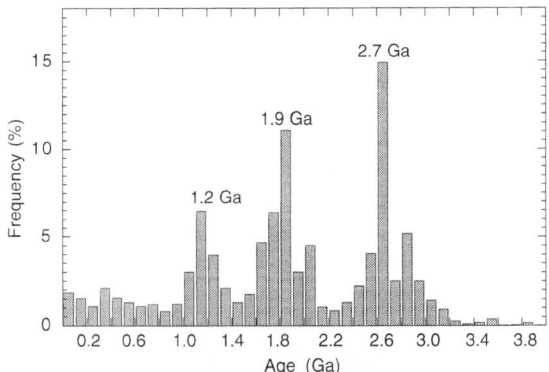

Fig. 1. Distribution of U/Pb zircon ages in juvenile continental crust. After Condie (1998a).

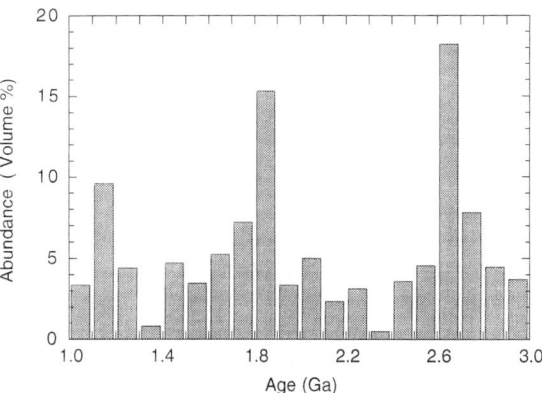

Fig. 2. Revised distribution of U/Pb zircon ages in juvenile continental crust between 3 and 1 Ga. Abundance is proportional to aerial distribution of juvenile age provinces scaled from an equal-area projection of the continents.

slab penetration as the mantle cools and the Rayleigh number decreases.

3. Discussion

3.1. Duration of superevents

Condie (1998a) suggested that each of the catastrophic superevents responsible for the three peaks in continental crust production rate may last for several hundred million years. Each superevent may comprise several subevents (Condie, 1998a, pp. 100–101), each lasting from 50 to 80 My. Since a superevent should happen rather rapidly in response to instability at the 660 km seismic discontinuity, a duration of 300–500 My seems unexplainably long, even if several 'breakthroughs' occur at the discontinuity at different locations as proposed by Condie (1998a).

Because new zircon ages have appeared since Condie (1998a), it is useful to re-examine the distribution of juvenile crustal provinces in light of these data. Zircon ages coupled with Nd isotopic data suggest that juvenile crust of 1.6–1.4 Ga may be much more widespread than suggested in Fig. 1. A potentially large belt of juvenile crust of this age has recently been recognized in western Amazonia (Condie and Melis, 1998; Geraldes et al., 1998), and several remnants of juvenile crust of this age are found in the Grenville province in Canada and in SW Scandinavia (Ahall et al., 1995; Rivers, 1997). Also, numerous zircon ages and Nd isotopic analyses suggest the existence of juvenile crustal belts of 2.3–2.1 Ga rocks in West Africa extending into the Guiana shield and in the basement of western Canada (Abouchami et al., 1990; Ross et al., 1991; Condie and Melis, 1998).

A revised histogram of the time interval from 3 to 1 Ga, based on zircon ages and Nd isotopic data and extrapolated to unexposed regions in the basement suggests that the abundance of juvenile continental crust between the three major peaks varies between about 3 and 7% for each 100 My age increment (Fig. 2). This could mean that the 'subevents' suggested by Condie (1998a) to accompany each major superevent may be part of the 'noise' in background growth rate of continental crust in the Precambrian. It is interesting that only two 100 My windows at 1.4–1.3 Ga and 2.4–2.3 Ga lack significant juvenile crust, and this could be due to inadequate sampling of rocks of these ages. The Late Archean spike straddles 2.7 Ga with ages falling between 2.75 and 2.65 Ga.

If the subevents associated with each of the three major events are not real, but part of the background crustal growth, each superevent is restricted in time to no more than about 100 My. Because rocks associated with each superevent are found on all of the continents today, the effects of superevents may have been very widespread in the mantle. Whether the superevents record slabs breaking through the 660 km discontinuity at one or at several locations at the same time is not yet clear.

3.2. Supercontinent and superevent cycles

One aspect of the catastrophic crustal growth models that has not been addressed is whether the supercontinent cycle can operate independently of the superevent (slab avalanche) cycle. Peltier et al. (1997) suggest that the supercontinent cycle is caused by slab avalanche events in the mantle. In their model, the avalanches produce mantle downwellings directly over the avalanches, which act as 'catchment basins' for an aggregating supercontinent. However, if supercontinents accumulate over geoid lows and break up over mantle upwellings (Anderson, 1982; Gurnis, 1988; Lowman and Jarvis, 1996), both of which are a consequence of the supercontinent cycle, slab avalanches in the mantle may not be a necessary part of supercontinent formation. The mantle upwellings that eventually break supercontinents result from thermal shielding of a large volume of mantle from subduction (and cooling) beneath supercontinents (Hager et al., 1985; Gurnis, 1988) (Fig. 3). If so, how is the superevent cycle (Condie, 1998a) related to the supercontinent cycle? Perhaps slab avalanches can be considered as an 'add-on' to the supercontinent cycle (Fig. 3). A slab avalanche, which initiates plume bombardment of the lithosphere and consequent production of continental crust, may occur at any time during the supercontinent cycle. In the model of Condie (1998a), supercontinent breakup involving increased rates of subduction may trigger a slab avalanche, and this may be the only common denominator to both cycles. This applies, however, only to the Mid and Early Proterozoic avalanches, since there may not have been any supercontinent to fragment for the Late Archean avalanche. In the model of Condie (1998a), the superevent cycle 'injects' juvenile crust into the supercontinent cycle, thus increasing the volume of each succeeding supercontinent.

If this model is correct, as the Earth gets older the superevent cycle will end before the supercontinent cycle because the 660 km seismic discontinuity becomes more permeable to descending slabs as the mantle cools and the Rayleigh number drops. With the exception of possible mini-events in the late Paleozoic and in the Cretaceous, superevents appear to have ended after 1.2 Ga (Fig. 1). The supercontinent cycle, however, may continue as long as convection continues in the mantle, in response to shielding of parts of the mantle by large lithospheric plates.

3.3. The first supercontinent

One of the intriguing yet puzzling questions of any of the episodic models is that of just how and why the first supercontinent formed. There are no robust data that support the existence of a supercontinent prior to the Late Archean. For a supercontinent to form requires a significant volume of continental crustal fragments that survive recycling into the mantle. Prior to the Late Archean, the high mantle temperatures and inferred large mantle convection rates in response to large Rayleigh numbers probably resulted in rapid recycling of continental crust, presumably before continental pieces had time to collide to make a supercontinent

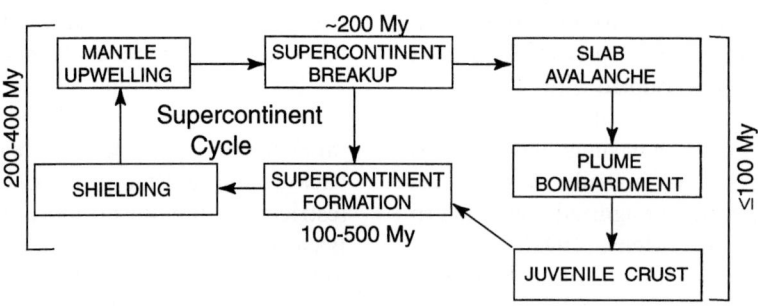

Fig. 3. The superevent cycle modified after Condie (1998a).

(Armstrong, 1991; Bowring and Housh, 1995). So what happened in the Late Archean that led to formation of the first supercontinent?

One possibility is that the first slab avalanche in the mantle at 2.7 Ga, which liberated mantle plumes from the D″ layer, led to the production of large volumes of continental crust in a relatively short period of time (≤100 My). If this is the case, unlike later supercontinents, the first supercontinent would form in response to the first slab avalanche. The mantle plumes resulting from the avalanche could produce juvenile crust in two ways: directly, by the production of oceanic plateaus, and indirectly by heating the upper mantle and increasing the production rate of ocean crust due to increased convection rates or/and increasing the total length of the ocean ridge system (Larson, 1991). The increased production rates of oceanic crust are accompanied by increased subduction rates, and hence increased rates of production of juvenile continental crust in arc systems. A supercontinent may form by collision of Archean oceanic plateaus (especially the thicker ones, which are negatively buoyant) (Cloos, 1993; Condie, 1997a), surviving fragments of continental crust older than 2.7 Ga, and oceanic arc systems. Also contributing to growth of the Late Archean supercontinent is a thick Archean subcontinental mantle lithosphere which is relatively buoyant (Griffin et al., 1998), thus resisting subduction during plate collisions.

If oceanic plateaus were a major contributor to the Late Archean supercontinent, the frequency of Late Archean greenstones with oceanic plateau geochemical affinities should be higher than their frequency in post-Archean greenstones. This seems to be the case from our limited database, using such incompatible element ratios as Th/Ta, La/Yb, and La/Nb (Rudnick, 1995; Condie, 1994, 1997a; Tomlinson and Condie, 1999). Plume-derived lavas contaminated by continental crust may acquire a pseudo-subduction zone geochemical signature (Ta–Nb depletion and Th enrichment). Hence, some Archean greenstone belts that have been described as arc-type based on geochemistry may represent crustally contaminated plume-type magmas. To distinguish these two cases, greenstone successions must be examined carefully, including

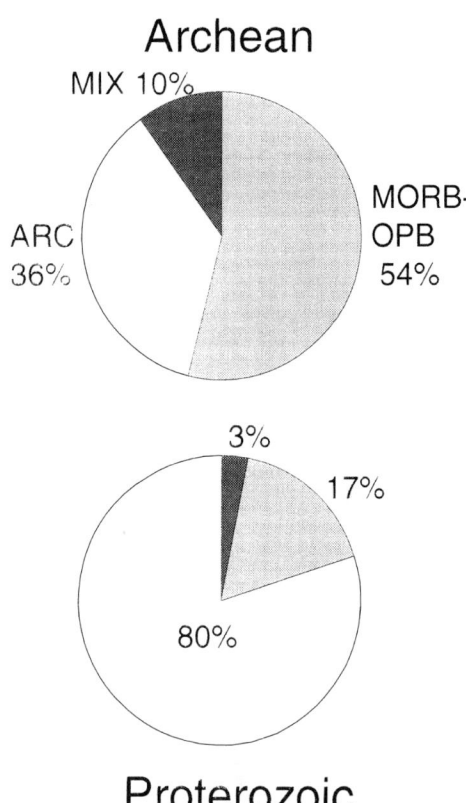

Fig. 4. Relative abundances of greenstones with arc and MORB-OPB affinities based on incompatible trace element distributions. Data sources given in Condie (1994). MORB, ocean ridge basalt; OPB, oceanic plateau basalt; MIX, greenstones with mixed tectonic affinities.

lithologic assemblage, trace element distributions (including Th/Ta, La/Nb and La/Yb relationships), and Nd isotopic studies to evaluate crustal contamination (Tomlinson and Condie, 1999). Because crustal contamination increases Th/Ta, La/Nb, and La/Yb ratios, the abundance of Archean greenstones with plume geochemical signatures represents a minimum.

Greenstones with MORB-OPB (oceanic plateau) affinities comprise about 50% of all Archean greenstones, whereas in the Proterozoic, only 17% have MORB-OPB affinities (Fig. 4). If a significant volume of Archean oceanic plateaus became incorporated in the lower crust (Condie, 1997a), we should see evidence for this in the trace element distributions of lower crustal and upper mantle

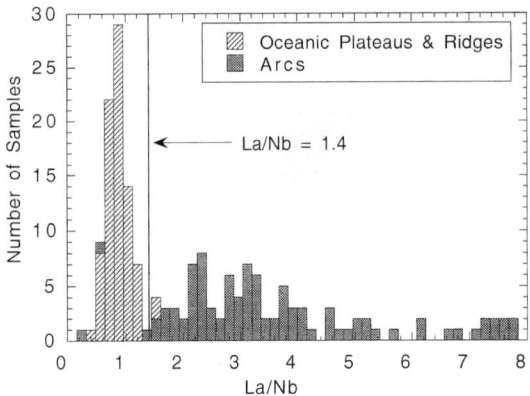

Fig. 5. La/Nb ratio in basalts from young oceanic plateaus and oceanic ridges compared with basalts from arcs. After Condie (1999).

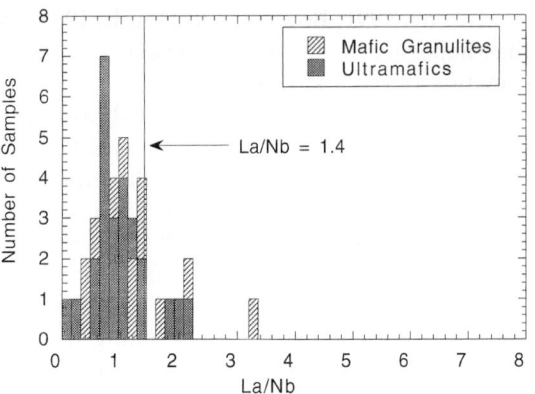

Fig. 6. La/Nb ratio in lower crustal and subcontinental lithospheric xenoliths (and exposed mantle fragments) of Archean age. References: Rogers (1977), Collerson et al. (1991), Huang et al. (1995) and Boyd et al. (1997).

xenoliths. The La/Nb ratio is a sensitive indicator of crust (and associated mantle lithosphere) with oceanic plateau and MORB affinities compared with crust with subduction affinities (Rudnick, 1995; Condie, 1999). Based on data from young basalts, Condie (1997a) suggested that OPB and MORB have La/Nb ratios < 1.4, whereas arc-related basalts have ratios > 1.4 (Fig. 5). One problem with using xenoliths to monitor Archean lower crust and mantle lithosphere composition is that these xenoliths (especially the mantle xenoliths) are often metasomatized, showing enrichment in LIL elements. Since LREE can be enriched during mantle metasomatism (Bodinier et al., 1990), only the lowest values of La/Nb may be useful in characterizing the original rocks. Although the database is small, most of the published La/Nb ratios for Archean lower crustal and mantle lithosphere xenoliths are low, consistent with dominantly mantle plume or MORB sources (Fig. 6). The Th/Ta ratio is an important index to distinguish between mantle plume and MORB sources, with oceanic plateau basalts derived from plumes typically having Th/Ta ratios > 1, whereas MORB sources have ratios generally < 1 (Tomlinson and Condie, 1999). An increasing database of Th and Ta data, especially from the Archean Superior Province in Canada, indicates that a significant proportion of Archean greenstones may have had mantle plume sources (Tomlinson and Condie, 1999). Although trace element distributions in greenstones and xenoliths seem to support a period of enhanced mantle plume activity in the Late Archean, which in turn could give rise to the first supercontinent, we need more geochemical data from greenstones and xenoliths of both pre- and post-2.7 Ga age to test this possibility further.

3.4. How do we make continental crust from mantle plumes?

In the model of Condie (1998a), mantle plumes produced by slab avalanches are responsible for producing juvenile continental crust during three superevents. Since mantle plumes give rise to basaltic magmas, which occur as oceanic plateaus, flood basalts, or mafic underplate of continental crust, how do plume-derived mafic rocks change into felsic continental crust? There are at least two ways of doing this (Condie and Chomiak, 1996; Condie, 1997a): (1) oceanic plateaus, being thick and buoyant, collide with and become part of the continental crust, and (2) mantle plumes heat up the upper mantle, increasing the production rate of both oceanic crust and arc-related crust, the latter of which becomes, at least in part, continental crust.

From a comparison of Th/Ta, La/Nb and Ni distributions in greenstones, Condie (1994, 1997a, 1999) suggested that oceanic plateaus may be

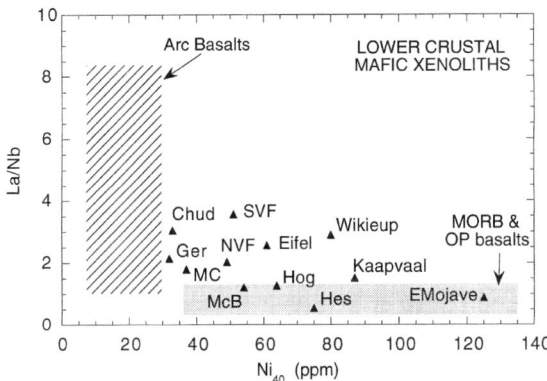

Fig. 7. La/Nb versus Ni_{40} diagram showing the distribution of lower crustal mafic xenoliths. References given in Condie (1999). Chud, Chudleigh, Australia; McB, McBride, Australia; Ger, Geronimo, Arizona; MC, Massif Central, France; SVF, San Francisco volcanic field, Arizona; NVF, Navajo volcanic field, Arizona/New Mexico; Hog, Hoggar, Algeria; Hes, Hessian, Germany; Eifel, Germany; Kaapvaal, South Africa; E Mojave, California. (For Eifel, Ni estimated from Ni/Co = 3; Nb from Nb/Ta = 17). Ni_{40} = Ni content in parts per million at Mg# of 40.

important in at least parts of the lower continental crust. For instance, Wrangellia in NW Laurentia may be an example of a relatively young oceanic plateau accreted to a continent (Condie, 1997a). The average composition of many mafic lower crustal xenoliths plot in or near the field of MORB-OPB compositions on a La/Nb versus Ni_{40} plot (Fig. 7). Because both crustal and host magma contamination should raise the La/Nb ratio in mafic xenoliths, the proportion of plume component in the lower crust deduced from xenolith compositions (approximately 30%; Condie, 1997a) should be considered a minimum value. Also supporting an important plume component in the subcontinental lithosphere are the compositions of spinel lherzolite xenoliths. These xenoliths typically have median Th/Ta (<1.0), La/Yb (2.85), and La/Nb (0.29) ratios similar to oceanic plateau basalts (McDonough, 1990), consistent with the subcontinental lithosphere containing a significant plume component, either as accreted oceanic plateaus or as a plume-derived underplate. In both cases, the spinel lherzolites represent restite remaining in the lithosphere after extraction of basaltic magma.

It has long been recognized that continental crust may also form in arcs from subduction processes (Taylor and McLennan, 1985; Pearcy et al., 1990). Arcs may become part of continents either directly as continental margin arc systems, or by terrane accretion of island arcs (Condie and Chomiak, 1996). Hence, higher rates of subduction during superevents will also enhance the rate of formation of continental crust.

Whether plume or subduction components are most important in the formation of continental crust is not yet clear. From our current geochemical database for greenstones and lower crustal xenoliths, however, it would appear that arc components exceed plume components in most post-Archean continental crust (Rudnick, 1995; Condie and Chomiak, 1996; Condie, 1999).

3.5. *Recycling and episodic growth of continental crust*

Higher temperatures in the Archean mantle should enhance both extraction and recycling rates of continental crust (Fig. 8). Large recycling rates in the Archean may be responsible for the very minor amount of continental crust older than 3 Ga preserved in the geologic record (Armstrong, 1991; Bowring and Housh, 1995). It would seem also

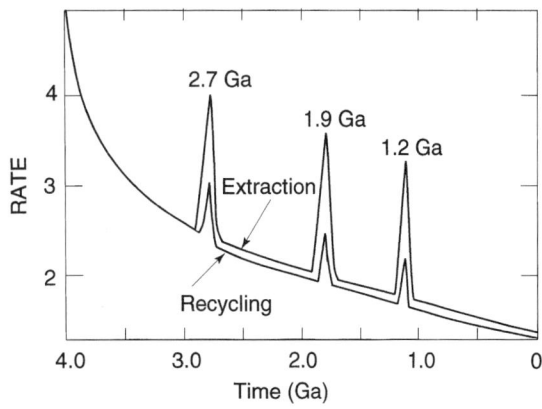

Fig. 8. Postulated rates of extraction and recycling of continental crust with time, relative to the present rates. Prior to about 3 Ga, extraction and recycling rates are assumed to be approximately equal. Background curves are assumed to follow the decrease of radiogenic heat in the mantle with time, and the relative height of peaks is arbitrary.

that during the three pulses of accelerated continental crustal growth rate at 2.7, 1.9, and 1.2 Ga, both extraction and recycling rates should increase. If this were the case, however, why do we have large volumes of continental crust of these ages that have survived?

Recycling of continental crust into the mantle can occur by at least two mechanisms, both of which appear to be potentially important (Kay and Kay, 1998; Plank and Langmuir, 1998): sediment subduction and delamination or slab break-off of lower crust during collisional events. If large amounts of continental crust are to be preserved during each of the three alleged superevents (Condie, 1998a), the extraction rate of continental crust must significantly exceed the recycling rate at these times (Fig. 8). Perhaps growth of a supercontinent during each superevent may be responsible for a *relative* decrease in mantle recycling rates. A large amount of juvenile continental lithosphere produced during each superevent may be 'trapped' and preserved in a growing supercontinent faster than it can be recycled. Because there is less perimeter per unit volume in a supercontinent than in the same volume of smaller continents, collisions should be less frequent, as should recycling by delamination and slab break-off. Also, internal river drainages should be more common in supercontinents and the total number of rivers entering the ocean per unit length of continent perimeter should be smaller in supercontinents than in small continents. Both of these features may lead to less sediment delivered to subduction zones for recycling and to less dissolved material from the continents being recycled into the mantle as altered oceanic crust.

If supercontinents are responsible for the preservation of large quantities of juvenile continental crust accompanying each superevent, then clearly they play a very important role in the growth of continental crust with time. If there were no supercontinents during episodes of enhanced juvenile crust formation, perhaps the total volume of continental crust today would be considerably less.

3.6. The appearance of enriched mantle

At least four, and perhaps as many as six compositional end members have been recognized in the mantle from radiogenic isotope and incompatible trace element distributions in basalts (Hart et al., 1992). The origin and location of these end members in the mantle is still a subject of controversy, although the depleted mantle component (DM) appears to comprise most or all of the mantle above the 660 km seismic discontinuity, and enriched mantle components (EM1, EM2, HIMU, FOZO) appear to reside in the lower mantle (Hart et al., 1992; Hofmann, 1997). The enriched components probably represent continental crust and lithosphere (EM) and oceanic crust (HIMU) that have been recycled into the mantle during subduction or during delamination or slab break-off. The enriched components are recognized chiefly in the Southern Hemisphere today, and are best represented in oceanic island basalts.

Only a few examples of Archean basalts with geochemical evidence for enriched mantle sources have been recognized (Xie et al., 1993; Francis et al., 1999; Tomlinson and Condie, 1999; Tomlinson et al., 1999), with the first major record for these sources appearing in basalts at about 1 Ga (Condie, 1994, 1997b). Why are these mantle components uncommon in most pre-1 Ga greenstone basalts that were erupted in oceanic tectonic settings? In particular, the HIMU component should be widespread in Archean greenstones, since the dominant crust in the Archean was probably oceanic crust and it was probably recycled rapidly into the mantle. Two reasons deserve consideration for the apparent absence of HIMU and EM components in Precambrian basalts. First, since these components are best developed in oceanic island basalts, and such basalts are rare in Precambrian greenstones, basalts with enriched mantle sources may have been overlooked. Equally plausible is the possibility that these components did not survive in the Precambrian mantle and especially in the Archean mantle due to more rigorous mixing in the mantle as suggested by Blichert-Toft and Albarede (1994) for the Archean. Also, during each of the three possible superevents the mantle may have been well mixed, thus losing much of the geochemical evidence for enriched and HIMU mantle domains. Mantle isochrons (Nd, Pb, Sr), however, suggest that geochemical domains began to be preserved in the mantle by about 2 Ga (Carlson, 1994; Hofmann, 1997), yet they are not recorded in mafic magmas until after this time. Why this is the case remains problematic, although it may be related to inadequate sampling of minor components in Proterozoic greenstones.

4. Conclusions

Consideration of the episodic distribution of ages of juvenile continental crust suggests that neither the 'big bang' nor the 'steady state' models of continental growth are acceptable. Instead, it would appear that there were three moderate 'bangs' in crustal growth at 2.7, 1.9, and 1.2 Ga. Although there are still problems that need to be resolved with episodic models for continental growth, a large number of seemingly unrelated geochemical and geophysical observations can be explained with such models. To evaluate the models more quantitatively we need more greenstone isotopic ages and incompatible element data, especially from greenstones that fall in the time windows between the three major age peaks at 2.7, 1.9, and 1.2 Ga (Figs. 1 and 2). Lower crustal and mantle xenolith ages and geochemical data are also needed to test the importance of plume components in the lower continental crust and subcontinental lithosphere. And finally, we need more Nd isotopic data, which together with zircon ages can better constrain the amount of juvenile continental crust formed during each of the three crustal growth episodes.

Geophysical data are needed to better constrain the slope to the 660 km seismic discontinuity reaction and more precisely relate stability of this discontinuity to increasing Rayleigh number with increasing mantle temperature in the Precambrian. Also, experimental data are needed to evaluate more precisely the role of change in composition at the 660 km seismic discontinuity. Increasingly sophisticated geophysical models, including three-dimensional models, will also help constrain the timing of slab avalanches and their relation to plume generation in the D'' layer.

Acknowledgements

This paper was significantly improved from comments by Paul Sylvester, Rob Kerrich, and Don Francis.

References

Abouchami, W., Bohler, M., Michard, A., Albarede, F., 1990. A major 2.1 Ga event of mafic magmatism in West Africa: an early stage of crust accretion. J. Geophys. Res. 95, 17 605–17 629.

Ahall, K., Persson, P., Skiold, T., 1995. Westward accretion of the Baltic shield: implications from the 1.6 Ga Amal–Horred belt, SW Sweden. Precambrian Res. 70, 235–251.

Anderson, D.L., 1982. Hotspots, polar wander, Mesozoic convection and the geoid. Nature 297, 391–393.

Armstrong, R.L., 1991. The persistent myth of crustal growth. Aust. J. Earth Sci. 38, 613–630.

Blichert-Toft, J., Albarede, F., 1994. Short-lived chemical heterogeneities in the Archean mantle with implications for mantle convection. Science 263, 1593–1596.

Bodinier, J.L., Vasseur, G., Vernieres, J., Dupuy, C., Fabries, J., 1990. Mechanisms of mantle metasomatism: geochemical evidence from the Lherz orogenic peridotite. J. Petrol. 31, 597–628.

Bowring, S.A., Housh, T., 1995. The Earth's early evolution. Nature 269, 1535–1540.

Boyd, F.R., Pokhilenko, N.P., Pearson, D.G., Mertzman, S.A., Sobolev, N.V., Finger, L.W., 1997. Composition of the Siberian cratonic mantle: evidence from Udachnaya peridotite xenoliths. Contrib. Mineral. Petrol. 128, 228–246.

Carlson, R.W., 1994. Mechanisms of earth differentiation: consequences for the chemical structure of the mantle. Rev. Geophys. 32, 337–361.

Cloos, M., 1993. Lithospheric buoyancy and collisional orogenesis: subduction of oceanic plateaus, continental margins, island arcs, spreading ridges, and seamounts. Geol. Soc. Am. Bull. 105, 715–737.

Collerson, K.D., Campbell, L.M., Weaver, B.L., Palacz, Z.A., 1991. Evidence for extreme mantle fractionation in the Early Archean ultramafic rocks from northern Labrador. Nature 349, 209–214.

Condie, K.C., 1994. Greenstones through time. In: Condie, K.C. (Ed.), Archean Crustal Evolution. Elsevier, Amsterdam, pp. 85–120. Chapter 3

Condie, K.C., 1997a. Contrasting sources for upper and lower continental crust: the greenstone connection. J. Geol. 105, 729–736.

Condie, K.C., 1997b. Sources of Proterozoic mafic dyke swarms: constraints from Th/Ta and La/Yb ratios. Precambrian Res. 81, 3–14.

Condie, K.C., 1998a. Episodic continental growth and supercontinents: a mantle avalanche connection? Earth Planet. Sci. Lett. 163, 97–108.

Condie, K.C., 1999. Mafic crustal xenoliths and the origin of the lower continental crust. Lithos. in press

Condie, K.C., Chomiak, B., 1996. Continental accretion: contrasting Mesozoic and Early Proterozoic tectonic regimes in North America. Tectonophysics 265, 101–126.

Condie, K.C., Melis, E.A., 1998. Filling the age gaps: a search for lost arcs. In: Proceedings of COPENA Conference, July 18–26, 1998, Montana State University, Bozeman, MT Part III., 10–11.

Davies, G.F., 1995. Punctuated tectonic evolution of the Earth. Earth Planet. Sci. Lett. 136, 363–379.

Francis, D., Ludden, J., Johnstone, R., Davis, W., 1999. Picrite evidence for more Fe in Archean mantle reservoirs. Earth Planet. Sci. Lett. 167, 197–213.

Geraldes, M.C., Van Schmus, W.R., Teixeira, W., 1998. Age of Proterozoic crust in SW Mato Grosso, Brazil: evidence for a 1450 Ma magmatic arc in SW Amazonia. Geol. Soc. Am. Abst. With Programs 30 (6), A96–A97.

Griffin, W.L., O'Reilly, S.Y., Ryan, C.G., Gaul, O., Ionov, D.A., 1998. Secular variation in the composition of subcontinental lithospheric mantle: geophysical and geodynamic implications. In: Geological Society of America and American Geophysical Union Geodynamics Series, 26., 1–25.

Gurnis, M., 1988. Large-scale mantle convection and the aggregation and dispersal of supercontinents. Nature 332, 695–699.

Hager, B.H., Clayton, R.W., Richards, M.A., Comer, R.P., Dziewonski, A.M., 1985. Lower mantle heterogeneity, dynamic topography and the geoid. Nature 313, 541–545.

Hart, S.R., Hauri, E.H., Oschmann, L.A., Whitehead, J.A., 1992. Mantle plumes and entrainment: isotopic evidence. Science 90, 273–296.

Hofmann, A.W., 1997. Mantle geochemistry: the message from oceanic volcanism. Nature 385, 219–229.

Huang, Y., Van Calsteren, P., Hawkesworth, C.J., 1995. The evolution of the lithosphere in southern Africa: a perspective on the basic granulite xenoliths form kimberlites in South Africa. Geochim. Cosmochim. Acta 59, 4905–4920.

Kay, R., Kay, S.M., 1998. Crustal recycling. Geol. Soc Am. Abst. With Programs 30 (6), A243–A244.

Larson, R.L., 1991. Latest pulse of Earth: evidence for a mid-Cretaceous superplume. Geology 19, 547–550.

Lowman, J.P., Jarvis, G.T., 1996. Continental collisions in wide aspect ration and high Rayleigh number two-dimensional mantle convection models. J. Geophys. Res. 101 (25), 485–497.

McCulloch, M.T., Bennett, V.C., 1994. Progressive growth of the Earth's continental crust and depleted mantle: geochemical constraints. Geochim. Cosmochim. Acta 58, 4717–4738.

McDonough, W.F., 1990. Constraints on the composition of the continental lithospheric mantle. Earth Planet. Sci. Lett. 101, 1–18.

Pearcy, L.G., DeBari, S.M., Sleep, N.H., 1990. Mass balance calculations for two sections of island arc crust and implications for the formation of continents. Earth Planet. Sci. Lett. 96, 427–442.

Peltier, W.R., Butler, S., Solheim, L.P., 1997. The influence of phase transformations on mantle mixing and plate tectonics. In: Crossley, D.J. (Ed.), Earth's Deep Interior. Gordon & Breach, Amsterdam, pp. 405–430.

Plank, T., Langmuir, C.H., 1998. The chemical composition of subducting sediment and its consequences for the crust and mantle. Chem. Geol. 145, 325–394.

Rivers, T., 1997. Lithotectonic elements of the Grenville province: review and tectonic implications. Precambrian Res. 86, 117–154.

Rogers, N.W., 1977. Granulite xenoliths from Lesotho kimberlites and the lower continental crust. Nature 270, 681–684.

Ross, G.M., Parrish, R.R., Villeneuve, M.E., Bowring, S.A., 1991. Geophysics and geochronology of the crystalline basement of the Alberta basin, western Canada. Can. J. Earth Sci. 28, 512–522.

Rudnick, R.L., 1995. Making continental crust. Nature 378, 571–578.

Stein, M., Hofmann, A.W., 1994. Mantle plumes and episodic crustal growth. Nature 372, 63–68.

Tackley, P.J., 1997. Effects of phase transitions on three-dimensional mantle convection. In: Crossley, D.J. (Ed.), Earth's Deep Interior. Gordon & Breach, Amsterdam, pp. 273–336.

Taylor, S.R., McLennan, S.M., 1985. The Continental Crust: its Composition and Evolution. Blackwell, London. 312 pp.

Tomlinson, K.Y., Condie, K.C., 1999. Archean mantle plumes: evidence from greenstone belt geochemistry. In: Geological Society of America Special Paper., in press.

Tomlinson, K.Y., Hughes, D.J., Thurston, P.C., Hall, R.P., 1999. Plume magmatism and crustal growth at 2.9 and 3.0 Ga in the Steep Rock and Lumby Lake area, western Superior Province. Lithos 46, 103–136.

Xie, Q., Kerrich, R., Fan, J., 1993. HFSE/REE fractionations recorded in three komatiite-basalt sequences, Archean Abitibi greenstone belt: implications for multiple plume sources and depths. Geochim. Cosmochim. Acta 57, 4111–4118.

Quantifying Precambrian crustal extraction: the root is the answer

Dallas Abbott [a,*], David Sparks [b], Claude Herzberg [c], Walter Mooney [d], Anatoly Nikishin [e], Yu Shen Zhang [f]

[a] *Lamont–Doherty Earth Observatory, Palisades, New York, NY 10964, USA*
[b] *Texas A and M, College Station, Texas, TX 77843, USA*
[c] *Rutgers University, New Brunswick, NJ 08903, USA*
[d] *US Geological Survey, Menlo Park, CA 94025, USA*
[e] *Geological Faculty, Moscow State University, Moscow 119899, Russia*
[f] *UC Santz Cruz, Santa Cruz, CA 95064, USA*

Received 7 June 1999; accepted for publication 19 October 1999

Abstract

We use two different methods to estimate the total amount of continental crust that was extracted by the end of the Archean and the Proterozoic. The first method uses the sum of the seismic thickness of the crust, the eroded thickness of the crust, and the trapped melt within the lithospheric root to estimate the total crustal volume. This summation method yields an average equivalent thickness of Archean crust of 49 ± 6 km and an average equivalent thickness of Proterozoic crust of 48 ± 9 km. Between 7 and 9% of this crust never reached the surface, but remained within the continental root as congealed, iron-rich komatiitic melt. The second method uses experimental models of melting, mantle xenolith compositions, and corrected lithospheric thickness to estimate the amount of crust extracted through time. This melt column method reveals that the average equivalent thickness of Archean crust was 65 ± 6 km, and the average equivalent thickness of Early Proterozoic crust was 60 ± 7 km. It is likely that some of this crust remained trapped within the lithospheric root. The discrepancy between the two estimates is attributed to uncertainties in estimates of the amount of trapped, congealed melt, overall crustal erosion, and crustal recycling. Overall, we find that between 29 and 45% of continental crust was extracted by the end of the Archean, most likely by 2.7 Ga. Between 51 and 79% of continental crust was extracted by the end of the Early Proterozoic, most likely by 1.8–2.0 Ga. Our results are most consistent with geochemical models that call upon moderate amounts of recycling of early extracted continental crust coupled with continuing crustal growth (e.g. McLennan, S.M., Taylor, S.R., 1982. Geochemical constraints on the growth of the continental crust. Journal of Geology, 90, 347–361; Veizer, J., Jansen, S.L., 1985. Basement and sedimentary recycling — 2: time dimension to global tectonics. Journal of Geology 93(6), 625–643). Trapped, congealed, iron-rich melt within the lithospheric root may represent some of the iron that is 'missing' from the lower crust. The lower crust within Archean cratons may also have an unexpectedly low iron content because it was extracted from more primitive, undepleted mantle. © 2000 Elsevier Science B.V. All rights reserved.

Keywords: Archean; crust; growth rates; lithosphere; mantle; Precambrian; xenoliths

---

* Corresponding author.
  E-mail address: dallas@ldeo.columbia.edu (D. Abbott)

1. Introduction

One of the outstanding problems in understanding the evolution of the Earth's crust is determining the rate of continental growth over time. It is known that about 40% of the Earth's surface is covered by continental crust, which is best defined as an unsubductable layer of rocks and sediments (Abbott et al., 1997). Areas of pre-1.8 Ga crust have overall lithospheric thicknesses of 220–400 km, much thicker than the 120–150 km thick lithosphere in regions of post-1.8 Ga crust (Jordan, 1978, 1988; Lerner-Lam and Jordan, 1987; Nolet et al., 1994). In contrast, pre-1.8 Ga continental crust has (within error) the same thickness as post-1.8 Ga continental crust (Table 1). These similarities in crustal thickness juxtaposed with large differences in lithospheric thickness could arise because the crust was modified over time or because continental crust formed differently over time (Abbott and Hoffman, 1984; Martin, 1986).

The processes by which continental crust is initially formed are still debatable. There are basically two models: the island arc model and the oceanic plateau model (Condie, 1999). The island arc model for the origin of the continental crust states that new continents are formed at subduction zones by the partial melting of the mantle wedge above the subducting plate (and sometimes by the melting of the subducting plate itself) (Kay and Kay, 1986; Reymer and Schubert, 1986; Plank and Langmuir, 1988). Because much of the upper continental crust appears to have chemical characteristics similar to convergent zone magmas, this model goes far toward explaining the observed composition of much of the continental crust. It is clear that most continental crust is affected by arc volcanism during its evolution. The question

Table 1
Seismic thickness and volume of crust and lithospheric mantle in the region between 80°N and 80° S

Name	Minimum	Mean	Maximum
Archean crust			
Crustal thickness (km)	36.4	39.4	42.4
Surface area (10^6 km^2)	36.8	40.9	45.0
Total crustal volume (10^9 km^3)	1.3	1.6	1.9
Thickness depleted mantle (km)	197	203	209
Lithosphere volume (10^9 km^3)	7.3	8.3	9.4
Early Proterozoic			
Crustal thickness (km)	35.0	41.6	48.3
Surface area (10^6 km^2)	29.3	132.5	35.7
Total crustal volume (10^9 km^3)	1.0	1.4	1.7
Thickness depleted mangle (km)	188	196	204
Lithosphere volume (10^9 km^3)	5.5	6.4	7.3
Post-Early Proterozoic			
Crustal thickness (km)	30.3	37.1	43.9
Surface area (10^6 km^2)	78.4	87.1	95.8
Total crustal volume (10^9 km^3)	2.5	3.2	4.2
Thickness depleted mantle (km)	93	97	101
Lithosphere volume (10^9 km^3)	7.3	8.4	9.6
Total crust			
Total surface area (80N–80S)	144.4	160.4	176.4
Total crustal volume (10^9 km^3)	4.7	6.2	7.8
Volume mantle root (10^9 km^3)	20.1	23.1	26.3
Percentage volume upper mantle (crust)	0.5	0.7	0.9
Percentage volume upper mantle (root)	2.2	2.5	2.9
Percentage volume depleted upper mantle needed to make the crust	26.5	40.6	44.8

is whether or not arc volcanism (unaided by plumes) can extract sufficient melt from the mantle wedge to make enough unsubductable material to form the volume of continental crust that is observed.

The oceanic plateau model states that new continents are formed by high degrees of partial melting within mantle plumes (Abouchami et al., 1990; Boher et al., 1992; Stein and Goldstein, 1996). When they reach the Earth's surface, mantle plumes erupt large quantities of basaltic lava. In the ocean basins, these lavas form oceanic plateaus. When oceanic plateaus reach a subduction zone, some are too thick and too buoyant to subduct, and therefore, they become part of a continent (Abbott et al., 1997; Mann et al., 1997; Petterson et al., 1997). The magmatic processes at convergent margins, through processes of partial melting, fractional crystallization, magma mixing, and so forth, cause the observed composition of the crust to change to the more silicic compositions that appear at the surface.

At the current time, much of the data appears to favor the oceanic plateau model. Boninites, a type of magma that forms only where the overriding plate is both oceanic and hot (Pearce et al., 1992), are rare in the early Precambrian geological record (Abbott and Mooney, 1995; Wyman, 1999). If most early continental crust formed at island arcs, there should be many more recognized boninites. The continental crust also has too much Ni and Cr to be explained by the island arc model (Taylor and McLennan, 1985). A recent re-evaluation of the average composition of the continental crust shows that it is lower in K_2O (0.9%) than previously thought (Rudnick et al., 1998). This, too, favors the oceanic plateau model.

The observed episodicity of continental growth has been used as evidence for the oceanic plateau model (Abbott and Mooney, 1995; Stein and Goldstein, 1996). However, episodic continental growth can result from the increased seafloor spreading rates caused by superplumes (Larson, 1991), or it can result from the addition of unsubductable oceanic plateaus produced by plumes. In the former case, the increased spreading rates are accompanied by more rapid subduction rates, thereby increasing the rate of production of new island arcs. In the latter case, the continental crust grows in surface area and volume by the addition of plume-generated oceanic plateaus. Thus, episodic continental growth (Gurnis and Davies, 1986; Stein and Hofmann, 1994) is largely a result of superplume events. Superplumes may produce continental growth by a combination of plateau and arc addition.

Another observation that is difficult to explain is that oceanic plateau basalts are much richer in iron than average lower continental crust (Mahoney et al., 1992; Rudnick, 1992). Because the lower part of the continental crust has a basaltic composition (Christensen and Mooney, 1995; Rudnick and Fountain, 1995), it may represent relatively unmodified parts of an originally basaltic oceanic plateau. However, lower crustal xenoliths have average FeO contents of about 9% (Rudnick and Presper, 1990). This is much lower than the FeO content of Ontong Java plateau basalts (Mahoney et al., 1992), which are typically about 11% FeO. If continents are the result of plume volcanism, why is the lower continental crust not rich in iron? A reasonable model of continental growth must explain why the observed lower crust has so little iron (Rudnick, 1995).

It is possible that the low iron content of the lower continental crust may be the result of large quantities of unerupted, congealed komatiitic melt. Experimental data suggest that komatiitic melt could become too dense to rise into the upper levels of the mantle root (Herzberg et al., 1983; Agee and Walker, 1993; Ohtani et al., 1995). If so, iron-rich komatiitic melt might rise some distance from its source to a level of neutral density before becoming trapped within the lithospheric root. This melt would react with the surrounding mantle as it solidified. If such trapped, congealed melt had a sufficient volume, it might provide a complementary iron reservoir to the lower continental crust.

There are basically two types of geochemical models for the rate of growth of the continental crust.. The first is that of Armstrong (1981) and can be called the 'early growth/recycling' model. Armstrong (1981) proposed that the bulk of continental material was separated from the mantle early in the Earth's history and that subsequent apparent growth is the result of the recycling of pre-existing crust through erosion and reactivation. However, many others have proposed that the addition of new material to the continents has been continually occurring over the history of the Earth (e.g. McLennan and Taylor, 1982; Veizer and Jansen, 1985). In these 'steady growth' models, the growth of continental crust by the addition of new mantle-derived material began early in Earth history and is ongoing. The 'steady growth' model and the 'early growth/recycling' model each have profoundly different implications for the origin of cratons and lithospheric roots.

These two types of geochemical models are subject to direct tests by observations of the presently existing volume of early Precambrian continental crust. Any viable geochemical model of crustal growth must be able to produce, as a minimum product, the presently existing volume of early Precambrian crust. The earliest such measurements were those of Hurley and Rand (1969). They calculated the amount of continental crust of various ages based on the surface distribution of isotopic ages, and they assumed that there were no age-dependent differences in crustal thickness and density. One of the problems with the data of Hurley and Rand (1969) is that subsequent improvements in isotopic analyses and new mapping have shown that there is much more very old crust than they had known (Graham et al., 1999; Tucker et al., 1999).

In this paper, we attempt to constrain the minimum rate of continental growth over time by taking advantage of newer estimates of crustal thickness, crustal erosion, lithospheric thickness, lithospheric composition, and geochronology. We are particularly interested in estimating the amount of continental crust that was present by 2.4 Ga and by 1.8 Ga. We also show how the observed structure and composition of the continental crust and the lithospheric mantle could arise from accreted oceanic plateaus that were later modified by arc volcanism.

2. Assembling the essential databases

In order to constrain the minimum rate of continental growth, it was first necessary to compile and organize the data for each of the variables that we intended to use in our calculations. These variables include the distribution of crustal thickness, the age distribution of continental crust, the erosion level of the crust, and the estimated amount of trapped, congealed melt. It was also necessary to calculate the total volume of the lithospheric roots of the continents. The methods for compiling these data and the reasoning behind them are discussed in the following sections.

3. Estimating the thickness of the lithosphere

The lithospheric root represents the most depleted part of the residual mantle remaining after the crust had been extracted from the mantle. As the crust increased in volume, the volume of severely depleted lithospheric mantle also increased. Thus, the volume of this depleted residue is directly related to the amount of the crust that was extracted from the mantle at each time in Earth history. Because the oldest roots are so refractory and buoyant, they are very difficult to recycle into the convecting mantle As a result, the oldest lithospheric roots may remain beneath their original crust while the upper crust moves laterally due to erosion by streams, thrusting during compressional events and block faulting during extensional events. Consequently, the volume of the lithospheric root beneath cratons may contain a complementary record of the amount of crust that was extracted during early Earth history.

Because the volume of lithospheric roots is an indirect measure of the amount of crustal extraction, we need to estimate the thickness and lateral extent of continental lithospheric mantle. For the primary data set on lithospheric thickness, we used the S-wave, tomographic model of (Zhang and Tanimoto, 1991). This tomographic model has a nominal 5 by 5° resolution, but was resampled at a 2 by 2° resolution. We used horizontal slices through the model that were spaced at 20 km depth intervals. These slices were stacked on top of one another to derive a vertical cross-section of the Earth at a given point. Because the average velocity changes as a function of pressure, we worked with velocity anomalies rather than the absolute velocity.

The Zhang and Tanimoto model (Zhang and Tanimoto, 1991) is not corrected for the effect of asthenospheric temperature changes. Because the lithospheric mantle is chemically depleted in comparison with the surrounding asthenosphere, it has a different seismic velocity. However, the seismic velocity of the lithospheric mantle also changes as its temperature changes.

In a tomographic model that has not been corrected for temperature variations, a given anomaly in travel time will not correspond uniformly to a given degree of mantle depletion. That is, an uncorrected tomographic model will overestimate the thickness of the lithospheric mantle in areas where the surrounding asthenosphere is unusually cold, and it will underestimate the lithospheric thickness where the surrounding asthenosphere is unusually warm. Consequently, it was necessary to correct the model of Zhang and Tanimoto (1991). In order to do this, we had to estimate the variations in temperature in the asthenosphere.

4. Temperature variability in the asthenosphere

It is known that the asthenosphere at depth has large-scale lateral temperature variations. The problem has been to identify the relative contribution from shallow and deep temperature variations using techniques where all observation points are at the surface of the Earth. Recent seismic studies get around this problem by measuring the differential travel time between two seismic discontinuities (Gossler and Kind, 1996).

In the upper mantle, there are two major seismic discontinuities: one at approximately 410 km depth, and the second at approximately 660 km depth. These zones are called discontinuities because they have a large contrast in seismic velocities across them, and this sharp contrast in seismic velocities causes seismic energy to be reflected by them. The seismic energy that is reflected by both of these discontinuities is received at the same seismic station at slightly different times. The differences in travel times for seismic waves reflected from the two discontinuities are a measure of the temperature of the mantle (Liu, 1994; Gossler and Kind, 1996).

These discontinuities are caused by mineralogical phase changes in the mantle, but not all phase changes produce discontinuities. At roughly 410 km, olivine changes to its beta (higher pressure) form, wadsleyite, and at roughly 660 km, spinel olivine changes to Mg-perovskite. Both of these phase changes are strong reflectors of seismic energy and have been well-mapped. However, at about 500 km depth, beta olivine (wadsleyite) converts to the spinel form of olivine, ringwoodite. For unknown reasons, this phase change does not often produce an observable seismic discontinuity.

What is important for our purposes in this paper is that the exact depths of the olivine:beta olivine (circa 410 km) and spinel olivine:Mg-perovskite (circa 660 km) transitions are known to vary as a function of the temperature of the local mantle. Also, these two phase changes have different signs on their Clausius–Clapeyron slopes. That is, as the temperature decreases, the olivine:beta olivine transition occurs at a shallower depth and the olivine:Mg perovskite transition occurs at a deeper depth in the mantle. This is illustrated in Fig. 1. Thus, the variations in the travel times between the two discontinuities are due to both the variation in temperature and the increasing or decreasing distance between the discontinuities as a result of these variations in temperature. The correction for travel time variations due to the temperature of the lithospheric mantle is very small, only about 15% (Revenaugh and Jordan, 1991). Therefore, the bulk of the variation in travel time is the result of the differential movement of the two discontinuities (Fig. 1).

Overall, the temperature in the transition zone between the olivine:beta olivine (circa 410 km) and spinel olivine:Mg-perovskite (circa 660 km) transitions is estimated by Gossler and Kind (1996) to vary by about 200°C. This is a significant difference, but as stated above, the change in seismic velocity due to this temperature variation is small. However, the variation in temperature causes the relative depths of the seismic discontinuities to shift, so that the seismic energy must travel either a shorter or a longer distance between the discontinuities. It is this change in the depths of the

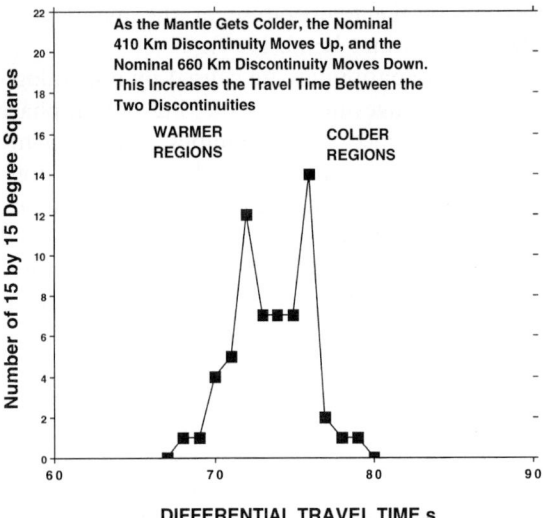

Fig. 1. Histogram of the travel time (in seconds) between the two major seismic discontinuities in the upper mantle. Both discontinuities are caused by phase changes in olivine. The phase changes cause a change in seismic velocity that results in an observable reflection of seismic waves. The travel time between the two discontinuities is a differential travel time derived by subtracting the arrival time of seismic energy reflected from the lower discontinuity from the arrival time of seismic energy reflected from the upper discontinuity. The differential travel times are from averages of data from 15 by 15° squares on the Earth.

discontinuities that causes almost all of the observed variations in the travel times between the two discontinuities.

Because lithospheric roots are highly depleted, their S-wave velocities are faster than those of comparable, undepleted mantle. The velocity changes are quite small, on the order of +0.5% (Jordan, 1978). Thus, they are in the same range as the velocity changes that are produced by temperature changes. We wish to correct for the effect of lateral temperature changes by finding some depth where there are no detectable velocity effects from lithospheric roots. However, the depth extent of lithospheric roots is debatable. Geochemists who study mantle xenoliths argue that they have a maximum thickness of only about 200 km (Boyd, 1987; Pearson et al., 1995b). Geophysicists who make tomographic models find that lithospheric roots go down to nearly 400 km depth (Grand, 1986; Lerner-Lam and Jordan,

1987). As a result, we needed to determine the maximum reasonable depth for lithospheric roots. We did this by looking at the correlation of slow seismic velocity anomalies with cratonic regions in successively deeper horizontal slices of the tomographic model. Once we reached a depth (350 km) where the slow seismic velocity anomalies were randomly correlated with cratonic regions, we concluded that we were below the lithospheric roots. Velocity anomalies at that depth are mainly the result of mantle temperature variations.

5. Correcting for lateral temperature variations

We used the velocity variations at 350 km depth as a proxy for temperature variations in the mantle. The velocity variations at 350 km depth have a relatively long wavelength. In order to calculate the variations in asthenospheric temperature, we must make two assumptions about the temperature structure of the mantle. Firstly, we assume that the lithospheric root is responsive to temperature changes in the underlying asthenosphere and that those temperature changes are observable by examining the velocity variations in the asthenospheric mantle. Secondly, we assume that heat transfer within the root involves linear thermal gradients (Rudnick et al., 1998).

The inferred horizontal length scale of variations in mantle temperature is quite broad, with average wavelengths of 1000–2000 km. This means that the overlying plate has millions of years to equilibrate its temperature profile with that of the underlying asthenosphere. The rate of heat transfer by conduction is on the order of 50 km in 10 Ma, equivalent to a plate tectonic velocity of 0.5 cm/yr. Continents typically move at velocities of 1–2 cm/yr (Stoddard and Abbott, 1996). However, radiative heat transfer goes as the absolute temperature to the fourth power, so there will also be a significant radiative component of heat transfer within the upper mantle. A recent re-evaluation of the temperature gradients within cratonic roots using improved geothermometry and geobarometry revealed linear, rather than curved, temperature–depth profiles (Rudnick et al., 1998; R. Rudnick, pers. commun.). If cratonic mantle roots were not in thermal equilibrium with the surrounding mantle, we would expect to find curved temperature–depth profiles using mantle xenoliths. This is not observed. Therefore, it is reasonable to assume that mantle roots are in thermal equilibrium with their surrounding asthenosphere.

To make the temperature corrections, the corrected tomographic model was merged into one file containing the vertical profiles of seismic velocity and of the velocity anomalies between 90 and 350 km depth. As explained above, we used the seismic data from the 350 km depth (the apparent maximum depth of compositional anomalies) as an index of the temperature of the asthenosphere. Then, we generated two sets of files: one in which the velocity data were not corrected for asthenospheric temperature variations, and one in which the data were corrected. The correction for the temperature of the asthenosphere was a linear conductive propagation of the velocity deviation at 350 km depth. We assumed that heat transfer between the asthenosphere and the overriding lithosphere produces linear variations in velocity anomalies with depth. That is, if the velocity deviation was $+1\%$ at 350 km depth, the velocity anomaly at 330 km was corrected downward by a quantity equal to $0.01 \times (330/350)$. Correspondingly, the velocity anomaly at 310 km was corrected downward by a quantity equal to $0.01 \times (310/350)$. We then used this result to correct the seismic velocities.

6. Computing the lithospheric thickness

We actually generated two models for lithospheric thickness: one using the uncorrected tomographic model of (Zhang and Tanimoto) 1991, and the other using the same tomographic model with corrections for lateral variations in asthenospheric temperature. Starting at 90 km depth, the calculation software is programmed to move down the velocity profile in a given tomographic slice of lithospheric mantle to a depth of 350 km. When the calculation reaches a depth where the measured seismic velocity is found to be less than $+1.005\%$ of the computed mean velocity of the Earth at that depth, that depth is registered as the lithospheric thickness for that slice. Because some lithosphere has seismic velocities between $+0.5\%$ and $+1.005\%$ of the mean velocity, our calculations provide a minimum estimate of lithospheric thickness. However, due to probable errors in the tomographic model, we decided to choose a conservative cut-off velocity between lithosphere and asthenosphere.

The results of the calculations using the uncorrected tomographic model are shown in Fig. 2a. The uncorrected tomographic model has a distribution of lithospheric thickness that is strongly bimodal between the continents and the ocean basins. Although some bimodality is expected, the uncorrected tomographic model shows too much thick lithosphere compared to what is known to exist from other studies. Furthermore, the uncorrected model shows an area of extremely thick lithosphere north of Australia that is partially due to subduction of buoyant continental lithosphere into the cold mantle beneath old, thick oceanic crust. Strictly speaking, this area does have a thick lithosphere, but the thickness of the continental lithosphere is overestimated.

The calculations using the corrected tomographic model produced much better results (Fig. 2b), although there are some lateral shifts in the location of thick lithosphere that we believe can be corrected with the better models that will eventually become available. Fig. 2b shows that the subduction zone artifact north of Australia is no longer present, and that the cratons in areas of recent hot spot activity in Africa and South America have a better resolution. Nevertheless, Africa and South America are much more poorly resolved compared to North America and Eurasia because there are fewer ray paths in Africa and South America. On average, the southern hemisphere has about half the number of ray paths of the northern hemisphere.

7. Comparing tomographic lithospheric thickness with a tectonic map

As a check on the results obtained with the corrected tomographic model (Fig. 2b), we made a tectonic map of lithospheric thickness (Fig. 3a). We used the following tectonic indicators of lith-

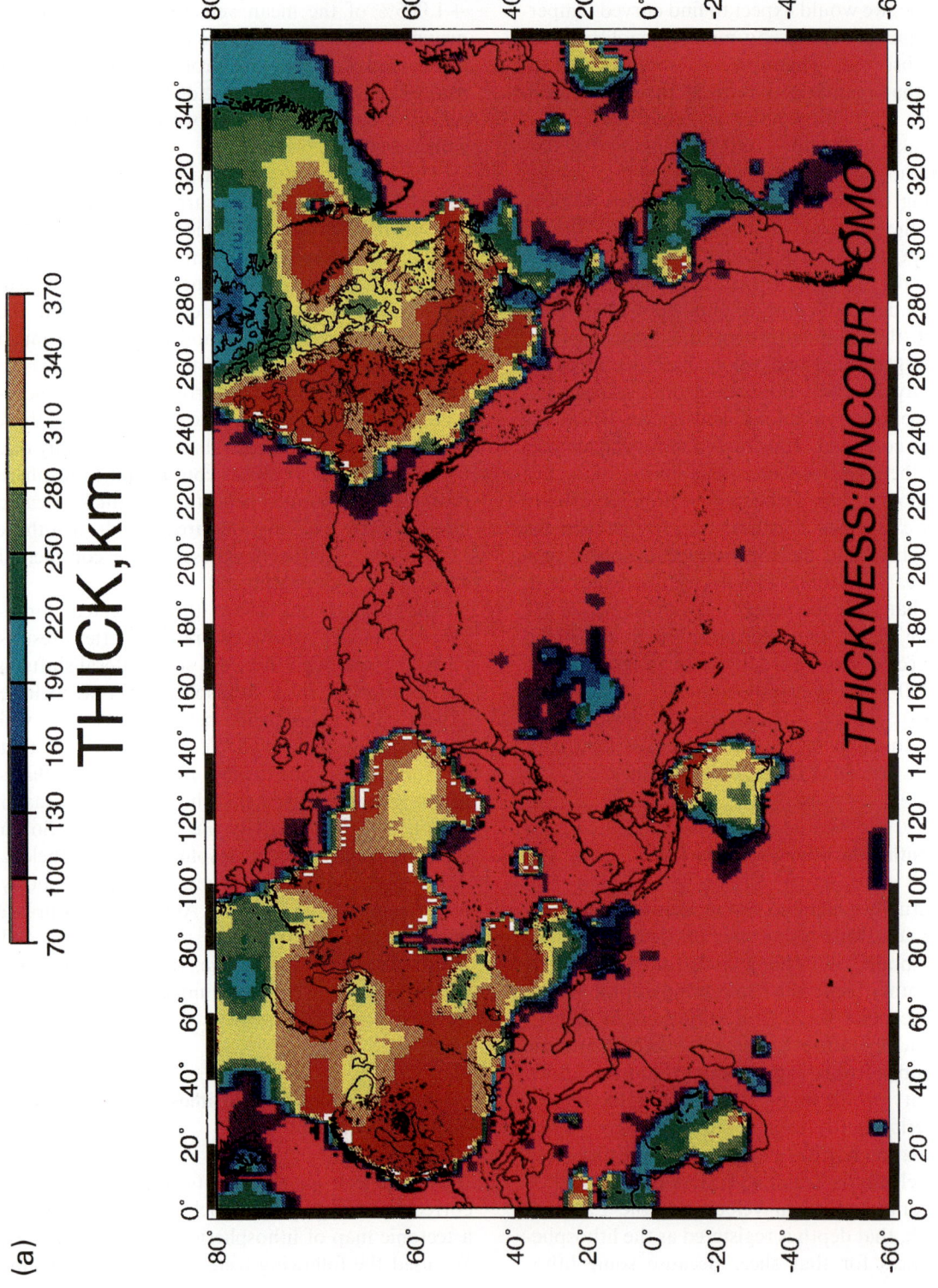

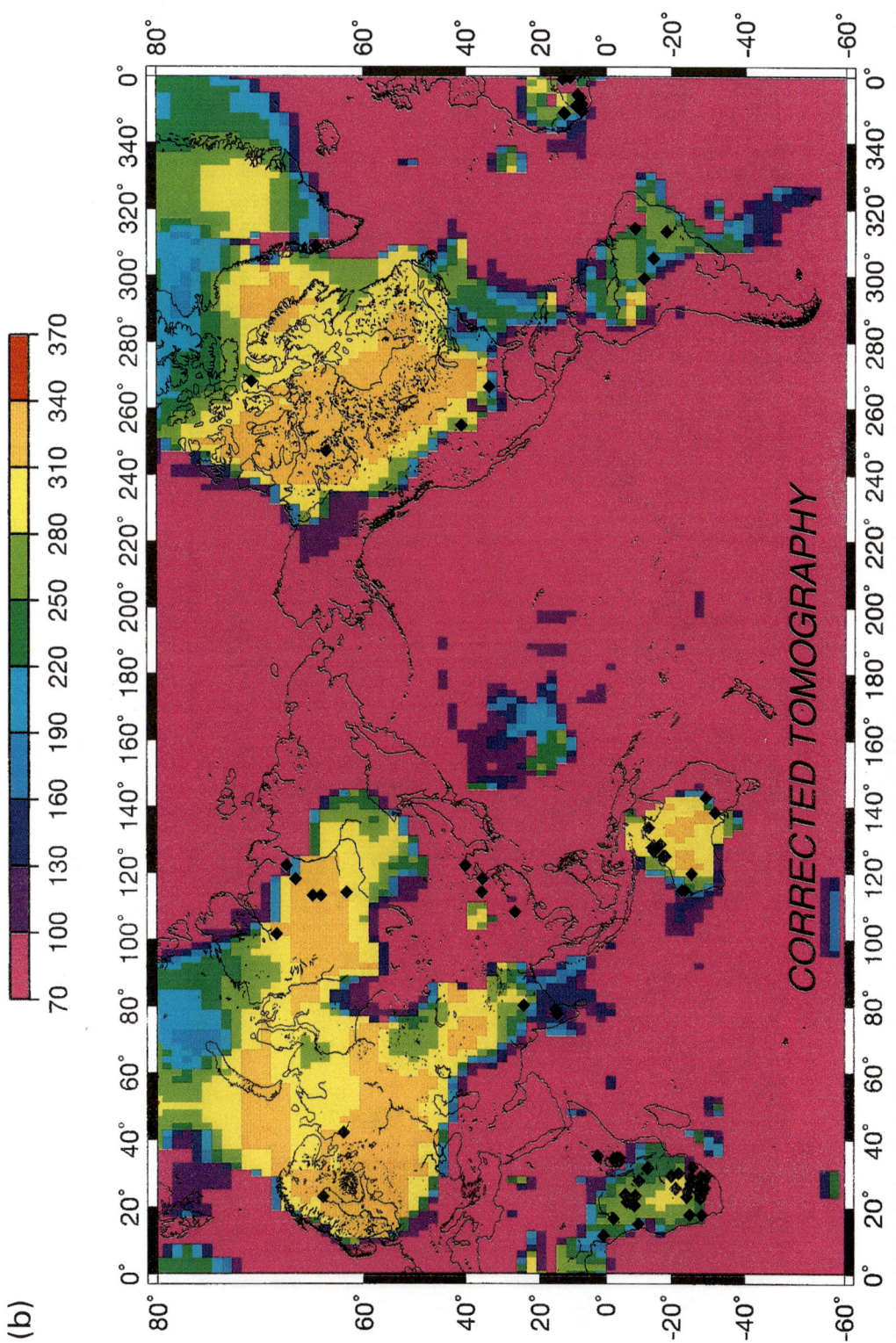

Fig. 2. (a) Global lithospheric thickness from a tomographic model that has not been corrected for temperature variations in the asthenosphere. The tomographic model has a typical resolution of 5 by 5°, but can be as poor as 8 by 8°. (b) Global lithospheric thickness from a corrected tomographic model. The tomographic model has a typical resolution of 5 by 5°, but can be as poor as 8 by 8°. Areas of thick lithosphere (>250 km) correspond broadly to cratonic areas. The resolution of the model is better in the northern hemisphere, where there are about twice as many ray paths. Areas of known diamondiferous kimberlites and lamproites appear as black diamonds.

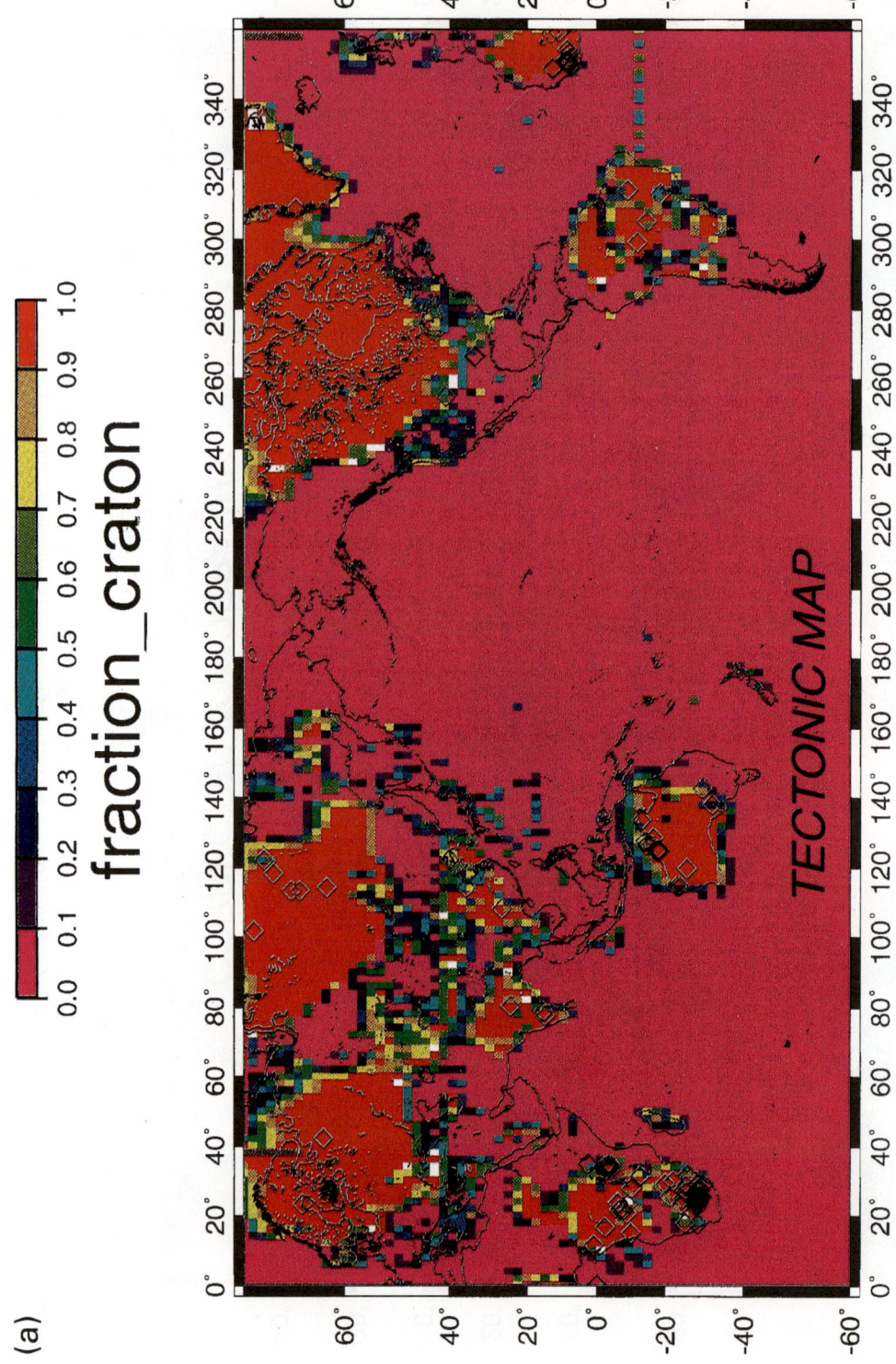

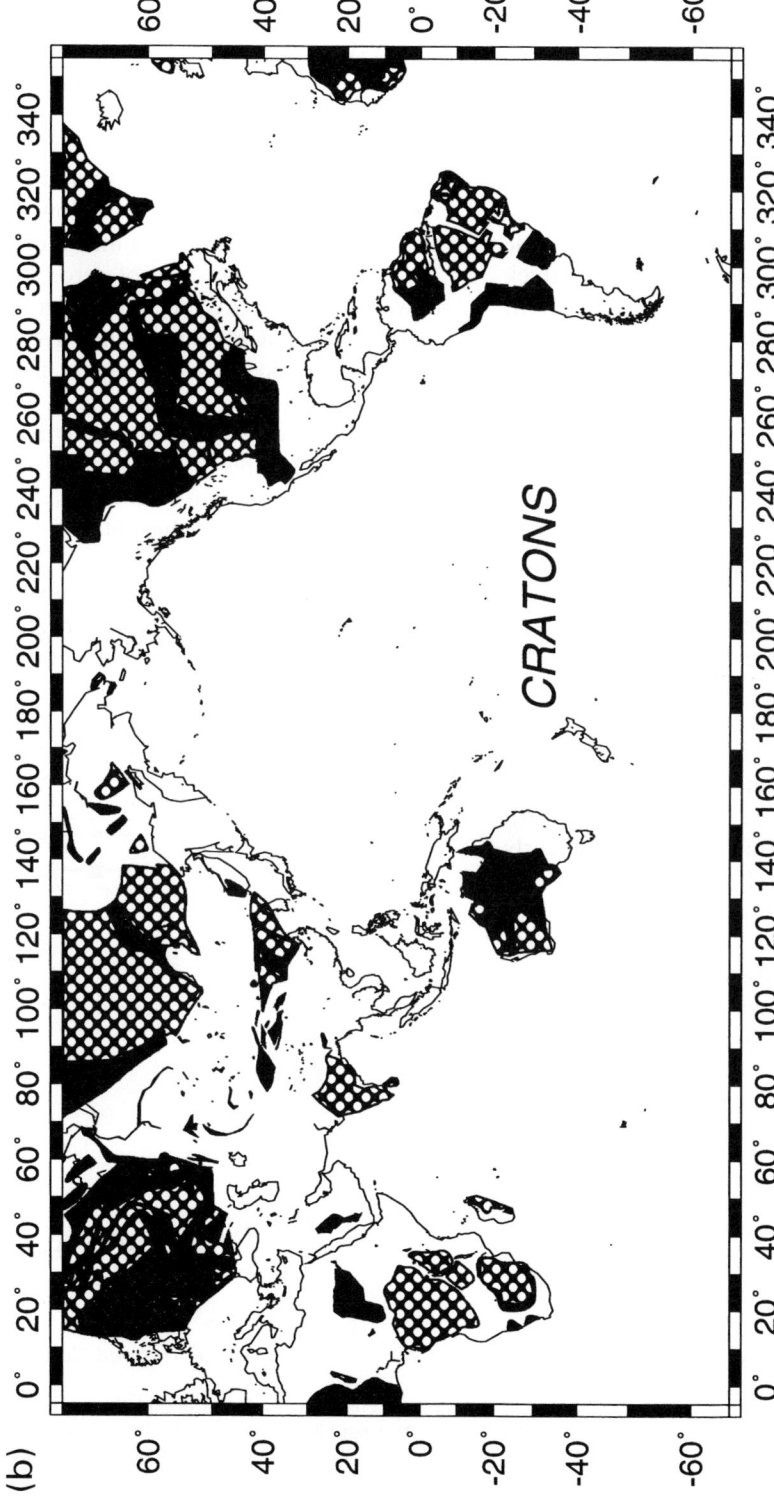

Fig. 3. (a) Fraction cratonic from cratonic indicators (age >1.8 Ga and heat flow <35 mW/m², and off craton indicators (age <1.8 Ga, heat flow >80 mW/m², and presence of basaltic volcanism). The data are averaged at a 2° grid spacing. Note that southern India and southern South America have cratonic areas that show up in this map that are poorly resolved in the tomographic model. Note also that the Madagascar shield is not well resolved on this map, due to the smoothing produced by a 2° grid spacing. Areas of known diamondiferous kimberlites and lamproites appear as white and black diamonds, respectively. (b) Estimated extent of Archean (black with white dots) and Early Proterozoic (black) continental crust.

ospheric thickness: heat flow, basement age, recent volcanism, and diamondiferous kimberlites or lamproites. Low heat flow and diamondiferous pipes are known to be characteristics of thick, cratonic lithosphere. Diamonds are not stable unless the lithospheric mantle is at least 150 km thick (Kennedy and Kennedy, 1969). Assuming average cratonic geotherms, the presence of diamondiferous mantle xenoliths is an indication that the underlying lithosphere is at least 175 km thick (Richter, 1988). Conversely, high heat flow, recent volcanism, and ages less than 1.8 Ga are usually correlated with lithosphere that is less than 150 km thick (Nixon, 1987; Nyblade and Pollack, 1993). The exceptions to these rules are some Archean areas that have had their thick lithosphere removed, for example, much of the Archean basement in China (Menzies et al., 1998).

The heat flow was derived from a database of reliable heat flow measurements, that is those from bore holes that are more than 100 m deep (Nyblade and Pollack, 1993). Because there are so few heat flow measurements within cratons, a simple gridding and smoothing of the data would bias the heat flow mean towards higher values.

For this reason, we divided the heat flow database into three groups: Archean (>2.4 Ga), early Proterozoic (1.8–2.4 Ga) and post-early Proterozoic (post-1.8 Ga) (Fig. 3b).

To prevent interpolation across craton boundaries, the boundaries of Archean and early Proterozoic cratons were assumed to have a heat flow value equal to the mean heat flow of these cratons. Where Archean and early Proterozoic cratons shared boundaries, they were assumed to have Archean values. The heat flow measurements were also truncated at values below 18 mW/m^2 (our estimated value of continental reduced heat flow) and above 250 mW/m^2 (estimated value of heat flow on 1 m.y. old crust) (Menke and Levin, 1994). These truncated heat flow values plus all reliable heat flow values were averaged within 1° squares. A smooth surface (GMT surface) was adjusted to fit the heat flow data exactly and to fill in all areas with no data.

After generating the heat-flow data set, we had two global data sets averaged over 1° blocks: basement age and heat flow. To make the tectonic map, we converted the two sets of gridded files (raster files) into ASCII files of points (lat, lon, value). We then assigned each set of points a value of 1 or zero based on the following strategy. If the area had a basement age greater than 1.8 Ga, it was given a a value of 1. If the area was younger, it was given a value of zero. In addition, areas of recent volcanic activity, heat flow higher than 80 mW/m^2, and mantle xenoliths erupted in alkali basalt were given a value of zero. Areas of heat flow less than 35 mW/m^2 were given a value of 1.

The ones and zeros were then averaged at a 2° grid spacing, like the tomographic model (Fig. 4). Two degree squares that have mean values of 0.9–1.0 plot as red, and are inferred to be cratonic. Two degree squares that have mean values of 0.0–0.1 plot as pink and are inferred to be non-cratonic. Areas with values of 0.9–0.1 are largely confined to craton boundaries.

When the tectonic map is compared with the tomographic map (Figs. 2b and 3b), we can evaluate the strengths and limitations of both methods

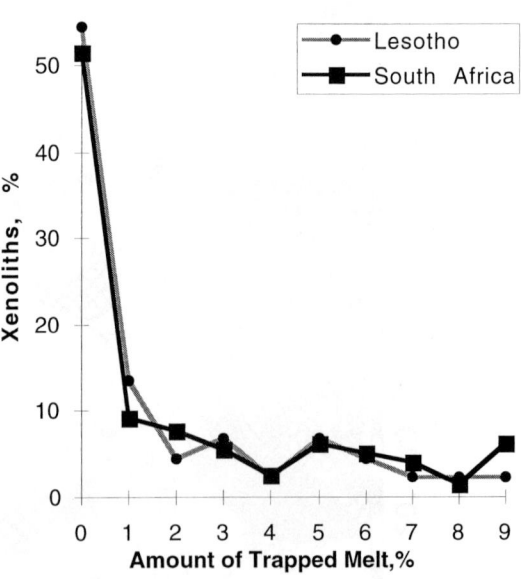

Fig. 4. Histograms of distribution of trapped, congealed melt in Archean (Kaapvaal) and Early Proterozoic (Lesotho) lithospheric mantle. As it cooled, this melt reacted with the surrounding, highly depleted mantle root.

for estimating lithospheric thickness. Remember that although the tomographic model was interpolated and resampled at a 2° grid spacing, the actual resolution of the tomographic model is just 5 by 5° in areas with larger numbers of ray paths. In areas with smaller numbers of ray paths (mostly in the southern hemisphere), the resolution is poorer, about 8 by 8°. In contrast, the tectonic map has a true 2° resolution. Thus, for the most part, the tectonic map is clearer, for example, in southern India. At the 5 by 5° resolution of the tomographic map, southern India would not appear as cratonic. On the tectonic map, southern India appears as cratonic because of a spatially restricted area of low heat flow (<35 mW/m^2) surrounding two areas with diamondiferous kimberlites (Nixon, 1987). For this reason, we infer that southern India has not lost its cratonic root. We attribute the fact that the tomographic model does not show the presence of this thick lithosphere to poor resolution in the mantle tomography of this area. The tomographic model does show a broad area of slightly thickened lithosphere that appears to surround India. We believe that this broad area is an artifact of the smearing of the data in the tomographic model due to its averaging of seismic travel times within 5 by 5° squares.

The limited resolution of the tomographic model also explains why some known narrow cratonic areas do not appear on the tomographic map. In general, most of the cratonic blocks in the southern hemisphere have a much smaller lateral extent than the cratonic blocks in the northern hemisphere. Because the southern hemisphere cratons are usually smaller, and because the tomographic model uses fewer rays in the southern hemisphere, the overall resolution of the southern hemisphere is much poorer in the tomographic model than in the tectonic model. The only exception to this is Australia, which is well resolved in the tomographic model. For example, at the 2° grid spacing on the tectonic map, the Archean/Early Proterozoic craton of Madagascar (Agrawal et al., 1992; Handke et al., 1997; Rambelson, 1998) appears as intermediate thickness lithosphere (Fig. 4). In other words, although the starting model contained the cratonic area in Madagascar, the 2° gridding interval averaged it with adjacent, non-cratonic crust. Because the Madagascar craton is only a few degrees wide, Madagascar lost its definition as a cratonic area. In South America, the tectonic map shows that the southernmost cratonic area constitutes a block of four separate 2 by 2° squares (Fig. 3a). In a tomographic model with an 8 by 8° resolution, these blocks would disappear.

8. Geological implications of the tomographic results

Although the tomographic results are uneven in quality, and there is some smearing due to the grid spacing of the model, Fig. 3a shows that the areas of thick lithosphere shown in the tomographic model correspond broadly to the areas of thick lithosphere that are known to exist. For example, the Kaapvaal craton (in Southern Africa), the Canadian shield, the east European platform, the Kenema-Man shield (in West Africa) and the Siberian craton all have inferred maximum thicknesses in excess of 250 km. All of these areas are clearly shown on the tomographic map.

The tomographic results also confirm what had been previously inferred from studies of mantle xenoliths and heat flow distributions. The lithospheric root has been removed in Wyoming (Eggler and Furlong, 1991) and in eastern China (Nixon, 1987; Nyblade and Pollack, 1993), where only relatively old diamondiferous kimberlites are found. The lithospheric root is still present is many areas where younger diamondiferous kimberlites are found, in Saskatchewan, at Prairie Creek, Arkansas, in the Slave Province, at Akhangelsk, northern Russia, in Siberia, on the Kaapvaal craton, on the west African craton, on the Kimberly craton, and in north central Australia (Nixon, 1987; Menzies et al., 1993; Lambert et al., 1995; Pearson et al., 1995a,b; Griffin et al., 1999). The tomographic results also show evidence of thick lithospheric roots in areas where placer diamonds are found, but the source kimberlites have not been located: in the Ukrainian shield region, in the Hoggar region of Africa (22N, 5E) and in the southern Urals (56N, 59E) (Bardet, 1973, 1974; Nixon, 1987).

The tomographic model also shows some unexpected results that are confirmed by more recent work. Fig. 3a shows that the largest area of thick lithosphere within the ocean basins is east of the Mariana–Izu–Bonin arc, in the western Pacific. Given the resolution of the model, this corresponds in part to the Ogasawara plateau. The Ogasawara plateau contains thickened crust whose origin is not known (Smoot and Richardson, 1988). The tomographic model also shows thick lithosphere in the model extending well south of the Ogasawara plateau to the area just east of the Mariana arc. Analyses of peridotites dredged from the Mariana forearc produced Re/Os model ages of 0.8–1.1 Ga (Parkinson et al., 1998). The old model ages may result from the incorporation of older depleted mantle into younger (Mesozoic) seamount magmas in the Pacific. The tomographic model also shows a small area of thickened lithosphere around the Seychelles islands, midway between India and Madagascar. The Seychelles islands have Precambrian granites (Yanagi et al., 1983). Plate reconstructions show that the Seychelles islands are continental fragments that remained behind when India was rifted from Madagascar. Thus, it is reasonable to conclude that the Seychelles might have thick lithospheric mantle beneath them.

9. Calculating the present-day volume of the continental crust and lithosphere

In order to calculate the volume of the continental crust and lithosphere at particular times in the past, it is necessary first to calculate the volumes of the continental lithosphere, lithospheric mantle, and continental crust that exist at the present time. We used a map of 1° squares of the surface of the Earth between 80°N and 80° S. Each 1° square was assigned to an age range based on the predominant age of the crust within the square. The age categories were: 0–1.8 Ga, 1.8–2.4 Ga, and pre-2.4 Ga (Stoddard and Abbott, 1996). We have updated this map using more recent estimates of crustal age in Northern Eurasia (Abbott and Nikishin, unpublished manuscript). Using spherical trigonometry, the surface area of each square was calculated from its average width measured in longitude and height measured in latitude. We derived the crustal thickness by sampling the global 5° grid of crustal thickness (Mooney et al., 1998) at the grid point closest to the center of each 1° square. The present-day crustal volume for each age group was calculated by summing the surface area for that age group and multiplying the total surface area by the mean crustal thickness for that age group. The standard error of the crustal thickness was subtracted or added to the mean crustal thickness to derive the probable range of crustal thicknesses in each age group. The minimum and maximum values of the crustal volume for each age group used the surface areas multiplied by the minimum and maximum values of the crustal thickness for that age group (Table 1).

The overall lithospheric thickness is derived by merging our 1° crustal thickness grid with the nearest point in the tomographic model. The tomographic velocity anomalies were calculated for 5 by 5° boxes and then resampled at a 2° grid spacing. The two grids are different in scale, and there are some Archean and Early Proterozoic blocks that are too small to be resolved by the tomographic model. These small blocks represent less than 9% of the overall surface area of Archean and Early Proterozoic crust, a small fraction of the total. Therefore, we did not introduce any large errors by using the average lithospheric thickness of the larger blocks as the average lithospheric thickness of all Archean and Early Proterozoic blocks. Then, we calculated the lithospheric volume by multiplying the average lithospheric thickness in each age range by the total surface area covered by continental crust within that age range (Table 1).

The mantle root is the most depleted part of the lithospheric mantle that was melted to make the crust. This part of the mantle was so intensely depleted by melting and its physical properties so profoundly changed, that it became isolated from the asthenosphere. This means that the average thickness of the mantle root supplies a conservative or minimum estimate of the volume of mantle that was depleted by extraction of the continental crust. We calculated the average thickness of the mantle root, T_r

$$T_r = \frac{\sum_i (T_1 - T_c)_i * (SA_i)}{SA_{total}}$$

where T_1 is the average lithospheric thickness of one region i, T_c is the average crustal thickness of one region i, SA_i is the surface area of one region i and SA_{total} is the total surface area of all regions of that age. We used the standard errors in average crustal and lithospheric thicknesses to derive estimates of the maximum and minimum thicknesses of the mantle root (Table 1).

10. Change in crustal volume through time: summation method

The crustal volume at any given time in Earth history is equal to the total volume of different parts of the crust: the volume of the surficial crust, the volume of trapped (i.e. unerupted) melt, and the volume of previously eroded crust. The volume of the surficial crust is equal to the surface area of the crust multiplied by the seismically determined crustal thickness. This is equal to the thickness of continental crust above a prominent seismic reflector, the Moho. Ideally, we would prefer to use the petrological thickness of the crust, that is, the total thickness of material that is geochemically and mineralogically continental crust rather than lithospheric mantle. Although, in some areas, the geochemical crustal thickness is not the same as the geophysical crustal thickness (Griffin and O'Reilly, 1987), in most cratonic areas, the two thicknesses are the same. One problem with this statement is that seismologists cannot distinguish between olivine cumulates derived from crystallizing komatiitic melts and olivine residues comprising the depleted lithospheric mantle (Durrheim and Mooney, 1991). If such olivine cumulates exist, it is not possible to account for them.

We determine the volume of crust that has been eroded by using recent estimates of the metamorphic grade of undeformed Archean granite-greenstone terranes (Galer and Metzger, 1998). The mean burial pressure of 1.5 ± 0.5 km corresponds to a mean depth of erosion of 5 ± 2 km since 3.0 Ga. We use the standard error of the mean depth of erosion to obtain a minimum depth of erosion of 3 km since 3.0 Ga. The maximum burial pressure of 3 ± 1 kb corresponds to a maximum depth of erosion of 10 ± 3 kb since 3.0 Ga. We use these numbers to calculate the minimum, average and maximum erosional levels of recent, Proterozoic, and Archean crust.

Although much erosion results from horizontal compression and deformation of terranes, thermal changes in the Earth's mantle also cause long-term uplift of terranes. This uplift is related to changes in the isostatic balance between the depleted mantle root underneath continents and the asthenosphere beneath the mantle root (Galer, 1991). As the asthenospheric temperature declined over time, the depleted mantle roots became relatively more buoyant. The increased buoyancy of their mantle roots produced uplift and erosion of the continental crust. Our models of erosion of the continental crust consider only the erosion due to isostatic uplift produced by the long-term decrease in the average temperature of the Earth's mantle (Abbott et al., 1994).

Because erosion due to isostatic uplift is produced by temperature changes in the Earth's mantle, it must change as the internal heat production of the Earth changes (Galer, 1991). Because the internal heat production of the Earth decays exponentially over time (Wasserburg et al., 1964), we argue that mantle temperatures must decay exponentially over time (Abbott et al., 1994). [The present data set of mantle temperatures over time is too sparse and could be fit equally well by a model of linearly decreasing mantle temperatures over time (Galer, 1991; Galer and Metzger, 1998).] However, if mantle temperatures do fall exponentially as the Earth ages, the total thickness of the layer of eroded crust must also decrease exponentially over time. We have made three best-fit exponentially decaying models for the total thickness of eroded continental crust versus time:

$$E_{min} = \exp(A \times 0.462098) - 1.0$$
$$E_{av} = \exp(A \times 0.597253) - 1.0$$
$$E_{max} = \exp(A \times 0.799298) - 1.0$$

where E_{min}, E_{av} and E_{max} are the minimum, average and maximum estimates of the thickness of eroded crust in kilometers, respectively, and A is the age of the Earth in Ga (10^9 years). At 3.0 Ga, these equations yield a total thickness of eroded crust of 3, 5 and 10 km, respectively. We then used these equations to calculate the minimum, average and mean thickness of crust eroded from continental areas with Archean, Early Proterozoic and post-Early Proterozoic ages. The results are shown in Table 2.

11. Volume of trapped, solidified melt

In addition to the existing surface volume of the crust and the amount of crust that has been eroded and removed, we must also take into account the amount of melt derived from depleted mantle that was never erupted. That is, there is some crustal material that was never successfully transported above the depleted mantle in the lithospheric root. It remained trapped at depth where it would have congealed and reacted with the surrounding lithospheric root (Ireland et al., 1994). Because this trapped melt caused depletion of the mantle from which it was removed and therefore represents a petrological complement to depleted mantle, it must be included in any calculation of the mass balance of the volume of continental crust.

We calculated an estimate of the volume of trapped melt using the composition of xenoliths from South Africa and Lesotho. Both sets of xenoliths were found in the Archean age basement of the Kaapvaal craton. Our modeling procedure relies upon having large numbers of xenoliths (>50) with whole rock compositions that are all from a single cratonic region. For this reason, only the xenolith suites from South Africa and Lesotho are suitable for this type of modeling calculation.

The oldest rocks on the Kaapvaal craton with an extensive surface exposure are about 3.5 Ga (de Wit and Armstrong, 1987; Riganti and Wilson, 1995) The craton was stabilized at about 3.1 Ga (de Wit et al., 1992). Lesotho lies at the southern edge of the Kaapvaal craton, and part of the lithospheric mantle beneath Lesotho is early Proterozoic in age (Olive and Ellam, 1997). Therefore, we use the xenoliths from South Africa as proxies for the amount of trapped melt beneath Archean basement and the xenoliths from Lesotho as proxies for the amount of trapped melt beneath early Proterozoic basement.

We model the compositions of melts formed in a one-dimensional melting column, using a parameterization of deep melting experiments (Herzberg et al., 1983; Herzberg and Hara, 1998; Herzberg, 1999). The melts are assumed to mix completely, so the melt composition reflects melts formed over a range of depths. The residual mantle composition at any depth in the melting column is found by subtracting the composition of the pooled melts from all greater depths in the column, from the initial mantle, taken to be represented by KLB-1. Because mantle xenoliths are enriched in incompatible elements by kimberlitic magmas, we perform our modeling using only five major elements: Si, Al, Ca, Mg and Fe. These elements represent over 98% of most mantle peridotites (O'Neill and Palme, 1988) This strategy avoids the problem of contamination of the xenolith by the kimberlite magma.

The compositions of the pooled melt and residual mantle are controlled by the final depth of the residual mantle and the depth of initial melting (determined by mantle potential temperature). The compositions of a xenolith would also depend on

Table 2
Thickness of the layer of crust that has been removed by erosion

Age of crust	Mean age (Ga)	Minimum (km)	Average (km)	Maximum (km)
Archean	2.7	2.5	4.1	7.8
Early Proterozoic	2.1	1.65	2.53	4.4
Post-1.8 Ga	0.9	0.56	0.8	1.3

the fraction of trapped melt. We varied the depth of initial and final melting, and the mixture of melt and residual mantle to obtain the potential compositions of xenoliths containing trapped melt. The best-fit model for each xenolith sample was determined by a least-squares fit to a combination of three parameters that were determined to be most diagnostic: Al_2O_3 content, MgO/FeO, and CaO/Al_2O_3.

The best-fit models exhibit a fairly wide range of parameters, but they all require high mantle temperatures and therefore high degrees of melting (most greater than about 35%). Most of the models allow only small amounts of trapped melt, <4%. The average of the best-fit compositions for the South African xenoliths and the Lesotho xenoliths is given in Table 3.

The average of the best-fit models in both cases had a mantle potential temperature of 1900–1925°C, and a final depth of melting of 80–85 km. These conditions result in about 40% melting, with the best fits for individual xenoliths mostly between 35 and 45%. The compositions of the trapped melts formed under these conditions are high-Mg komatiites and are consistent with the primitive lava compositions of Archean komatiites, such as Belingwe.

For South Africa, the average composition of the depleted mantle resembles a peridotite that has been depleted by the extraction of 38.1% melt. The average amount of trapped melt in the best-fit model is 2.1% (Fig. 4). For Lesotho,. the average composition of the depleted mantle resembles a peridotite that has been depleted by the extraction of 40.0% melt. The best-fit model for Lesotho has an average of 1.7% trapped melt.

12. Checking our models of melting

We check that the melt does not represent unextracted melt rather than trapped melt by plotting Mg number of the xenolith versus our modeled per cent trapped melt. If the melt was simply unextracted primitive mantle, we would expect to see higher amounts of trapped melt in the xenoliths with the lowest Mg numbers. We do see a slight hint of such a trend, but Fig. 5 is an almost perfect scatter plot, consistent with our assertion that the xenoliths contain trapped melt rather than unextracted melt.

The compositions of the modeled trapped melt are extremely interesting. Both of the best-fit models of trapped melt have major element compositions that resemble those of komatiites (Table 3). The best-fit composition of both trapped melts resembles a Belingwe komatiite. Both of the trapped melt compositions are high in iron, with 10.8% FeO.

As a further check, we have calculated the mean composition of all xenoliths with an Mg number of less than 89, the Mg number of primitive undepleted mantle. These xenoliths must have high percentages of trapped melt. We find that their average composition is roughly komatiitic. The mean composition of the three xenoliths with the lowest Mg numbers is the most compelling. These

Table 3
Modeled composition of trapped melt and xenoliths plus trapped melt compared to the observed mean xenolith composition and the mean composition of xenoliths with Mg numbers less than 89

Rock type	SiO_2	CO	Al_2O_3	FeO	MgO
Mean South African xenoliths	46.5	1.2	1.6	6.6	44.1
Best-fit model South African xenoliths + melt	43.6	1.0	1.5	6.7	46.9
Trapped melt composition	47.4	7.5	7.3	10.8	27.0
Average all South African xenoliths with Mg#<89	47.9±2.3	4.9±2.0	4.3±0.89	10.1±2.0	32.9±3.5
Three lowest Mg # xenoliths (Mg#=82.9)	46.8±0.9	4.0±0.8	4.6±0.7	12.0±0.7	32.6±1.8
Mean lesotho xenoliths	46.4	0.8	0.9	6.4	45.6
Best-fit model Les. xenoliths + melt	43.6	0.9	1.3	6.6	48.2
Trapped melt composition	47.4	7.6	7.4	10.8	26.8

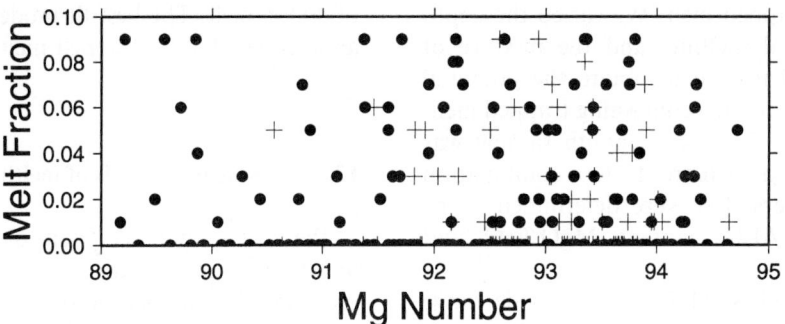

Fig. 5. Mg number of peridotites versus fraction of melt within the peridotite. Black dots: Archean peridotites. Crosses: Early Proterozoic peridotites. If the melt fraction represented unextracted melt, the plot would show a trend between melt fraction and Mg number. There is no trend, and the plot is an almost perfect scatter plot. This lack of a trend suggests that the melt fraction represents trapped melt rather than unextracted melt.

xenoliths have a composition that is clearly komatiitic and is very high in iron (Table 3).

We also estimate the reasonableness of these results by using the upper bound of the K_2O continent of the mantle root (Rudnick et al., 1998). Rudnick et al. (1998) estimated that mantle roots could have no more than 0.03% K_2O before the internal heat production would be too high to fit known crustal and cratonic geotherms. If we take the average K_2O content of lavas from the Ontong Java plateau ($0.326 \pm 0.02\%$) (Mahoney et al., 1992), a peridotite can have an average of 9.2% trapped melt before these limits are exceeded. This upper bound on the percentage of trapped melt is much higher than the percentages that we obtain.

If, however, we take a komatiitic trapped melt, the peridotite can have over 40% trapped melt before these limits on radiogenic heat production are exceeded. Because both of our model trapped melts were komatiitic, we conclude that our estimates for the percentage of trapped melt are reasonable for Early Proterozoic and Archean age lithospheric mantle. For younger lithosphere, we use the Early Proterozoic values of trapped melt (Table 4).

The amount of crust that is potentially trapped within the lithospheric mantle is significant.

Table 4
Volume of trapped melt within lithospheric mantle of different ages

Source	Melt minimum	Melt average	Melt maximum
South Africa	0.019	0.0211	0.0232
Lesotho	0.014	0.0166	0.0192
Rudnick (Ontong Java)	0.081	0.092	0.106
	Root minimum	Root average	Root maximum
Archean	197.1	203	208.9
Early Proterozoic	188.2	196	203.8
Post-1.8 Ga	93.3	97	100.7
	Crust minimum	Crust average	Crust maximum
Archean	3.74	4.28	4.85
Early Proterozoic	2.63	3.25	3.91
Post-1.8 Ga	1.31	1.61	1.93

Proportionately, Archean areas have the most trapped melt, equivalent to a crustal thickness of 3.7–4.9 km (Table 4). Early Proterozoic areas also have considerable amounts of trapped melt, equivalent to a crustal thickness of 2.6–3.9 km. For post-1.8 Ga crust, we estimate that the maximum amount of trapped melt is equivalent to an extra 1.3–1.9 km of crust. Thus, trapped melt may represent a significant portion of the crust extracted from the mantle, particularly in cratonic regions.

13. Constraining the rate of continental growth

The first technique, which we call the summation method, uses the sum of the following quantities to estimate the minimum volume of crust at different times in Earth history: distribution of crustal thickness, erosion level of the crust, and the amount of trapped melt. The surface areas are taken from our compilations of crustal ages. The crustal thicknesses are from the recent model of (Mooney et al., 1998). The amount of crust removed by erosion is taken from the work of Galer and Metzger (1998). Finally, we estimate the amount of unextracted melt in the lithospheric mantle using the compositions of mantle xenoliths. The amount of unextracted melt is multiplied by the total volume of the lithospheric root to derive an estimate of the amount of crust trapped within the lithospheric mantle. The sum of these three parameters, crustal thickness, crustal erosion, and trapped melt, equals the original amount of crust that was extracted from the mantle.

The second technique uses the corrected thickness of the continental lithosphere to estimate the overall crustal volume present at the end of the Archean, the Early Proterozoic and at the present time. Lithospheric roots are derived from partial melting of the mantle to make the crust. Thus, the overall thickness of the lithosphere of a given age is directly related to the amount of crust that was extracted from the mantle at that time in Earth history. Because the oldest Re/Os model ages either match or exceed the age of the oldest crust within a craton (Pearson et al., 1995a,b; Olive and Ellam, 1997; Graham et al., 1999), we believe that the root volume may be recycled much more slowly than the crustal volume. Thus, the lithospheric thickness may preserve a less biased record of the total amount of crust extracted from the mantle at a given time in the Earth's history.

14. Crustal extraction — the summation method

We computed the sum of the total amount of crust created for each of
our three periods in Earth history. As shown in Table 5, after combining the seismic thickness, the eroded thickness, and the trapped melt, we found that the volume of crust produced in the Archean was considerable. If no melt had been trapped within the lithospheric mantle during the Archean, it would have been equivalent to an overall thickness of 43–55 km of crust. Similarly, when we included the melt that was trapped and frozen within the lithospheric mantle, Early Proterozoic crustal volume was equivalent to a crustal layer with a thickness of 39–57 km. The total equivalent thickness of post-Early Proterozoic crust was somewhat less, equivalent to a layer that is 32–47 km thick (Table 5).

As Table 5 shows, the summation method, which calculates the overall volume of crust that was extracted from the mantle, gives values for the volume of the continental crust over time that are much closer to the isotopic and geochemical estimates of total crustal volume (McLennan and Taylor, 1982; Taylor and McLennan, 1985; Veizer and Jansen, 1985; McCulloch and Bennett, 1994) than previous summations of crustal volume (Hurley and Rand, 1969). From the summation results, we estimate that at least 27–30% of the continental crust was extracted by the end of the Archean. By the end of the Early Proterozoic (1.8 Ga), between 50 and 52% of the continental crust had been extracted from the mantle (Fig. 5). However, the results of the summation method do not require large amounts of recycling of early formed continental crust.

15. Crustal volume through time — the melt column method

Our second method for estimating the volume of crust that was extracted relied on the overall

Table 5
Total thickness of crust through time derived from the summation method

	Minimum	Average	Maximum
Archean			
Seismic (km)	36.4	39.4	42.4
Eroded (km)	2.5	4.1	7.8
Melt (km)	3.7	4.3	4.8
Equivalent crustal thickness	42.6	47.8	55.1
Fraction trapped melt	0.0190	0.0211	0.0232
Root thickness (lith-crust)	197	203	209
Surface area (10^6 km^2)	36.8	40.9	45.0
Total volume (10^9 km^3)	1.6	2.0	2.5
Percentage total volume of all crust	29.9	28.2	27.5
Early Proterozoic			
Seismic (km)	35.0	41.6	48.3
Eroded (km)	1.7	2.5	4.4
Melt (km)	2.6	3.3	3.9
Equivalent crustal thickness	39.2	47.4	56.6
Fraction trapped melt	0.0140	0.0166	0.0192
Root thickness (lith-crust)	188	196	204
Surface area (10^6 km^2)	29.3	32.5	35.7
Total volume (10^9 km^3)	1.1	1.5	2.0
Percentage total volume of all crust	21.9	22.2	22.4
Post-1.8 Ga			
Seismic (km)	30.3	37.1	43.9
Eroded (km)	0.6	0.8	1.3
Melt (km)	1.3	1.6	1.9
Equivalent crustal thickness	32.2	39.5	47.1
Fraction trapped melt	0.0140	0.0166	0.0192
Root thickness (lith-crust)	93	97	101
Surface area (10^6 km^2)	78.4	87.1	95.8
Total volume (10^9 km^3)	2.5	3.4	4.5
Percentage total volume of all crust	48.1	49.6	50.1
All crust			
Total volume (10^9 km^3)	5.2	6.9	9.0

thickness of the lithospheric mantle and theoretical modeling of melt extraction. We assumed that present-day lithospheric thickness, after a correction for erosion, provided a valid estimate of the depth at which melt began to be extracted to make the crust. We then used models based on experimental results to estimate the amount of melt that would have been extracted. This value was multiplied by the thickness of the depleted lithospheric root to obtain an estimate of the amount of crust that had been extracted by a given time in Earth history. The results of these calculations are shown in Tables 6 and 7 and in Fig. 6.

16. Checking our melt column results using average xenolith composition

A check on the melt column method of estimating the volume of crustal extraction is based on theoretical models of xenolith composition. These models are the same models that were used to estimate the amount of trapped, congealed melt using xenolith suites from South Africa and Lesotho. Because mantle xenoliths sample only the uppermost part of the lithospheric root, the average degree of melting represented by the xenolith suite is larger than the average degree of melting within the lithospheric root.

Table 6
Total thickness of crust through time derived from the melt column method

	Minimum	Average	Maximum
Archean			
Seismic (km)	36.4	39.4	42.4
Eroded (km)	2.5	4.1	7.8
Root thickness (km)	197	203	209
Present lithospheric thickness (km)	236	247	259
Past lithospheric thickness (km)	238	251	267
Surface area (10^6 km^2)	36.8	40.9	45.0
Average fraction melted (melt column)	0.300	0.317	0.337
Average fraction melted (xenolith composition)	0.311	0.381	0.451
Average fraction melted (upper 160 km)	0.344	0.381	0.423
Crustal thickness (km) (melt column)	59.1	64.4	70.4
Crustal volume (10^9 km^3) (melt column)	2.2	2.6	3.2
Percentage total volume (melt column)	46.3	45.1	44.1
Early Proterozoic			
Seismic (km)	35.0	41.6	48.3
Eroded (km)	1.7	2.5	4.4
Root thickness (km)	188	196	204
Present lithospheric thickness (km)	225	240	256
Past lithospheric thickness (km)	226	243	261
Surface area (10^6 km^2)	29.3	32.5	35.7
Average fraction melted (melt column)	0.284	0.306	0.330
Average fraction melted (xenolith composition)	0.350	0.400	0.450
Average fraction melted (upper 160 km)	0.313	0.359	0.407
Crustal thickness (km) (melt column)	53.4	60.0	67.3
Crustal volume (10^9 km^3) (melt column)	1.6	1.9	2.4
Percentage total volume (melt column)	33.3	33.4	33.4
Post-1.8 Ga			
Seismic (km)	30.3	37.1	43.9
Eroded (km)	0.6	0.8	1.3
Root thickness (km)	93	97	101
Present lithospheric thickness (km)	124	135	146
Past lithospheric thickness (km)	125	136	147
Surface area (10^6 km^2)	78.4	87.1	95.8
Average fraction melted (melt column)	0.131	0.149	0.167
Crustal thickness (km) (melt column)	12.2	14.5	16.8
Crustal volume (10^9 km^3) (melt column)	1.0	1.3	1.6
Percentage total volume (melt column)	20.4	21.6	22.4
All crust			
Crustal volume (10^9 km^3) (melt column)	4.7	5.8	7.2

We model the average degree of melting in the xenoliths, assuming that the overall lithospheric thickness is the same as our estimated past lithospheric thickness, but that the xenoliths sample only the upper 160 km of the melt column. This value of 160 km is based on a maximum depth of origin of the xenoliths of about 200 km (Finnerty and Boyd, 1987; Pearson et al., 1995a,b) and assumes a 40 km thick crust. We call this a truncated melt column model. The truncated melt column model of the average fraction of melting in the xenoliths matches our theoretical estimates of the average fraction of melting in the xenoliths (Table 6). We take this as a confirmation that our

Table 7
Comparison of the cumulative volume of continental crust extracted through time from the melt column and summation methods

Age	Percentage total crust	Error
Method 1: Trapped melt + erosion + seismic		
2.7	28.7	1.2
2.1	50.9	0.9
0	100.0	0.3
Method 2: Melt column		
2.7	45.2	1.1
2.1	78.6	1.1
0	100.0	4.8

seismic estimate of the overall length of the melt column is basically correct.

17. Volume of crust from xenolith composition and melt column models

The average fraction of melting in the melt column is multiplied by the thickness of the depleted lithospheric root to obtain the amount of crust extracted by a given time in Earth history (Table 6, Fig. 5). From the melt column model, we obtain an equivalent of 59–70 km of crust extracted from the mantle by the end of the

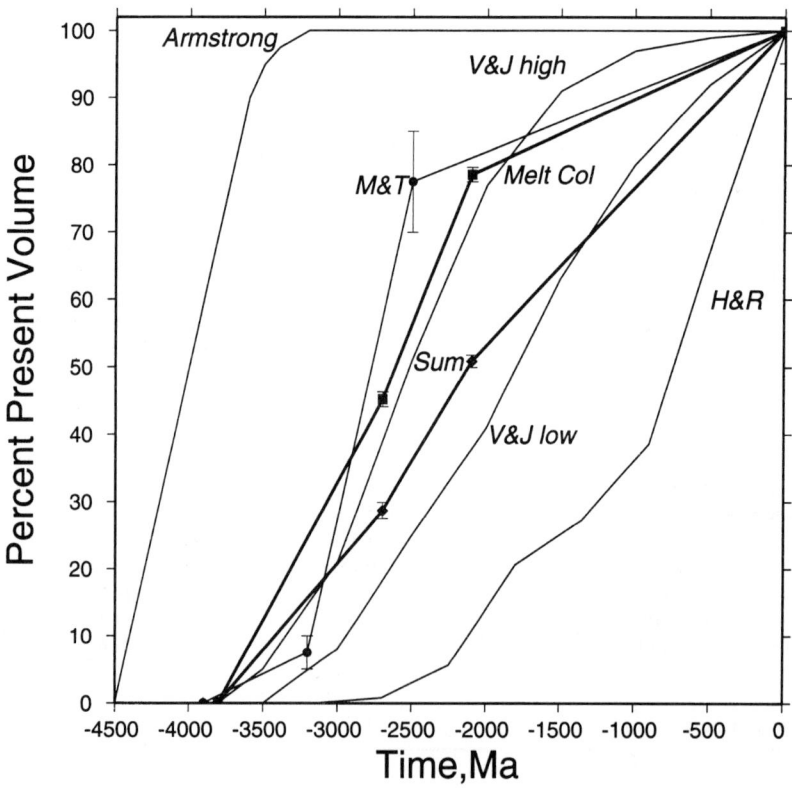

Fig. 6. Cumulative volume of continental crust extracted through time from the two methods compared to some other models of the volume of continental crust extracted through time. In order to keep the figure legible, the models are not exhaustive. H&R: Hurley and Rand (1969); M&T: McLennan and Taylor (1982); V&J: Veizer and Jansen (1985); Armstrong: Armstrong (1981). Sum: our results from summing the overall volume of the seismic crust, eroded crust, and trapped melt in the lithospheric mantle. Melt Col: Our results from assuming that the lithospheric root is the residue of a melt column. Thus, the overall lithospheric thickness is an estimate of where melting starts. This type of melting model is consistent with the average amount of melting within mantle xenoliths from Archean and Early Proterozoic age mantle roots. Note that our estimate of crustal volume through time does not include the areas north and south of 80°.

Archean. For Early Proterozoic time, the equivalent thickness of crust is between 53 and 67 km. For post-Early Proterozoic time, the equivalent thickness of crust is only 12–17 km.

When we consider the distribution of crustal extraction over time from the melt column models, we obtain a crustal growth history that closely approaches the estimates of McLennan and Taylor (1982). Overall, the melt column results imply that a minimum of 44–46% of continental crust was extracted by the end of the Archean. By the end of the Early Proterozoic (1.8 Ga), between 77 and 79% of the continental crust was extracted from the mantle (Fig. 6).

18. Discussion: crustal thicknesses and volumes

The melt column models of melt extraction predict that large amounts of crust were extracted from the lithospheric mantle in the Archean and Early Proterozoic. For pre-1.8 Ga, the composite crustal thicknesses from the melt column models are much larger than those from the summation method. However, the melt column results for post-1.8 Ga crust are much smaller than the results from the summation method. The experimentally based models of melt extraction predict that only about 15 ± 2 km of crust has been extracted from the mantle in areas of post-1.8 Ga crust (Table 6). This number contrasts quite markedly with the estimated crustal thickness of post-1.8 Ga crust from seismic observations, estimated erosion, and estimated trapped melt fraction, which averages about 40 km (Table 5).

There are several possible explanations for this discrepancy. Some workers have found that the seismic thickness overestimates the true geochemical crustal thickness in areas of high heat flow (Griffin and O'Reilly, 1987). However, the continental areas of high heat flow are relatively spatially restricted, and thus, this cannot explain the overall pattern.

The second possibility is that the use of the metamorphic grade of undeformed Archean greenstone belts to estimate the degree of erosion of the crust underestimates the true amount of crustal erosion since 3.0 Ga. By picking undeformed greenstone belts, we are, by definition, excluding greenstone belts that have had no erosion due to tectonic compression. We know that large areas of the Earth's crust have undergone tectonic compression since the crust was initially extracted from the mantle. Recent work also suggests that some greenstone belts represent thrust slices of the upper few kilometers of Archean oceanic plateaus (Kusky and Kidd, 1992; Kusky and Vearncombe, 1997). If this is generally true of undeformed greenstone belts, this means that undeformed greenstone belts record only the erosion that occurred after their emplacement. Therefore, we need some other complementary method of estimating the amount of erosion experienced by the continental crust through time.

A third problem is that we know that not all depleted mantle is sufficiently buoyant to resist subduction. As a result, the basal part of the melt column that melts to form the continental crust is reincorporated into the convecting mantle. Unfortunately, we do not really know exactly how much of the basal part of the melt column is remixed into the mantle. We do know that some depleted mantle is remixed: this is the origin of the depleted mantle signature of MORB (mid-ocean ridge basalts). Consequently, our estimates of melt extraction from experimental data combined with the overall thickness of the lithospheric root can only be minimum values.

Finally, there is the possibility that these differences in the estimated amount of crustal extraction are basically a reflection of the relative importance of recycling and new crustal extraction. In that case, the difference in estimated crustal thickness derived from lithospheric thickness and observed seismic thickness is an estimate of the difference between the amount of new crustal extraction and the amount of recycled continental crust. This is a reasonable inference because our estimates of the rate of crustal growth from experimental constraints and lithospheric thickness are much closer to geochemical models than any previous, geophysically based mass-balance calculations.

19. Importance of the trapped melt

The theoretically based estimates of the amount and composition of trapped melt suggest that the continental lithospheric mantle contains large quantities of iron-rich, komatiitic melt, which, according to our models, has an average FeO content of about 10.8%. This result is important because it can help to explain the discrepancy between the FeO content of lower crustal xenoliths within Archean cratons (about 9%) and the known FeO content of komatiitic melts. Furthermore, it has been suggested that melting of a primitive, previously unmelted mantle would produce high degree melts with an average FeO content of 9% (O'Neill and Palme, 1988). As a result, there is no need for a mechanism to delaminate the lower crust in Archean times. These ideas are much more compatible with the evidence that thick Archean continental lithosphere formed very quickly and cooled very rapidly (Nisbet, 1987; Abbott, 1991; Pearson et al., 1995a).

Overall, our estimates of volumes of trapped melt within Archean and Early Proterozoic cratonic roots indicate that much of the komatiitic melt that was generated within the mantle was trapped within the root. Only relatively small proportions made it to the surface. The residual melt that did reach the surface was, in most cases, basaltic, with much less iron. This can also explain why komatiites are so rare in greenstone belts, constituting at most a few per cent of the overall volume of extrusive rocks (de Wit and Ashwal, 1997).

This model of the origin of the moderate iron content of lower continental crust is testable in several ways. This model suggests that the iron content of oceanic plateau basalts may be higher in shallower layers than in deeper layers of the crust. This inference can be tested by looking at crustal xenolith suites within oceanic plateaus that are too young to be modified by arc volcanism. The xenolith localities on the island of Malaita in the Solomons tap the lithosphere of the Ontong Java Plateau. If the crustal xenoliths could be characterized by FeO content and depth of origin, they would constitute an important test of this model.

The melt column model of crustal extraction also suggests that areas of thick lithospheric mantle are dominantly very old. In some cases, there are proposed areas of thick lithosphere that lie entirely within the ocean basins. One notable candidate is the Mascarene bank, south of Madagascar (Fig. 3a). If Madagascar is returned to its former position in north Africa, this proposed area of thick lithosphere lies next to the Zimbabwe and Kaapvaal cratons. This supposition about a thick lithosphere could be tested by a seismic experiment with stations on Reunion, Madagascar and South Africa. Other areas where this assumption could be tested include the Seychelles and the Ogasawara plateau.

20. Conclusions

The results of a geophysically based compilation of the volume of continental crust generated through time are most consistent with isotopic and geochemical studies that indicate that relatively large volumes of continental crust were extracted early in Earth's history (McLennan and Taylor, 1982; Taylor and McLennan, 1985; Veizer and Jansen, 1985; Calderwood, 1998; McCulloch and Bennett, 1998; Sylvester, 1998). Although our present results do not match the models with the greatest amount of early extraction, our technique is biased towards younger ages of crustal extraction by the selection of 2.4 Ga as our first age interval. Our technique is also biased towards smaller lithospheric thicknesses in areas of Archean basement by the smearing effects of the tomographic model. As a result, it is likely that a better tomographic model combined with smaller age divisions would produce higher estimates of the amount of early crustal growth.

We suggest that the greater lithospheric thickness of Archean cratons is a direct result of the large amounts of crust extracted during that era. Some of this crust never made it to the surface, but was stored as congealed trapped melt within the lithospheric mantle. This trapped melt is quite komatiitic in mean composition, meaning that it is rich in iron. In addition, melting of previously undepleted mantle may produce melts that are lower in iron than melts of mantle containing recycled crustal material. Thus, the low iron content of Archean crust is not the result of lithospheric delamination, but rather the result of sequestration of iron within Archean lithospheric roots and the result of deriving continental crust from melting within a truly primitive mantle.

The discrepancies between the volumes of crust inferred from melt column models and xenolith composition models and those inferred from summations of the geophysical crust, eroded crust, and trapped melt volumes are partially the result of a transfer of mass by recycling (Veizer and Jansen, 1985). Within Archean and Proterozoic cratons, the equivalent thickness of crust from the summation method is less than the equivalent thickness of crust from the melt column method. We suggest that these differences reflect the lateral transfer of mass from Archean cratons to Proterozoic cratons by erosion and recycling within subduction zones.

Our studies also suggest that there are some areas of unrecognized cratonic basement that lie within the ocean basins. In some cases, the cratonic basement has been rifted off from other cratons and is covered by volcanic rocks (e.g. the Mascarene bank, the Ogasawara plateau) (Ishii, 1985; Bassias et al., 1993). However, these volcanic rocks may retain a cratonic signature within their isotopes, in particular in Re/Os. The chemistry of these volcanic rocks should be studied with this hypothesis in mind. Once we have a full inventory of the areas of Archean and Early Proterozoic age cratonic roots, we will be able to make a more accurate assessment of the amount of early crustal growth.

Acknowledgements

We thank Martin Menzies, Stephen Galer, and Steve Grand for helpful, constructive reviews of the manuscript. D. Sparks thanks NSF contract #EAR-9614178 for support of this work. LDEO contribution number 6064.

References

Abbott, D.H., Hoffman, S.E., 1984. Archean plate tectonics revisited 1: Heat flow, spreading rate, and the age of subducting oceanic lithosphere and their effects on the origin and evolution of continents. Tectonics 3, 429–448.

Abbott, D.H., 1991. The case for accretion of the tectosphere by buoyant subduction. Geophys. Res. Lett. 18, 585–588.

Abbott, D., Burgess, L., Longhi, J., Smith, W.H.F., 1994. An empirical thermal history of the Earth's upper mantle. J. Geophys. Res. 99, 13 835–13 850.

Abbott, D.H., Mooney, W.D., 1995. Crustal Structure and Evolution: Support for the Oceanic Plateau Model of Continental Growth. Reviews of Geophysics, Supplement, US National Report to the IUGG. pp. 231–242

Abbott, D.H., Drury, R., Mooney, W., 1997. Continents as lithological icebergs: the importance of buoyant lithospheric roots. Earth Planet. Sci. Lett. 149, 15–27.

Abouchami, W., Boher, M., Michard, A., Albarede, F., 1990. A major 2.1 Ga event of mafic magmatism in West Africa: an early stage of crustal accretion. J. Geophys. Res. 95, 17 605–17 629.

Agee, C.B., Walker, D., 1993. Olivine flotation in mantle melt. Earth Planet. Sci. Lett. 114, 315–324.

Agrawal, P.K., Pandey, O.P., Negi, J.G., 1992. Madagascar: A continental fragment of the paleo-super Dharwar craton of India. Geology 20, 543–546.

Armstrong, R.L., 1981. Radiogenic isotopes: the case for crustal recycling on a near-steady-state no-continental growth Earth. Philos. Trans. R. Soc. London Ser. A 301, 443–472.

Bardet, M.G., 1973. Geologie du diamont: Paris — France. Bureau de Recherches Geologiques et Minieres, Paris.

Bardet, M.G., 1974. Geologie du diamont: deuxieme partie: Gisements de diamant d'Afrique: Paris — France. Bureau de Recherches Geologiques et Minieres, Paris.

Bassias, Y., Denis-Clocchiatti, M., Leclaire, L., 1993. Le Plateau des Mascareignes, evolution d'une plate-forme neritique en mileau oceanique. C. R. Acad. Sci. 317, 507–514.

Boher, M., Abouchami, W., Michard, A., Albarede, F., Arndt, N.T., 1992. Crustal growth in West Africa at 2.1 Ga. J. Geophys. Res. 97, 345–369.

Boyd, F.R., 1987. High and low temperature garnet peridotite xenoliths and their possible relation to the lithosphere–asthenosphere boundary beneath southern Africa. In: Nixon, P.H. (Ed.), Mantle Xenoliths. Wiley, Chichester, UK, pp. 403–412.

Calderwood, A.R., 1998. Sm–Nd isotopic modeling of the evolution of the Earth's depleted mantle and crust, estimates of continental recycling and accretion rates. Geol. Soc. Am. Abstr. Programs, 30, 207

Christensen, N.I., Mooney, W.D., 1995. Seismic velocity structure and composition of the continental crust: a global view. J. Geophys. Res. 100, 9761–9788.

Condie, K.C., 1999. Mafic crustal xenoliths and the origin of lower continental crust. Lithos 46, 95–102.

de Wit, M.J., Armstrong, R., 1987. Felsic igneous rocks within the 3.3 to 3.5-Ga Barberton Greenstone Belt: High crustal level equivalents of the surrounding tonalite–trondhjemite terrain, emplaced during thrusting. Tectonics 6 (5), 529–549.

de Wit, M.J., Jones, M.G., Buchanan, D.L., 1992. The geology

and tectonic evolution of the Pietersburg Greenstone Belt, South Africa. Precambrian Res. 1, 123–153.
de Wit, M.J., Ashwal, L.D., 1997. Greenstone Belts. Clarendon Press, Oxford.
Durrheim, R.J., Mooney, W.D., 1991. Archean and Proterozoic crustal evolution: evidence from crustal seismology. Geology 19, 606–609.
Eggler, D., Furlong, K.P., 1991. Destruction of subcratonic mantle keel: the Wyoming Province. Fifth International Kimberlite Conference, Extended Abstracts, Araxa, Brazil, United States.
Finnerty, A.A., Boyd, F.R., 1987. Thermobarometry for garnet peridotites: basis for the determination of thermal and compositional structure of the upper mantle. In: Nixon, P.H. (Ed.), Mantle Xenoliths. Wiley, Chichester, UK, pp. 381–402.
Galer, S.J.G., 1991. Interrelationships between continental freeboard, tectonics and mantle temperatures. Earth Planet. Sci. Lett. 105, 214–228.
Galer, S.J.G., Metzger, K., 1998. Metamorphism, denudation and sea level in the Archean and cooling of the Earth. Precambrian Res. 92, 389–412.
Gossler, J., Kind, R., 1996. Seismic evidence for very deep roots of continents. Earth Planet. Sci. Lett. 138, 1–13.
Graham, S., Lambert, D.D., Shee, S.R., Smith, C.B., Reeves, S., 1999. Re–Os isotopic evidence for Archean lithospheric mantle beneath the Kimberly block — Western Australia. Geology 27, 385–480.
Grand, S., 1986. Shear Velocity Structure of the Mantle Beneath the North American Plate. California Institute of Technology.
Griffin, W.L., O'Reilly, S.Y., 1987. The composition of the lower crust and the nature of the continental Moho — xenolith evidence. In: Nixon, P.H. (Ed.), Mantle Xenoliths. Wiley, Chichester, UK, pp. 413–432.
Griffin, W.L., et al., 1999. Layered mantle lithosphere in the Lac de Gras area, Slave craton, Composition, structure and origin. J. Petrol. 40, 727–795.
Gurnis, M., Davies, G.F., 1986. Apparent episodic crustal growth arising from a smoothly evolving mantle. Geology 14, 396–399.
Handke, M.J., Tucker, R.D., Hamilton, M.A., 1997. Early Neoproterozoic (800–790 Ma) intrusive rocks in central Madagascar, geochemistry and petrogenesis. Geol. Soc. Am., Abstr. Programs 29, 468.
Herzberg, C.T., Fyfe, W.S., Carr, M.J., 1983. Density contraints on the formation of the continental Moho and crust. Contrib. Mineral. Petrol. 84, 1–5.
Herzberg, C., 1999. Phase equilibrium constraints on the formation of cratonic mantle. Mantle petrology: field observations and high pressure experimentation: A tribute to Francis R. (Joe) Boyd. In: Fei, Y., Bertka, C., Mysen, B.O. (Eds.), Geochemical Society Spec. Publ. 6, 241–257.
Herzberg, C., Hara, M.J.O., 1998. Phase equilibrium constraints on the origin of basalts, picrites, and komatiites. Earth-Sci. Rev. 44, 39–79.
Hurley, P.M., Rand, J.M., 1969. Pre-drift continental nuclei. Science 164, 1229–1242.
Ireland, T.R., Rudnick, R.L., Spetsius, Z., 1994. Trace elements in diamond inclusions from eclogites reveal link to Archean granites. Earth Planet. Sci. Lett. 128, 199–213.
Ishii, T., 1985. Dredged samples from the Osagawara fore-arc seamount or 'Osagawara Paleoland'–'fore-arc ophiolite'. In: Nasu, N., et al. (Eds.), Formation of Active Ocean Margins, pp. 307–342.
Jordan, T.H., 1978. Composition and development of the continental tectosphere. Nature 274, 544–548.
Jordan, T.H., 1988. Structure and formation of the continental tectosphere. Special Lithosphere Issue. J. Petrol., 11–37.
Kay, R.W., Kay, S.M., 1986. Petrology and geochemistry of the lower continental crust: an overview. In: Dawson, J.B., Carswell, D.A., Hall, J., Wedepohl, K.H. (Eds.), The Nature of the Lower Continental Crust, pp. 147–159.
Kennedy, C.S., Kennedy, G.C., 1969. The equilibrium boundary between graphite and diamond. J. Geophys. Res. 81, 2467–2470.
Kusky, T.M., Kidd, W.S.F., 1992. Remnants of an Archean oceanic plateau, Belingwe greenstone belt, Zimbabwe. Geology 20, 43–46.
Kusky, T.M., Vearncombe, J.R., 1997. Structure of Archean greenstone belts. Tectonic evolution of greenstone belts. In: de Wit, M., Ashwal, L.D. (Eds.), Greenstone Belts. Clarendon Press, Oxford, pp. 91–124.
Lambert, D.D., Shirey, S.B., Shirey, S.C., 1995. Proterozoic lithospheric mantle source for the Prairie creek lamproites: Re–Os and Sm–Nd isotopic evidence. Geology 23, 273–276.
Larson, R.L., 1991. Geological consequences of superplumes. Geology 19, 963–966.
Lerner-Lam, A.L., Jordan, T.H., 1987. How thick are the continents? J. Geophys. Res. 92, 14 007–14 040.
Liu, M., 1994. Asymmetric phase effects and mantle convection patterns. Science 264, 904–1907.
McCulloch, M.T., Bennett, V.C., 1994. Progressive growth of the Earth's continental crust and depleted mantle: Geochemical constraints. Geochim. Cosmochim. Acta 58, 4717–4738.
McCulloch, M.T., Bennett, V.C., 1998. Early Differentiation of the Earth: An Isotopic Perspective. In: Jackson, I. (Ed.), The Earth's Mantle. Cambridge University Press, Cambridge, pp. 127–158.
McLennan, S.M., Taylor, S.R., 1982. Geochemical constraints on the growth of the continental crust. J. Geol. 90, 347–361.
Mahoney, J.J., Storey, M., Duncan, R.A., Spencer, K.J., Pringle, M., 1992. Geochemistry and geochronology of leg 130 basement lavas: Nature and origin of the Ontong Java plateau. In: Berger, W.H., Kroenke, L.W., Mayer, L.A. (Eds.), Proceedings of the Ocean Drill. Progr. Sci. Results. College Station, TX, Ocean Driiling Program, 3–22.
Mann, P., Gahagan, L., Coffin, M., Shipley, T., Cowley, S., Phinney, E., 1997. Regional tectonic effects resulting from the progressive east-to-west collision of the Ontong Java Plateau with the Melanesian Arc system. EOS Trans. Am. Geophys. Union 77, 712

Martin, H., 1986. Effect of steeper Archean geothermal gradient on geochemistry of subduction-zone magmas. Geology 14, 753–756.

Menke, W., Levin, V., 1994. Cold crust in a hot spot. Geophys. Res. Lett. 21, 1967–1970.

Menzies, M.A., Zhang, M., Weiming, F., 1993. Paleozoic and Cenozoic lithoprobes and the loss of >120k of Archean Lithosphere from the Sino-Korean craton China. In: Pritchard, H. (Ed.), Magmatic Processes and Plate Tectonics. Geological Society of London, London, pp. 71–81.

Menzies, M.A., Xu, Y., Flower, M.F.J., Chung, S.L., Lo, H., Lee, T.Y., 1998. Geodynamics of the North China Craton. In: Chung, S.L., Lo, C.H., Lee, T.Y. (Eds.), Mantle Dynamics and Plate Interactions in East Asia. American Geophysical Union, Washington, DC, pp. 155–165.

Mooney, W.D., Laske, G., Masters, T.G., 1998. Crust 5.1, a global crustal model at 5 by 5 degrees square. J. Geophys. Res. 103, 727–747.

Nisbet, E.G., 1987. The Young Earth: An Introduction to Archean Geology. Allen and Unwin, Boston, MA.

Nixon, P.H., 1987. Mantle Xenoliths. Wiley, Chichester, UK.

Nolet, G., Grand, S.P., Kennett, B.L.N., 1994. Seismic heterogeneity in the upper mantle. J. Geophys. Res. 99, 23 753–23 766.

Nyblade, A.A., Pollack, H.N., 1993. A global analysis of heat flow from Precambrian terrains: Implications for the thermal structure of Archean and Proterozoic lithosphere. J. Geophys. Res. 98, 12 207–12 218.

Ohtani, E., Nagata, Y., Suzuki, A., Kato, T., 1995. Melting relations of peridotite and the density crossover in planetary mantles. In: McDonough, W.F., Arndt, N.T., Shirey, S. (Eds.), Chemical Evolution of the Mantle. Elsevier, Amsterdam, pp. 207–221.

Olive, V., Ellam, R.M., 1997. A Re–Os isotope study of ultramafic xenoliths from the Matsoku kimberlite. Earth Planet. Sci. Lett. 150, 129–140.

O'Neill, S.C., Palme, H., 1988. Composition of the silicate Earth: implications for accretion and core formation. In: Jackson, I. (Ed.), The Earth's Mantle: Composition Structure and Evolution. Cambridge University Press, Cambridge, pp. 3–126.

Parkinson, I.J., Hawkesworth, C.J., Cohen, A.S., 1998. Ancient mantle in a modern arc: Osmium isotopes in Izu–Bonin–Mariana forearc peridotites. Science 281, 2011–2013.

Pearce, J.A., et al., 1992. Boninite and harzburgite from Leg 125 (Bonin–Mariana forearc): A case study of magma genesis during the initial stages of subduction. In: Fryer, P., Pearce, J.A., Stokking, L.B. (Eds.), Proc. Ocean Drill. Progr. Sci. Results. National Science Foundation, Washington, DC, pp. 623–659.

Pearson, D.G., Carlson, R.W., Shirey, S.B., Boyd, F.R., Nixon, P.H., 1995a. Stabilization of Archean lithospheric mantle: a Re–Os isotope study of peridotite xenoliths from the Kaapvaal craton. Earth Planet. Sci. Lett. 134, 341–357.

Pearson, D.G., Shirey, S.B., Carlson, R.W., Boyd, F.R., Pokhilenko, N.P., Shimuzu, N., 1995b. Re–Os, Sm–Nd, and Rb–Sr isotopic evidence for thick Archean lithospheric mantle beneath the Siberian craton modified by multistage metasomatism. Geochim. Cosmochim. Acta 59, 959–977.

Petterson, M.G., et al., 1997. Structure and deformation of north and central Malaita, Solomon Islands: tectonic implications for the Ontong Java Plateau–Solomon arc collision, and for the fate of oceanic plateaus. Tectonophysics 283, 1–33.

Plank, T., Langmuir, C.W., 1988. An evaluation of the global variations in the major element chemistry of arc basalts. Earth Planet. Sci. Lett. 90, 349–370.

Rambelson, R.A., 1998. The Malagasy Shield. In: de Wit, M., Ashwal, L.D. (Eds.), Greenstone Belts. Clarendon Press, Oxford, pp. 636–639.

Revenaugh, J., Jordan, T.H., 1991. Mantle layering from ScS reverberations: 2, the transition zone. J. Geophys. Res. 96, 19 763–19 780.

Reymer, A., Schubert, G., 1986. Rapid growth of some major segments of continental crust. Geology 14, 299–302.

Richter, F.M., 1988. A major change in the thermal state of the Earth at the Archean Proterozoic boundary: consequences for the nature and preservation of continental lithosphere. Special Lithosphere Issue. J. Petrol. 1, 39–52.

Riganti, A., Wilson, A.H., 1995. Geochemistry of the mafic/ultramafic volcanic associations of the Nondweni greenstone belt, South Africa, and constraints on their petrogenesis. Lithos 34, 235–252.

Rudnick, R., Presper, T., 1990. Geochemistry of intermediate to high pressure granulites. In: Vielzeuf, D., Vidal, P. (Eds.), Granulites and Crustal Differentiation. Kluwer, Amsterdam, pp. 523–550.

Rudnick, R.L., 1992. Xenoliths — Samples of the lower continental crust. In: Fountain, D., Arculus, R., Kay, R.W. (Eds.), Continental Lower Crust. Elsevier, New York, pp. 269–316.

Rudnick, R., 1995. Making Continental Crust. Nature 378, 571–578.

Rudnick, R.L., Fountain, D.M., 1995. Nature and composition of the continental crust: a lower crustal perspective. Rev. Geophys. 33, 267–309.

Rudnick, R.L., McDonough, W.P., O'Connell, R.J., 1998. Thermal structure, thickness, and composition of continental lithosphere. Chem. Geol. 145, 395–411.

Smoot, N.C., Richardson, D.B., 1988. Tectonic and geomorphic interpretation of the Osagawara Plateau by multibeam-based 3D methods. Mar. Geol. 79, 141–147.

Stein, M., Hofmann, A.W., 1994. Mantle plumes and episodic crustal growth. Nature 372, 63–68.

Stein, M., Goldstein, S.L., 1996. From plume head to continental lithosphere in the Arabian Nubian Shield. Nature 382, 773–777.

Stoddard, P.R., Abbott, D.H., 1996. The influence of the tectosphere upon plate motion. J. Geophys. Res. 101, 5425–5433.

Sylvester, P.J., 1998. Formation of the continents, dribble or big bang? The Geochemical News 94, 12–25.

Taylor, S.R., McLennan, S.M., 1985. The Continental Crust: Its Composition and Evolution. Blackwell Scientific, Oxford.

Tucker, R.D., Ashwal, L.J., Handke, J., Hamilton, M.A., LeGrange, M., Rambleson, R.A., 1999. U–Pb geochronology and isotope geochemistry of the Archean and Proterozoic gneiss belts of Madagascar. J. Geol. 107, 135–154.

Veizer, J., Jansen, S.L., 1985. Basement and sedimentary recycling — 2: time dimension to global tectonics. J. Geol. 93 (6), 625–643.

Wasserburg, G.J., MacDonald, G.J., Hoyle, F., Fowler, W.A., 1964. Relative contributions of uranium, thorium, and potassium to heat production in the Earth. Science 143, 465.

Wyman, D.A., 1999. Paleoproterozoic boninites in an ophiolite-like setting. Geology 27, 455–458.

Yanagi, T., Wakizaka, Y., Suwa, K., 1983. Rb–Sr whole rock ages of granitic rocks from the Seychelles Islands. In: Eighth Preliminary Report of African Studies. University of Nagoya, Nagoya, pp. 23–26.

Zhang, Y.S., Tanimoto, T., 1991. Global love wave phase velocity variation and its significance to plate tectonics. Phys. Earth Planet. Inter. 66, 160–202.

Continental emergence and growth on a cooling earth

N.J. Vlaar *

Department of Theoretical Geophysics, University of Utrecht, PO Box 80021, 3508 TA, Utrecht, The Netherlands

Received 27 January 1999; accepted for publication 16 August 1999

Abstract

Isostasy considerations are connected to a 1-D model of mantle differentiation due to pressure release partial melting to obtain a model for the evolution of the relative sea level with respect to the continent during the earth secular cooling. In this context, a new mechanism is derived for the selective exhumation of exposed ancient cratons. The model results in a quantitative scenario for sea-level fall due to the changing thicknesses of the oceanic basaltic crust and its harzburgite residual layer as a function of falling mantle temperature. It is also shown that the buoyancy of the harzburgite root of a stabilized continental craton has an important effect on sea-level and on the isostatic readjustment and exhumation of exposed continental surface during the earth's secular cooling.

The model does not depend on the usual assumption of constant continental freeboard and crustal thickness and its application is not restricted to the post-Archaean. It predicts large-scale continental emergence near the end of the Archaean and the early Proterozoic. This provides an explanation for reported late Archaean emergence and the subsequent formation of late Archaean cratonic platforms and early Proterozoic sedimentary basins.

For a period of secular cooling of 3.8 Ga, corresponding to the length of the geological record, the model predicts a fall of the ocean floor of some 4 km or more. For a constant ocean depth, this implies a sea-level fall of the same magnitude. A formula is derived that allows for an increasing ocean depth due to either the changing ratio of continental with respect to oceanic area, or to a possible increase of the oceanic volume during the geological history. Increasing ocean depth results in a later emergence of submarine ancient geological formations compared to the case when ocean depth is constant. Selective exhumation is studied for the case of constant ocean depth. It is shown that for this case, early exposed continental crust can be exhumed to a lower crustal depth, which explains the relative vertical displacement of low-grade- with respect to high-grade terrain. Increasing ocean depth is not expected to result in diminished exhumation. © 2000 Elsevier Science B.V. All rights reserved.

Keywords: continental emergence and growth; cooling earth; isostasy

1. Introduction

Isotopic geochemistry of the present-day sedimentary record reveals that most of the continental upper crust had been fractionated from the mantle at the start of the Proterozoic (2.5 Ga). Estimates range from 60 to 80% or more (Moorbath, 1975;

* Fax: +31-30-253-5030.
E-mail address: vlaar@geof.ruu.nl (N.J. Vlaar)

Taylor and McLennon, 1985; Patchett and Arndt, 1986; Rudnick, 1995). Since then, most of the upper continental crust has been recycled by erosion and deposition. Continental growth was low during the early Archaean, but it experienced a strong growth pulse during the later Archaean, which generated most, probably some 5% or more, of the upper continental crust (Windley, 1977; Taylor and McLennon, 1985; Condie, 1989; Rudnick, 1995). At 2.5 Ga ago, probably most of the cratonic continental crust and its deep mantle root had been fractionated from the mantle and had been stabilized. The formation of late Archaean cratonic sedimentary platforms and vast early Proterozoic sedimentary basins requires part of the continental crust to have emerged above sea level during the middle to late Archean to be eroded and exhumed subsequently (Windley, 1977).

Continental freeboard has been considered to be a crucial parameter in the discussion of continental growth. Phanerozoic changes in continental freeboard have been the subject of discussion in terms of marine transgressions and regressions since the 1930's (Kuenen, 1939; Umbgrove, 1939). The causes of 'the pulse of the earth' of eustatic changes were ascribed to unknown workings of the interior of the earth. Only after the advent of the hypothesis of plate tectonics could Phanerozoic changes of continental freeboard be related to plate tectonic cyclic activity (Hays and Pitman, 1973; Hallam, 1977; Vail et al., 1978). Oscillating sea-level changes with an amplitude of about 200 m and a wavelength of about 200 Ma are recorded to be superposed on a small secular fall during the Phanerozoic. As no direct observation of seafloor older than 200 Ma is available, the Precambrian record of eustatic sea-level changes connected to plate tectonic activity is unknown. Therefore, only secular changes of continental growth and freeboard have been the subject of investigation (Wise, 1973; Schubert and Reymer, 1985; Galer, 1991; Galer and Mezger, 1998) based on model calculations constrained by assumptions about model parameters. The volume of the ocean, continental freeboard, and continental thickness have been constrained to be constant. Also, the condition of Airy isostasy for continental and oceanic crusts has been assumed in these studies.

Wise (1973) considers a constant volume post-Archaean continental crust that is maintained by stationary recycling of continental material through the mantle. Airy isostatic equilibrium is required to hold for a constant thickness continental and oceanic crusts.

Schubert and Reymer (1985), in order to accommodate post-Archaean additions from the mantle to the continental crust to compensate for losses by erosion of the continental surface require secular deepening of the oceanfloor in order to meet the constraints of constant freeboard and crustal thicknesses of continental and oceanic crusts. They ascribe the secular foundering of the ocean bottom to the decline of radioactive heat production in the earth's interior. However, they do not present a rationale connecting deepening of the ocean basin to the decline of heat sources. Moreover, neither do they clarify why the continental surface does not founder as a consequence of the same decline of heat production. Galer (1991) and Galer and Mezger (1998) try to overcome these difficulties by proposing a decreasing thickness of the ocean crust due to the secular cooling of the earth. Following McKenzie (1984), they assume that, like at present, new oceanic crust is generated at oceanic spreading centres by isentropic decompression melting. Moreover, he takes Airy isostatic equilibrium to be governed by the density contrast between crustal rocks end fertile mantle peridotite (fertile mantle peridotite means upper mantle rock from which no basalt has been segregated). However, it has been demonstrated by Vlaar and Van den Berg (1991) that on a cooling earth, the low-density residual harzburgite layer also has to be taken into account when comparing isostatic conditions at different mantle temperatures. Moreover, they showed that the segregated basaltic layer has a strongly reduced thickness compared to that which McKenzie (1984) deduced. This is due to the circumstance that decompression melting only proceeds to the base of the formed crust.

Hence, the oceanic crustal thickness used by Galer (1991) and Galer and Mezger (1998) is about a factor of two in excess of a more realistic scenario involving decompression melting at an oceanic spreading ridge (Vlaar and Van den Berg, 1991). As the two effects (neglecting the buoyancy effect of the low density harzburgite layer and taking the segregated basaltic layer to be largely overestimated, both by about a factor of two for the Archaean) oppose each other, by fortuity, the resulting sea-level fall does not deviate strongly from that derived in the present paper.

However, by strongly overestimating the thickness of the oceanic basaltic layer and assuming a constancy of the thickness and freeboard of continental crust from the middle Archaean onward,

Galer's conclusion (Galer, 1991; Galer and Mezger, 1998) that the mantle could not have been more than 150 K hotter than at present, cannot be validated as Airy isostasy of an oceanic crust, which is taken to be overestimated by about a factor of two, is required to be in isostatic equilibrium with respect to a constant-thickness continental crust.

In the present study, continental crustal thickness and freeboard are not constrained a priori. Instead, from alternative isostatic considerations, it is shown that the early Archaean proto-continental crust was probably mainly submerged and that the emergence of the continents above sea level from the middle Archaean onward was highly selective and caused by a falling mantle temperature. Late Precambrian emergence of larger parts of the Proterozoic continent cannot be excluded a priori (Hargraves, 1976).

Apart from sporadic findings of Proterozoic ophiolites, no clearly recognizable remnants of Precambrian oceanic crust have been observed. No direct evidence for modern style plate tectonic activity is available for the latter period, nor is the timing known of the advent of plate tectonics in the earth's history. Vlaar (1985, 1986a,b) presented theoretical arguments that the conditions under which plate tectonics operates at present are violated at higher mantle temperatures. The strongest argument put forward concerns the strongly increasing gravitational stability of the oceanic lithospheric layering with increasing mantle temperature. The conditions under which subduction of Precambrian and particularly Archaean oceanic lithosphere could have been initiated could not be easily reached when mantle temperatures were higher than at present.

In the present paper, a basic assumption regarding the ancient oceanic lithosphere is that it is formed by differentiation and segregation of mantle diapirs rising to the base of an existing oceanic crust like at present at oceanic spreading centres. However, in this study, spreading is not required. The basic mechanism is the same as that for the creation of a new lithosphere at a modern spreading ridge, that is, isentropic decompression melting to the pressure of the base of an existing crust. A scenario for recycling this lithosphere at higher mantle temperatures is described by Vlaar et al. (1993). As no Precambrian oceanic crust has survived to the present, it must be assumed that this crust and its harzburgite residue have been destroyed and reabsorbed and remixed in the mantle at the same rate as they were formed. In a cooling earth, this results in evolving decreasing thicknesses of the oceanic crust and its residual harzburgite layer, and — as will be shown in the following — changing isostatic conditions and fall of sea level during the earth's history.

The presence of high-melting-temperature komatiites in Archaean greenstone belts suggests that the earth was some 300 K hotter at the beginning of the Archaean (3.8 Ga) than at present. Some authors suggest (e.g. Abbott et al., 1994) that the temperature indicated by komatiites is not representative for the temperature of the upper mantle from which oceanic crust is fractionated. They assume that komatiites found in greenstone belts represent the higher temperature of mantle hot spots and that this temperature is not representative for the middle Archaean average upper mantle. However, it still remains to be seen whether greenstone belts represent oceanic crust. Here, they are viewed to be the ancient precursors of modern intracontinental basins (Hunter, 1974; McKenzie et al., 1980). From petrological arguments, Abbott et al. (1994) conclude that the mean mantle potential temperature, T_p, has decreased from about 1600°C at the middle Archaean to 1380°C at present. This appears not to be much at odds with the values for T_p used in this paper: $T_p = 1338$°C at present and 1667°C at 3.8 Ga. Rapid cooling of the earth during the (early) Archaean (Vlaar et al., 1993) brings these values within the range considered by Abbott et al. (1994).

A separate argument is given by the mere existence of the deep and strongly depleted cratonic lithospheric root as such. This suggests high upper mantle temperatures when the crust was segregated from the mantle during the early geological evolution.

2. Decompression melting

Decompression melting of undepleted mantle peridotite is the most important crust forming process in the earth. It is held to be the cause of the present-day generation of basaltic oceanic crust at mid-ocean ridges. It is generally agreed upon that in a hotter mantle, the melting of a rising diapir starts at a deeper level and that a larger volume of basaltic magma is formed. The thickness of the formed basaltic crust is a function of the potential temperature, T_p, of the mantle. T_p is defined as the temperature obtained when a material volume in the mantle is decompressed under isentropic and metastable conditions to the pressure at the earth's surface. When, under supersolidus conditions, partial melting takes place, the temperature of the diapir decreases as a consequence of latent heat consumption. McKenzie (1984), Vlaar (1986b), and McKenzie and Bickle (1988) derive the thickness of the basaltic crust as a function of mantle potential temperature. An important effect that has not been implemented by these authors is that the amount of melting is strongly influenced by the pressure exerted by the crust already formed. This leads to a considerably reduced thickness of the segregated crust (Vlaar and Van den Berg, 1991).

In the following, it is assumed that throughout the geological evolution, oceanic crust is generated by 1-D isentropic decompression melting of a fertile mantle diapirs that rise to the base of the crust. At present, this process is associated with the spreading of lithospheric plates at oceanic ridges. The formalism developed by McKenzie (1984) is used but is corrected for the crustal pressure effect (Vlaar and Van den Berg, 1991). Calculations are based on the experimentally determined solidus and liquidus of fertile garnet peridotite by Takahashi (1980), and partial melting behaviour by Jaques and Green (1980).

Fig. 1 shows the thicknesses as a function of mantle potential temperature of the segregated basaltic crustal layer for two cases. In the lower curve, the pressure effect of the crust already formed has been accounted for (Vlaar and Van den Berg, 1991). The upper curve represents the case when all expelled melt or formed crust is removed and the harzburgitic residue is allowed to rise to the surface (McKenzie, 1984). The latter condition is not met in practice as the segregated oceanic crust is taken to be subaqueous and in

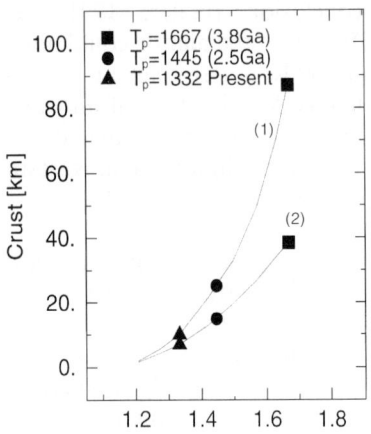

Fig. 1. Thickness of basaltic melt layer for two cases: (1) isentropic decompression melting proceeds to surface where melt is removed instantaneously until the top of the depleted layer reaches the surface; (2) melting proceeds only to the base of the crust. The Mantle potential temperature, T_p, is anchored at the early Archaean (3.8 Ga) (melting temperature of komatiite), an assumed temperature at the early Proterozoic (2.5 Ga) based on an high rate of cooling during the Archaean, and slow thereafter until the present.

general cannot be eroded before it is recycled in the mantle again. In the following, this condition will be simulated by replacing the basaltic crust by a lower density continental crust that emerges above sea level and is subject to erosion.

3. Cooling of the earth, falling sea level, and continental growth

At a higher mantle potential temperature, T_p, partial melting at the solidus starts at a deeper level. The thicknesses of the segregated and residual layers increase strongly with increasing T_p. In the 1-D model used in this paper, the degree of depletion of the residual harzburgite layer increases continuously from zero at the solidus to a maximum value at the base of the crust. This causes the composition of the harzburgitic root to decrease continuously relative to the underlying undepleted peridotite mantle and therefore compositional density too. Following Jordan (1997), his normalized density as a function of depletion is used in the present study. The densities of the basalt and the harzburgite are smaller than that of unde-

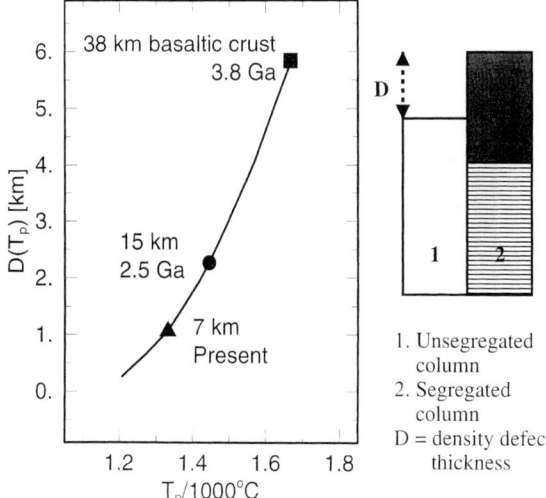

Fig. 2. The density of a segregated column is lower than that of a column of undepleted peridotite of equal mass and cross-section from which it has been derived. The lengthening of the column, the density defect thickness $D(T_p)$, is a strongly increasing function of mantle potential temperature. Segregation does not change the pressure at the base of the column.

pleted peridotite. Upon segregating an undepleted column, its length becomes larger (Fig. 2). This lengthening is termed density defect thickness (Oxburgh and Parmentier, 1977) and is obtained by integration of the density defect (Vlaar and Van den Berg, 1991). The density defect thickness, D, also increases strongly as a function of T_p (Fig. 2). In Fig. 3, the states of segregation at different mantle potential temperatures are compared ($T_{p1} > T_p$). As reference depths, the depth to the solidus at T_{p1}, $Z_{sol}(T_{p1})$, and the constant depth to the top of the non-segregated column, Z_{undepl}, are taken. The length of the undepleted column $[Z_{sol}(T_{p1}) - Z_{undepl}]$ is constant for all T_p. When comparing at the temperature, T_p, the states of segregation at T_{p1} and T_p ($T_p < T_{p1}$), the effect of the fall in T_p on the shrinking of the earth has to be considered. As at $T_p = T_{p2}$, this effect is the same for all columns with $T_p > T_{p2}$, it can be neglected.

For $T_p < T_{p1}$, only part of the undepleted column $[Z_{sol}(T_{p1}) - Z_{undepl}]$ above the solidus $Z_{sol}(T_p)$ is subject to segregation. For all $T_p < T_{p1}$, this results in a lengthening of the total undepleted column by the amount of lengthening of only the segregated part. This is equal to $D(T_p)$, the density defect thickness at T_p. The ocean bottom for $T_p < T_{p1}$ is falling, therefore, as $D(T_p)$. Taking also oceanic water depth to be a function of T_p, $d(T_p)$, the falling ocean bottom results in a fall of sea level relative to a continent stabilized at T_{p1} (or to the earth's centre) by:

$$D(T_{p1} - D(T_p) + d(T_{p1}) - d(T_p) = D_1 - D + d_1 - d \quad (1)$$

where $D_1 = D(T_{p1})$, $D = D(T_p)$, $d_1 = d(T_{p1})$ and $d = d(T_p)$.

The secular change in depth of the ocean depends on the change in the ratio of continental versus oceanic area, thus on a lateral continental growth, or on the change of ocean volume. As only a secular change of sea level is considered here, the effect on sea level due to short-term geodynamic processes like present-day plate tectonics is neglected. In the following calculations, ocean depth is taken either to be constant or to increase by 1 km during the earth's evolution. The latter case results in later emergence of a once submerged crust and is compatible with lateral growth of the continent. Still, later emergence results if an additional increase in ocean volume had taken place. For the sake of simplicity, it is assumed here that the earth's volatile content had been outgassed prior to 3.8 Ga and that therefore the ocean water volume can be taken to be constant in time. The depth of the ocean then is mainly determined by the area of the ocean.

Lateral growth of the ancient continent and consequent decrease of the ocean area therefore results in deepening of the ocean with time. As the evolution of continental growth and the shrinking of the ocean area is not known in detail, only schematic values for ocean depth are used.

Taking $d_1 = d$, in Eq. (1), hence constant ocean depth over time for $T_p < T_{p1}$, Eq. (1) reduces to $(D_1 - D)$. The fall in sea level is then equal to the fall in density defect thickness $D(T_p)$.

For $T_{p1} = 1667°C$ at 3.8 Ga, and $T_p = 1332°C$ at present, Fig. 2 gives a sea-level fall of $(5.8 - 1.1)$ km = 4.7 km over this period. Throughout the Archaean, when T_p is assumed to drop from 1667 to 1445°C, the sea-level fall is 3.5 km. For the

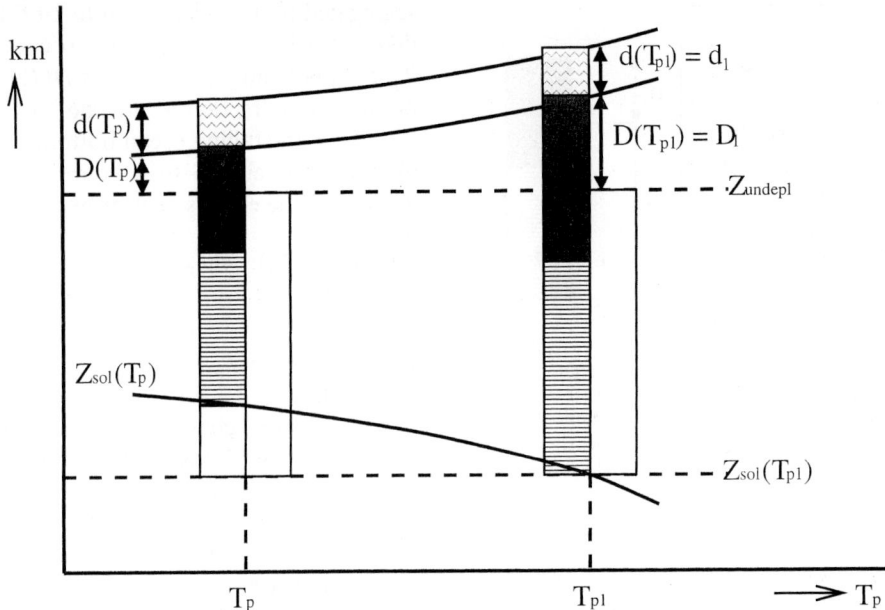

Fig. 3. (Not to scale) Depth to the solidus $Z_{sol}(T_p)$, density defect thickness $D(T_p)$, and ocean depth $d(T_p)$. The state at T_{p1} is taken as the state of reference for $T_p < T_{p1}$. $[Z_{undepl} - Z_{sol}(T_{p1})]$ is the length of the column before segregation, subject to a decreasing degree of depletion for $T_p < T_{p1}$. After segregation, the pressure at depth $Z_{sol}(T_{p1})$ for $T_p < T_{p1}$ is constant and can be taken as the depth of isostatic compensation. The length of the segregated column above the constant reference depth, Z_{undepl}, including the ocean depth is $D(T_p) + d(T_p)$. The small isostatic effect of variable ocean depth is not taken into account. The situation can be viewed as one in which columns of variable length with their base on the solidus and their top at $D(T_p)$ at variable T_p are floating in a sea of undepleted mantle.

post-Archaean until the present, the sea-level fall is 1.2 km. When the schematic ocean depth is increased by 1 km at the Archaean–Proterozoic boundary, the sea-level fall for the post-Archaean is decreased to 0.2 km over this period. When considering matters related to continental free-board in connection with continental growth, a post-Archaean (gradual) increase in sea level due to shrinking of the oceanic area, even on a cooling earth, should be taken into account.

In the present paper, it is assumed that oceanic lithosphere is continuously recycled through the upper mantle. Cratons and their mantle roots, which have been stabilized early in the continental evolution, have not been destroyed and recycled. Their stabilization can be shown to be due to thermal blanketing by the low-density continental crust laying on top of the mantle root (de Smet et al., 1999). As the oceanic lithosphere exhibits a strongly increased gravitational stability with increasing T_p (Vlaar, 1986b), its stability can become permanent when it is covered by a thick layer of sediments causing thermal blanketing. I suggest that a substantial contribution to continental growth of volume and areal extent is due to stabilization of oceanic crust, which has been subject to thermal blanketing by sediments in epicontinental basins. In this case, fractionation of a new continental crust from the mantle is effected by incorporating oceanic crust into the continent.

A sedimentary basin formed in this manner is subaqueous at the time of its formation and stabilization. A sea-level fall may cause the newly formed addition to the continent to emerge above sea level at some time later during the geological history.

As an example, a sedimentary basin formed at 2.5 Ga is considered. If its surface was at a depth of h km below sea surface, its emergence above sea level takes place [Eq. (1)] when $D_1 - D = h$, if the ocean depth, d, does not change in time, or $D_1 - D = h + 1$ when the ocean depth increases by 1 km. Taking $T_{p1} = 1445°C$, $D_1 = 2.3$ km at 2.5 Ga, and $h=2$ km, this geological structure would emerge at $D=1.9$ or 0.9 km, depending on the constant or increasing ocean depth. This means that emergence took place at some point during the first half of the Proterozoic or that the basin would still be submerged at present. In the present model, a late Precambrian emergence of late Archaean and early or later Proterozoic platforms cannot therefore be excluded (Hargraves, 1976) unless T_{p1} at 2.5 Ga has been strongly overestimated. In this case, it must be assumed that since 2.5 Ga, the mantle has cooled considerably less than is indicated in the present model. It has been shown by Vlaar (1986b) that even a slight increase in mantle potential temperature results in a strongly increased stability of the oceanic lithospheric layering. An 50 K increase gives an increase in the age of transition from stable to unstable oceanic lithosphere from the present 30 Ma to about double this value. Therefore, the possibility of creating stable epicontinental basins, even at moderately elevated mantle temperatures compared to the present, is plausible. Even at present, this may be the case in large deltaic sediment fans.

4. Selective exhumation of the continental crust

The existence of the deep depleted harzburgite mantle root of the Arechaean craton must be the result of extensive depletion and the segregation of a thick (basaltic) crust. It, appears, therefore to be a paradox that the cratonic crust is continental. Consequently, it must be inferred that the present cratonic crust is allochtonous with respect to the mantle root. This felsic continental crust therefore must have replaced an earlier oceanic mafic crust, and its function has been to stabilize the cratonic structure in its history by thermal blanketing and its low density (de Smet et al., 1999). Lateral translation of floating proto-continental fragments must have been facilitated by the low strength of the lower part of a thick basaltic crust (Hoffman and Ranalli, 1988).

Early Archaean growth and addition of continental rocks to the crust have probably occurred at a slow rate (Taylor and McLennon, 1985; Rudnick, 1995). Small proto-continental crustal fragments must, therefore, have been floating in a thick basaltic layer. Isostatic conditions were determined by the small density contrast between felsic and mafic rocks. Vlaar (1986a) has shown that under these conditions, protracted loading of the continental surface with felsic rocks or their detritus results in protracted burial of continental crustal material. The small density difference and resulting isostatic conditions favour replacing the basaltic layer by felsic rocks under subaqueous conditions. The early continental proto-crust therefore must have remained subaqueous until it emerged above sea level due to crustal growth and to the sea-level fall described in this paper.

Sediments in the oldest terrains, however, are evidence that small tonalite islands must have been subaerial to provide sediments for high- and low-grade terrains.

The Archaean is characterized by the existence of high-grade gneissic terrains consisting of rocks of the T.T.G. suite, low-grade granite-greenstone belts, and late Archaean cratonic platforms.

At present, the high-grade terrains are outcropping at the surface, and exhibit deep crustal metamorphism. They appear to have been buried to lower crustal depths (20–40 km) in the earlier part of the Archaean and subsequently eroded, resulting in crustal exhumation subsequently, and appear to have stabilized in the late Archaean. This process can be understood in terms of a protracted increase in emerging continental mass above sea level and its denudation to deep crustal depths. In the present paper, this is attributed to a rapid fall in sea level due to strong cooling of the earth during the Archaean.

Granite-greenstone belts are assumed here to be basin-like structures with alternating layering of mainly basaltic magmatic outflows and volcanoclastic sediments. The sediments display a deep- to shallow-water deposition. This layering is intruded by later granitic intrusive plutons. The greenstone belts are emplaced on the continental basement and can probably be seen as ancient analogues of modern graben-like basins (Hamilton, 1998). In the present paper, they are supposed to be generated by mantle plumes impinging from below on an already existing subaqueous continental (proto)crust. This gives rise to magmatism leading to volcanic outflows and lower crustal anatexis in order to generate granitic plutons. The low-pressure metamorphic grade of granite greenstone belts ranges from greenschist to amphibolite facies. This implies that they have been buried to pressures of up to 2–3 kb or depths of up to 8–12 km after their formation under shallow- to deep-water conditions (Condie, 1989; Galer and Mezger, 1998). As, in the present paper, they are assumed to be caused by mantle thermal events, thermal relaxation after their formation is assumed to have caused subsidence to deeper water levels and coverage by sedimentary layers that have now disappeared.

Adjacent high- and low-grade terrains have undergone considerably different exhumation histories indicating large relative vertical displacements prior to their cratonic stabilization in the later Archaean. In the following, it will be shown that selective exhumation can be explained by protracted subaerial erosion and exhumation of the high-grade terrains, and a subaqueous position of the granite greenstone belts until the late Archaean.

In order to simulate erosion and exhumation of the high-grade terrains, the basaltic oceanic crust is replaced by a continental crust of lower density in order to create a positive continental freeboard. For the sake of simplicity, the ocean depth is kept constant in time at 3 km.

The increase in ocean depth over time is not expected to result in largely different conclusions as exhumation is mainly governed by the buoyant harzburgite residual layer.

At T_{p1}, the oceanic basaltic crust and water layer are replaced by a lower density continental crust, resulting in a continental freeboard, h_1, without changing the depth of isostatic compensation at $Z_{sol}(T_{p1})$ (Fig. 4).

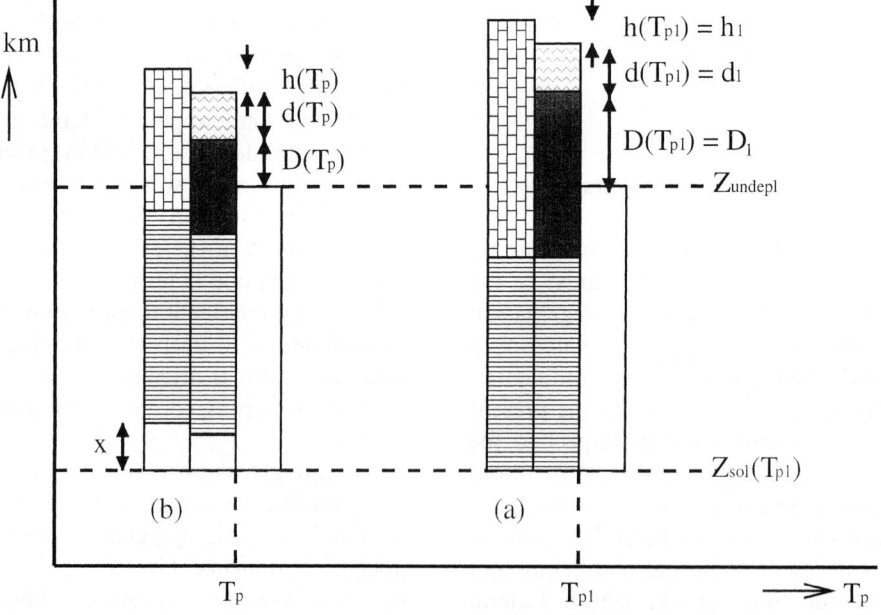

Fig. 4. (Not to scale) At T_{p1}, the oceanic basaltic crust and water layer are replaced by a lower density continental crust, resulting in a continental freeboard, h_1, without changing the depth of isostatic compensation: $d_1\rho_w + d_{b1}\rho_b = (D_1 + d_1 + h_1)\rho_c$. d_1 and d_{b1} are the layer thicknesses of the ocean and oceanic crust at T_{p1}, and ρ_w, ρ_b, and ρ_c are the densities of water, basalt, and continental rocks, respectively. At $T_p < T_{p1}$, the sea level has fallen by $(D_1 + d_1 - D - d)$. If no erosion takes place, the continental surface would stand at $(D_1 + d_1 + h_1 - D - d)$ above sea level at T_p. If erosion results in a continental freeboard, h, at T_p, isostatic readjustment displaces the continental column upward over the distance, x. The depth of crustal exhumation then is $(D_1 + d_1 + h_1 - D - d - h + x)$ and is compensated at the base by replacing depleted by undepleted peridotite. Isostatic balance is governed by $x = (D_1 - D + d_1 - d + h_1 - h)\rho_c/(\rho_p - \rho_c)$. ρ_p is the density of undepleted peridotite.

Isostasy pertains to the base of the oceanic and the continental crust and requires:

$$d_1 \rho_w + d_{b1} \rho_b = (D_1 + d_1 + h_1) \rho_c \quad (2)$$

where d_1 and d_{b1} are the thicknesses of the oceanic layer and the oceanic crust at T_{p1}, and ρ_w, ρ_b, and ρ_c are the densities of water, basalt, and continental rocks, respectively.

Taking $\rho_w = 1$ g cm^{-3}, $\rho = 3$ g cm^{-3}, $d_1 = 3$ km, and requiring the continental freeboard to be $h_1 = 0.75$ km at T_{p1}, Eq. (2) gives:

$$3(1 + d_{b1}) = (d_{b1} + 3.75) \rho_c \quad (3)$$

d_{b1} and ρ are therefore interdependent due to the assumption that continental and oceanic crusts have their base at the same depth.

For $T_{p1} = 1667°C$, and $d_{b1} = 38$ km at 3 Ga, the density of continental rocks becomes $\rho_c = 2.8$ g cm^{-3}. This value appears to be reasonable for a crust of intermediate composition.

For $T_{p1} = 1445°C$, and $d_{b1} = 15$ km at 2.5 Ga, $\rho_c = 2.56$ g cm^{-3}. Though marginal, this is still an acceptable value for a late Archaean, more felsic crust. Therefore, with acceptable values for continental density, a freeboard of 0.75 km can be obtained, departing from the specific type of isostasy used here. At a lower post-Archaean T_{p1}, however, a continental freeboard of 0.75 km or more can only be realized by assuming crustal thickening subject to Airy isostasy. Airy isostasy is then superposed on the isostatic state presented in this paper. In the following, only the effects of the latter, in particular for the Archaean exposed continental surface, are investigated.

If no erosion had taken place, the continental surface with freeboard h_1 at T_{p1} would stand at $(D_1 + d_1 + h_1 - D - d)$ above sea level at T_p. Taking the constant depth of the ocean with time, $T_{p1} = 1667°C$, and $h_1 = 0.75$ km, in agreement with $\rho_c = 2.8$ g cm^{-3}, the early Archaean exposed continent would stand at $(D_1 + h_1 - D)$ above ambient sea level at T_p, 4.25/5.45 km above sea level at 2.5 Ga/present day.

As erosion did in fact take place, the continental surface at T_{p1} with freeboard, h_1, is assumed to have been eroded to h above sea level at T_p. Erosion and exhumation proceed from T_{p1} to T_p under conditions of isostatic balance in a cooling and changing mantle. The base of the column formed at T_{p1} thereby is shifted upward over the distance, x. Isostatic balance at T_p requires:

$$x = (D_1 - D + d_1 - d + h_1 - h) \rho_c / (\rho_p - \rho_c) \quad (4)$$

where ρ_p is the density of undepleted mantle rock.

Erosion then has resulted in a crustal exhumation of the amount:

$$x + (D_1 - D + d_1 - d + h_1 - h). \quad (5)$$

For the entire Archaean period (3.8–2.5 Ga), constant ocean depth, ($d_1 = d$), and erosion to sea level ($h = 0$), by substituting $D_1 - D = 3.5$ km, $h_1 = 0.75$ km, $\rho_c = 2.8$ g cm^{-3}, and $\rho_p = 3.33$ g cm^{-3} in Eqs. (4) and (5), this results in $x = 22.88$ km, and a crustal exhumation of 27.13 km. If erosion of the 3.8 Ga continental surface had proceeded to the present sea level, the crustal exhumation would be 34.24 km. These values are consistent with the observed erosion of the early Archaean high-grade terrains to a lower crustal depth.

In order to explain the large relative vertical displacement of high-grade terrains and low-grade granite-greenstone belts, in the present model, greenstone belts must have become aerially exposed only later in the geological history and should not have been subject to the large vertical displacements that high-grade terrains underwent. These conditions can only be met when, after their formation, greenstone belts that were formed under shallow to deep water conditions had foundered to a deeper, but low-grade, crustal depth below the sea surface. The greenschist to amphibolite metamorphic grade indicates shallow crustal burial. One way to achieve this is by thermal relaxation of the lithosphere after formation of the greenstone belts by a thermal anomaly underneath the crust.

It is now assumed that at 3.8 Ga, a greenstone belt existed side by side to a future high-grade terrain that had not yet been exhumed. The high-grade terrain had a positive freeboard, and the greenstone belt was buried at a low metamorphic grade in a crust whose top was 3 km below sea level.

Therefore, from the above, a scenario can be imagined whereby, at 2.5 Ga, the greenstone belt would just have emerged above sea level, and the high-grade terrain would have been exhumed to a crustal depth of 27 km. Selective exhumation of high-grade terrain probably took place mostly during the early Archaean when the earth's cooling was most rapid. A secular sea-level fall, however, also continued thereafter, and protracted erosion of the stabilized craton proceeded. This gave rise to the formation of early Archaean cratonic platforms and early Proterozoic epicontinental sedimentary basins.

Exhumation causes thinning of the continental crust. A 3.8 Ga crust of 41.75 km is reduced to about 15 km at 2.5 Ga. Therefore, 26.75 km of crust have been eroded away.

The present Archaean crustal thickness is about 40 km, and this had been in place when the craton had stabilized in the late Archaean. In order to compensate the loss of the upper crust caused by its exhumation, new mantle-derived mafic material must have been added to the crust before its stabilization. Exhumation and isostatic readjustment as such cause decompression and renewed melting in the harzburgite mantle root. If this additional decompression follows shortly after isentropic segregation of the crust from the mantle, the mantle root can be regarded to generate melt under the same isentropic conditions. The generated volume of extra melt is an upper bound. If decompression takes place at a later time, the amount of melt is less because of conductive cooling of the shallower mantle layers. A rough estimate of the maximum amount of additional melt can be made from Fig. 1. The difference between the upper and lower curves is the additional melt generated when all pre-existing crust is removed. This volume of melt is roughly equal to the crustal volume. For example, at 3.8 Ga, some 25 km of newly added crust have to be accounted for. This is approximately half the difference between curves (a) and (b). This implies that renewed isentropic decompression of the mantle root can easily satisfy the required crustal growth. After conductive cooling of the mantle root, however, additional melt generation takes place at a deeper level in the harzburgite root. The amount of melt generated is then smaller. However, it has been demonstrated by de Smet et al. (1999) that convection in the hot Archaean upper mantle also leads to protracted melt generation. It could be assumed, therefore, that sufficient additional melt can be generated to make up for the eroded crust.

5. Discussion and conclusions

Primary basaltic crust is assumed to have been generated by supersolidus isentropic decompression melting in a convective shallow upper mantle. When a low-density, more felsic, continental secondary crust replaces the basaltic crust, it advances stabilization of the underlying harzburgite root by thermal blanketing (de Smet et al., 1999). The oceanic basaltic crust and its harzburgite residue, not being shielded in this way, are assumed to be continuously generated and destroyed.

On a secularly cooling earth, the thickness of the oceanic compositional lithospheric layering decreases as the mantle temperature falls. This causes changing isostatic conditions and a considerable decrease in sea level. The Archaean craton, however, has escaped destruction, and has been stabilized during the later Archaean and early Proterozoic. The regularly falling sea level caused parts of the stabilized continent to emerge above water. For the Archaean craton, this took place mostly during the later or latest Archaean, whereas this may have occurred in early Proterozoic basins during the later Precambrian. Erosion of the exposed continental surface leads to its exhumation by isostatic readjustment. The particular circumstances that determine isostasy on a cooling earth — the decreasing thickness of the oceanic compositional layering, secular deepening of the ocean floor and a corresponding fall in sea level, and also the substantial buoyancy of the low-density harzburgite root — favour exhumation of the Archaean high-grade terrain to deep crustal levels.

However, the model presented in this paper requires also that granite-greenstone belts, which have experienced low-grade metamorphism only, should have been buried at a shallow depth in a crust that did not emerge above sea level until the late stages of high-grade terrain exhumation. The large relative vertical displacements between high- and low-grade terrains must have been settled before the final stabilization of the craton.

Buick et al. (1995) observed that greenstone belts can be buried to deeper crustal levels. They reported a 3.5 Ga old greenstone belt that had been formed on top of an existing older basement of greenstone belt. Their inference was that this structure emerged repeatedly around 3.5 Ga. The present model, however, requires that shortly after its emplacement, the structure must have foundered to larger crustal depths and remained subaqueous until their late Archaean emergence.

The model relates the geological history to the falling temperature of the mantle. As such, it can be considered to be robust under the suppositions stated. Mantle temperature is secularly falling. As the decrease in mantle temperature, however, is dependent on the geodynamic mode of cooling, the cooling history of the earth as a function of time is strongly dependent on the cooling mechanism. As the geodynamic processes by which the earth cools are not known for the entire Precambrian, a time record for secular sea-level fall and its consequences are not available. Cooling of the earth could have been episodic and concentrated in shorter periods between longer periods of stagnancy and stabilization. Therefore, the cooling of the earth as a function of time, and therefore its effect on sea level and selective exhumation, cannot be directly related to the smooth fall of T_p used in this investigation. In particular, the temperature at 2.5 Ga as used here may have been over- or underestimated. This can affect considerations concerning post-Archaean continental emergence and freeboard and growth, and should be taken into account in future research.

Short-term non-secular events comparable to the modern plate tectonic cycle, orogenic processes, magmatic additions to the crust by plume activity, and associated crustal thickening together with Airy isostatic readjustment and resulting emergence have not been taken into consideration. Only the dominant effects of the earth's secular cooling on crustal segregation, sea-level change and their consequences have been treated here.

Acknowledgement

The author acknowledges partial funding of research by the Dr. Schürmann foundation.

References

Abbott, D., Burgess, L., Longhi, J., 1994. An empirical thermal history of the Earth's upper mantle. J. Geoph. Res. 99, 13835–13850.

Buick, R., Thornett, J.R., McNaughton, N.J., Smith, J.B., Barley, M.E., Savage, M., 1995. Record of emergent continental crust ~3.5 billion years ago in the Pilbara craton of Australia. Nature 375, 574

Condie, K.C., 1989. Plate Tectonics and Crustal Evolution, third ed. Pergamon Press, Oxford. 476 pp.

de Smet, J.H., van den Berg, A.P., Vlaar, N.J., 1999. The evolution of continental roots in numerical thermo-chemical mantle convection models. Lithos 48, 153–170.

Galer, S.J.G., 1991. Interrelationships between continental freeboard, tectonics and mantle temperature. Earth Planet. Sci. Lett. 105 (214), 228–1991.

Galer, S.J.G., Mezger, K., 1998. Metamorphism, denudation and sealevel in the Archaean and cooling of the earth. Precambrian Res. 92, 389–412.

Hallam, A., 1977. Secular changes in marine inundations of USSR and North America through the Phanerozoic. Nature 269, 769–772.

Hamilton, W.B., 1998. Archean magmatism and deformation were not products of plate tectonics. Precambrian Res. 91, 143–179.

Hargraves, R.B., 1976. Precambrium Geologic History. Science 193, 363–371.

Hays, J.D., Pitman, W.C., 1973. Lithospheric plate motion, sealevel changes and climateand ecological consequences. Nature 246, 18–22.

Hoffman, P.F., Ranalli, G., 1988. Archaean oceanic flake tectonics. Geophys. Res. Lett. 15, 1077–1080.

Hunter, D.R., 1974. Crustal development in the Kaapvaal Craton, I. The Archaean. Precambrian Res. 1, 259–294.

Jaques, A.L., Green, D.H., 1980. Anhydrous melting of peridotite at 0–15 kb pressure. Contrib. Mineral. Petrol. B73, 287–310.

Jordan, T.H., 1997. Mineralogies, densities and seismic velocities of garnet lherzolites and their geophysical implications. In: Boyd, F.R., Meyer, H.O.A. (Eds.), The Mantle Sample: Inclusions in Kimberlites and Other Volcanics. Am. Geophys. Union, Washington, DC, pp. 1–14.

Kuenen, P.H., 1939. Quantitave estimations relating to eustatic movements. Geol. Mijnbouw 18, 194–201.

McKenzie, D., Nisbet, E., Sclater, J.G., 1980. Sedimentary basin development in the Archaean. Earth Planet. Sci. Lett. 48, 35–41.

McKenzie, D., 1984. The generation and compaction of partially molten rock. J. Petrol. 25, 713–765.

McKenzie, D., Bickle, M.J., 1988. The volume and composition of melt generated by the extension of the lithosphere. J. Petrol., 625–679.

Moorbath, S., 1975. Evolution of Precambrian crust from strontium isotope evidence. Nature 254, 395.

Oxburgh, E.R., Parmentier, E.M., 1977. Compositional and density stratification in oceanic lithosphere: causes and consequences. J. Geol. Soc. London 133, 343–355.

Patchett, J.P., Arndt, N.T., 1986. Nd isotopes and tectonics of 1.9–1.7 Ga crustal genesis. Earth Planet. Sci. Lett. 78 (329), 328–1986.

Rudnick, R.L., 1995. Making continental crust. Nature 378, 591

Schubert, G., Reymer, P.S., 1985. Continental volume and freeboard through geological time. Nature 316, 336–339.

Takahashi, E., 1980. Speculations on the Archaean mantle: missing link between komatiite and depleted garnet peridotite. J. Geophys. Res. 95, 15941–15945.

Taylor, S.R., McLennon, S.M., 1985. The Continental Crust: Its Composion and Evolution. Blackwell Scientific, Oxford. 312 pp.

Umbgrove, J.H.F., 1939. On rhythms in the history of the earth. Geol. Mag. 76, 116–129.

Vail, P.R., Mitchum, R.M., Thompson, S., 1978. Seismic stratigraphy and global changes of sealevel 4. Global cycles of relative changes of sealevel, Payton, E. (Ed.), Seismic Stratigraphy: Applications to Hydrocarbon Exploration. Am. Assoc. Petrol. Geol. Mem. Vol. 26., 83–97.

Vlaar, N.J., 1985. Precambrian geodynamical constraints. In: Tobi, C., Touret, J.L.R. (Eds.), The Deep Proterozoic Crust in the North Atlantic Provinces. pp. 3–20.

Vlaar, N.J., 1986a. Archaean global dynamics. Geol. Mijnbouw 65, 91–101.

Vlaar, N.J., 1986b. Geodynamic evolution since the Archaean. Proc. Kon. Ned. Akad. Wetensch. Ser. B 89, 387–406.

Vlaar, N.J., van den Berg, A.P., 1991. Continental evolution and archeo-sealevels. In: Sabadini, R., Lambeck, K., Boschi, E. (Eds.), Glacial Isostasy, Sealevel and Mantle Rheology. Kluwer, Dordrecht.

Vlaar, N.J., van Keken, P.E., van den Berg, A.P., 1993. Cooling of the Earth in the Archaean: Consequences of pressure release melting in a hotter mantle. Earth Planet. Sci. Lett. 121, 1–18.

Windley, B.F., 1977. Timing of continental growth and emergence. Nature 270, 426.

Wise, D.U., 1973. Freeboard of continents through time. Geol. Soc. Am. Mem. 132, 87–100.